Nitrile Oxides, Nitrones, and Nitronates in Organic Synthesis
Novel Strategies in Synthesis

ORGANIC NITRO CHEMISTRY SERIES

Managing Editor
Dr. Henry Feuer
Purdue University
West Lafayette, Indiana 47907
(USA)

EDITORIAL BOARD

Hans H. Baer
Ottowa, Canada

Robert G. Coombes
London, England

Leonid T. Eremenko
Chernogolovka, USSR

Milton B. Frankel
Canoga Park, CA, USA

Mortimer J. Kamlet
Silver Springs, MD, USA

Nathan Kornblum
West Lafayette, IN, USA

Philip C. Myhre
Clairmont, CA, USA

Arnold T. Nielsen
China Lake, CA, USA

Wayland E. Noland
Minneapolis, MN, USA

George A. Olah
Los Angeles, CA, USA

Noboru Ono
Kyoto, Japan

C.N.R. Rao
Bangalore, India

John H. Ridd
London, England

Glen A. Russell
Ames, IA, USA

Dieter Seebach
Zurich, Switzerland

François Terrier
Rouen, France

Heinz G. Viehe
Lowain-la-Neuve, Belgium

Zhou Fa-qi
Peking, China

Also in the Series
NITROAZOLES
The C-Nitro Derivatives of Five-Membered
N- and N,O-Heterocycles
by Joseph H. Boyer

Nitrile Oxides, Nitrones, and Nitronates in Organic Synthesis

Novel Strategies in Synthesis

Kurt B. G. Torssell

Kurt B. G. Torssell
Department of Organic Chemistry
University of Aarhus
8000 Aarhus C, Denmark

Library of Congress Cataloging-in-Publication Data

Torssell, Kurt, 1926-
 Nitrile oxides, nitrones, and nitronates
in organic synthesis.

 (Organic nitro chemistry ; 2)
 Bibliography: p.
 Includes index.
  1. Nitrogen Oxides. 2. Chemistry, Organic--
Synthesis. I. Title. II. Series
QD305.N8T67 1987   547'.2   87-23010
ISBN 0-89573-304-8

©1988 VCH Publishers, Inc.

This work is subject to copyright.

All rights are reserved, whether the whole or part of the material is concerned, specifically those of translation, reprinting, re-use of illustrations, broadcasting, reproduction by photocopying machine or similar means, and storage in data banks.

Registered names, trademarks, etc. used in this book, even when not specifically marked as such, are not to be considered unprotected by law.

Printed in the United States of America.

ISBN 0-89573-304-8 VCH Publishers
ISBN 3-527-26641-0 VCH Verlagsgesellschaft

Distributed in North America by:

VCH Publishers, Inc.
220 East 23rd Street, Suite 909
New York, New York 10010

Distributed Worldwide by:

VCH Verlagsgesellschaft mbH
P.O. Box 1260/1280
D-6940 Weinheim
Federal Republic of Germany

Series Foreword

In the organic nitro chemistry era of the fifties and early sixties, a great emphasis of the research was directed towards the synthesis of new compounds that would be useful as potential ingredients in explosives and propellants.

In recent years, the emphasis of research has been directed more and more toward utilizing nitro compounds as reactive intermediates in organic synthesis. The activating effect of the nitro group is exploited in carrying out many organic reactions, and its facile transformation into various functional groups has broadened the importance of nitro compounds in the synthesis of complex molecules.

It is the purpose of the Series to review the field of organic nitro chemistry in its broadest sense by including structurally related classes of compounds such as nitroamines, nitrates, nitrones, and nitrile oxides. It is intended that the contributors who are active investigators in the various facets of the field will provide a concise presentation of recent advances that have generated a renaissance in nitro chemistry research.

Henry Feuer
Purdue University

Preface

1,3-Dipolar addition of nitrile oxides, nitrones, and nitronates to alkenes or acetylenes is a well-established carbon-carbon bond coupling reaction leading to isoxazole derivatives that occasionally have been used synthetically. The basic chemistry of the nitrile oxide and nitrone dipoles is also well known as well as the chemistry of the third member of the dipole family: the silyl and alkyl nitronates, which have been investigated quite recently. They are all easily available compounds. The interest in the isoxazoles, isoxazolines, and isoxazolidines increased considerably when it became evident that they possessed significant masked functionalities that, on unmasking, give rise to several other heterocycles and functionalized alkanes. A network of useful reactions was developed with interesting and broad applications. We have here a valuable instrument in organic synthesis. Common, inexpensive chemicals (aldehydes, aliphatic nitro compounds, alkenes, acetylenes, hydroxylamines) serve as starting material, and the experimental conditions are simple.

This volume collects the contributions found in the earlier literature on the synthetic use of the isoxazole nucleus together with the results from recent, extensive, systematic research by several groups.

Chapter 1 is an introductory discussion of the general principles for using heterocycles in synthesis, and special attention is devoted to the chemistry of isoxazole derivatives and the problem of why these derivatives formed by 1,3-dipolar addition are especially suited as intermediates. The mechanistic aspects of 1,3-dipolar additions are discussed jointly for the three dipoles in this chapter. The nitrile oxides and nitrones are already excellently covered in recent reviews and monographs, but the essential parts of the chemistry of nitrile oxides and nitrones are highlighted in a condensed version in Chapters 2 and 3 for the sake of coherency and for their importance as immediate reactants. Therefore, references given in these sections are not comprehensive and may sometimes, regrettably, be chosen by chance. Chapter 4 reviews the chemistry of the nitronates. These sections function as background and support to the main part, Chapter 5, of the book, which is a contribution to the art of carbon-carbon coupling, the essence of organic synthesis. I hope this will inspire other workers to venture into pyrixizole chemistry. The literature has been searched up to the end of 1985. Articles published in the period December 1985 to August 1987 have been reviewed in the Addendum.

I wish to express my thanks to Mrs. Ella M. Larsen for the excellent typing and art work, and to Dr. John R. Cashman for linguistic corrections.

Aarhus 1987
KURT B. G. TORSSELL

Contents

SERIES FORWARD v
PREFACE vii

CHAPTER 1. INTRODUCTION AND GENERAL CONSIDERATIONS 1

1.1 THE DIPOLES: NITRILE OXIDES, NITRONES, AND SILYL NITRONATES. THE REVIEW LITERATURE 1
1.2 1,3-DIPOLAR ADDITION—A PRINCIPLE FOR CARBON-CARBON COUPLING. HETEROCYCLES IN ORGANIC SYNTHESIS 3
1.3 STARTING MATERIAL. REACTION CONDITIONS 8
1.4 ISOXAZOLES. REACTIONS AND PHYSICOCHEMICAL PROPERTIES. THE MASKED FUNCTIONALITY 9
1.5 ISOXAZOLINES. REACTIONS AND PHYSICOCHEMICAL PROPERTIES. THE MASKED FUNCTIONALITY AND THE ALDOL CONCEPT 14
1.6 ISOXAZOLIDINES. REACTIONS AND PHYSICO-CHEMICAL PROPERTIES. CYCLOREVERSIONS AND THE MASKED FUNCTIONALITY 20
1.7 MECHANISMS OF THE 1,3-DIPOLAR CYCLOADDITION. STEREOCHEMICAL AND REGIOCHEMICAL CONSIDERATIONS. ASYMMETRIC INDUCTION 25
1.8 RELATIVE REACTIVITIES OF REACTANTS 44
REFERENCES 47

CHAPTER 2. THE NITRILE OXIDES 55

2.1 REACTIONS AND PHYSICOCHEMICAL PROPERTIES 55
2.2 SYNTHESIS OF FULMINIC ACID (FORMONITRILE OXIDE) 60
2.3 SYNTHESIS OF HALOGEN AND SULFUR-SUBSTITUTED FULMINIC ACID 62
2.4 SYNTHESIS OF NITRILE OXIDES 64

ix

2.5	NUCLEOPHILIC ADDITION TO NITRILE OXIDES	69
2.6	THE 1,3-DIPOLAR CYCLOADDITION	70
2.7	ALLERGENIC PROPERTIES OF NITRILE OXIDES	70
REFERENCES		71

CHAPTER 3. THE NITRONES **75**

3.1	PHYSICOCHEMICAL PROPERTIES	75
3.2	REACTIONS OF NITRONES	79
	A. THE NITRONE-AMIDE REARRANGEMENT	79
	B. THE BEHREND REARRANGEMENT	80
	C. THE NITRONE-OXIME O-ETHER REARRANGEMENT	80
	D. THERMOLYTIC ALKENE ELIMINATION	81
	E. REACTIONS OF NITRONES WITH NUCLEOPHILES	81
	F. OXIDATIONS OF NITRONES	83
	G. REDUCTIONS OF NITRONES	84
	H. NITRONES AS RADICAL SCAVENGERS	84
	I. 1,3-DIPOLAR CYCLOADDITION	85
	J. Z-E-ISOMERIZATION	85
3.3	SYNTHESIS OF NITRONES	86
	A. CONDENSATION OF N-MONOSUBSTITUTED HYDROXYLAMINES WITH CARBONYL COMPOUNDS	86
	B. DEHYDROGENATION OF N,N-DI-SUBSTITUTED HYDROXYLAMINES	88
	C. N-ALKYLATION OF OXIMES	88
	D. REACTION OF AROMATIC NITROSO COMPOUNDS WITH ACTIVE METHYLENE COMPOUNDS: THE KRÖHNKE REACTION	89
	E. N-OXIDATION OF IMINES (SCHIFF'S BASES)	89
REFERENCES		89

CHAPTER 4. ALKYL AND SILYL NITRONATES **95**

4.1	INTRODUCTION. THE NITRONIC ACIDS	95
4.2	ALKYLATION OF NITRO COMPOUNDS. SYNTHESIS OF ALKYL NITRONATES (NITRONIC ESTERS)	101
4.3	PHYSICOCHEMICAL PROPERTIES OF ALKYL NITRONATES. REACTIONS OF ALKYL NITRONATES	109
4.4	SYNTHESIS OF SILYL NITRONATES	113

4.5	PHYSICOCHEMICAL PROPERTIES OF SILYL NITRONATES	114
4.6	REACTIONS OF SILYL NITRONATES	119
REFERENCES		122

CHAPTER 5. APPLICATIONS OF NITRILE OXIDES, NITRONES, NITRONATES, AND INTERMEDIATE ISOXAZOLES, ISOXAZOLINES, AND ISOXAZOLIDINES IN SYNTHESIS 129

5.1	2-NITROALCOHOLS AND 2-AMINOALCOHOLS	129
5.2	α-CYANOKETONES AND 2-CYANOALCOHOLS. CARBOXY-HYDROXYLATION	131
5.3	WOODWARD-OLOFSON'S PEPTIDE SYNTHESIS. CLEAVAGE OF ISOXAZOLIUM SALTS	137
5.4	KETONE ANNELATION. REACTIONS VIA α-DEPROTONATION OF ISOXAZOLES	138
5.5	1,3-CARBONYL TRANSPOSITION	142
5.6	β-POLYKETONES, PHENOLS, ANILINES	145
5.7	2-ISOXAZOLINES, THE MASKED ALDOLS. SYNTHESIS OF β-HYDROXYKETONES	147
5.8	α,β-UNSATURATED CARBONYL COMPOUNDS, β-ACYLATED ACRYLIC ESTERS AND ACROLEIN, 2-ENE-1,4-DIONES, 1,4-DIONES γ- AND δ-KETOESTERS, α-HYDROXYESTERS, AND KETONES	151
5.9	FURANS, 3-(2H)-FURANONES, AND PYRONES	157
5.10	PYRROLE AND INDOLE DERIVATIVES	160
5.11	AZIRIDINES AND 1-AZIRINES	167
5.12	PYRAZOLES, IMIDAZOLES, OXAZOLES, AND ISOTHIAZOLES	169
5.13	SYNTHESIS OF HETEROCYCLES BY RING TRANSFORMATION INVOLVING THE SIDE CHAIN OF ISOXAZOLES	173
5.14	PYRIDINES, PYRIMIDINES, PYRIDAZINES, OXAZINES, QUINOLINES, AND QUINOXALINES	176
5.15	AZETIDINES, β-LACTAMS, AND β-LACTONES	180
5.16	MISCELLANEOUS HETEROCYCLES. CONDENSED SYSTEMS. CYCLOADDITIONS TO HETERO DIPOLAROPHILES	183
5.17	REACTIONS WITH α-CHLORONITRONES AND α,β-EPOXYNITRONES	189

5.18	CYCLOADDITION OF OXIMES WITH OLEFINS	191
5.19	CYCLOPENTANE DERIVATIVES, PROSTAGLANDIN PRECURSORS, AND TERPENOIDS. RING ANNULATION, INTRAMOLECULAR CYCLIZATION, BRIDGED CYCLOALKANES	192
5.20	1,3-AMINOALCOHOLS. ALKALOIDS. INTRAMOLECULAR CYCLIZATION	209
5.21	CARBOHYDRATES	232
5.22	AMINO ACIDS	239
5.23	MACROCYCLES, BIOTIN, MISCELLANEOUS NATURAL PRODUCTS	246
CONCLUDING REMARKS		255
REFERENCES		255
ADDENDUM		**271**
CHAPTER 1		271
CHAPTER 2		277
CHAPTER 3		278
CHAPTER 4		280
CHAPTER 5		283
REFERENCES		297
AUTHOR INDEX		**301**
SUBJECT INDEX		**327**

Chapter 1

Introduction and General Considerations

1.1 THE DIPOLES: NITRILE OXIDES, NITRONES, AND SILYL NITRONATES. THE REVIEW LITERATURE

Breakthrough and advancement in science rest always on the foundations laid by earlier workers. The history of nitrile oxides goes back to 1800, when E. Howard prepared the explosive mercury fulminate. The correct structure of fulminic acid and its salts remained unknown for a long time and was the subject of much speculation. It had to wait for nearly 100 years for its elucidation. In 1899 H. Ley suggested that fulminic acid was the nitrile oxide of formic acid, $HC\equiv N^+-O^-$, i.e., the parent compound of nitrile oxides. The present status of the chemistry of the dipolar nitrile oxides acknowledges particularly the contributions by H. Wieland in the early 1900s, A. Quilico and associates in the 1940s, and R. Huisgen, who, in the 1960s, systematized comprehensively the 1,3-dipolar reactions and arrived at a better understanding of their mechanism based on molecular orbital theory.

The nitrones were discovered by E. Beckmann in the 1880s by *N*-alkylation of oximes. There has been a constant interest in the chemistry of nitrones ever since. In the 1960s the high-yielding addition of this 1,3-dipole to olefins with formation of isoxazolidines was discovered independently by Cope, LeBel, Brown, Delpierre, and Huisgen.

The silyl nitronates are the youngest members of this dipole family. V. A. Tartakovskii and his group prepared the first silyl nitronates and investigated their properties and reactions in the 1970s. This work was extended by us and other workers, and we have undertaken an exploration of the utilization of isoxazoles, isoxazolines, and isoxazolidines as intermediates in organic synthesis.

Various aspects of the subject of this book have been reviewed earlier in journals, comprehensive series, and a few monographs. A bibliographic list of reviews follows.

BIBLIOGRAPHY

Silyl Nitronates and Alkyl Nitronates

E. Breuer In "The Chemistry of Amino, Nitroso and Nitro Compounds, and Their Derivatives" (S. Patai, Ed.), Wiley, Chichester, 1982, p. 459.
S. L. Joffe; L. M. Leonteva; V. A. Tartakovskii, *Russ. Chem. Revs.* (Engl. Transl.) *46* (1977), 872.
A. T. Nielsen In "The Chemistry of the Nitro and Nitroso Compounds", Part 1 (H. Feuer, Ed.), Wiley-Interscience, New York, 1969, p. 349.

Nitrones

N. Balasubramanian, *Org. Prep. Proc. Int. 17* (1985), 25.
D. St. C. Black; R. F. Crozier; V. C. Davis, *Synthesis, 1975*, 205.
E. Breuer In "The Chemistry of Amino, Nitroso and Nitro Compounds, and Their Derivatives" (S. Patai, Ed.), Wiley, Chichester, 1982, p. 459.
G. R. Delpierre; M. Lamchen, *Quart. Revs. 19* (1965), 329.
J. Hamer; A. Macaluso, *Chem. Revs. 64* (1964), 473.
A. R. Katritzky; J. M. Lagowski, "Chemistry of the Heterocyclic *N*-Oxides, Academic Press, London, 1971.
W. Oppolzer, *Angew. Chem. 89* (1977), 10.
A. Padwa, *Angew. Chem. 88* (1976), 131.
W. Rundel In Houben-Weyl: "Methoden der Organischen Chemie", 10:4, G. Thieme Verlag, Stuttgart, 1968, p. 309.
J. J. Tufariello, *Accts. Chem. Res. 12* (1979), 396.
J. J. Tufariello In "1,3-Dipolar Cycloaddition Chemistry" (A. Padwa, Ed.), Vol. 2, Wiley, New York, 1984, p. 83.

Nitrile Oxides

P. Caramella; P. Grünanger In "1,3-Dipolar Cycloaddition Chemistry" (A. Padwa, Ed.), Vol. 1, Wiley, New York, 1984, p. 291.
C. Grundmann; P. Grünanger, "The Nitrile Oxides", Springer-Verlag, Berlin, 1971.
C. Grundmann In Houben-Weyl: "Methoden der Organischen Chemie", 10:3 (E. Müller, Ed.). G. Thieme Verlag, Stuttgart, 1965, 837; E5:2, (J. Falbe, Ed.), 1985, 1585.
A. Quilico, *Experientia 26* (1970), 1169.

Isoxazoles, Isoxazolines, and Isoxazolidines

J. P. Freeman, *Chem. Revs. 83* (1983), 241.
V. Jäger; I. Müller; R. Schohe; M. Frey; R. Ehrler; B. Häfele; D. Schröter In "Lectures in Heterocyclic Chemistry", Vol. 8, Hetero Corp., Tampa, *1985*, p. 79.
N. K. Kochetkov; S. D. Sokolov In "Advances in Heterocyclic Chemistry, Vol. 2, 1963, p. 365.

S. A. Lang, Jr.; Y.-i Lin In "Comprehensive Heterocyclic Chemistry (A. Katritzky and C. Rees, Eds.), Vol. 6, Pergamon, 1984, p. 1.
A. Quilico In "The Chemistry of Heterocyclic Compounds" (A. Weissberger and E. C. Taylor, Eds.), Vol. 17, Wiley-Interscience, New York, 1962, p. 1.
Y. Takeuchi; F. Furusaki In "Advances in Heterocyclic Chemistry, Vol. 21, 1977, 208.
B. J. Wakefield; D. J. Wright In "Advances in Heterocyclic Chemistry, Vol. 25, 1979, p. 147.

Heterocycles in Synthesis

A. A. Akhrem; F. A. Lakhvich; V. A. Khripach, *Khim. Geterosikl. Soed. 1981*, 1155. Engl. Transl., p. 853.
A. P. Kozikowski, *Accts. Chem. Res. 17* (1984), 410.
A. P. Kozikowski In "Comprehensive Heterocyclic Chemistry" (A. Katritzky and C. Rees, Eds.), Vol. 1, Pergamon, 1984, p. 413.
A. I. Meyers, "Heterocycles in Organic Synthesis", Wiley, New York, 1974.
T. Nishiwaki, *Synthesis 1975*, 20.
H. C. Van der Plas, "Ring Transformation of Heterocycles", Academic Press, London, 1973.

1,3-Dipolar Reactions

G. Bianchi; C. De Micheli; R. Gandolfi In "The Chemistry of Double-Bonded Functional Groups" (S. Patai, Ed.), Wiley, London, 1977, p. 369.
R. Huisgen, *Angew. Chem. 75* (1963), 604.
A. Padwa, Ed. "1,3-Dipolar Cycloaddition Chemistry", Wiley, New York, 1984.

1.2   1,3-DIPOLAR ADDITION—A PRINCIPLE FOR CARBON-CARBON COUPLING. HETEROCYCLES IN ORGANIC SYNTHESIS

Numerous carbon-carbon coupling reactions of general applicability have been worked out during the last century. Most of the reactions can be explained and predicted by the polarity of bonds involved, by Coulomb forces between charged species, or by pairing of radicals. A few, such as the concerted Diels-Alder and 1,3-dipolar reactions, are best rationalized in terms of frontier orbital interactions.[1-3] In this case the naive polar model fails to explain the regioselectivities and reactivities of the addition. The Diels-Alder reaction has secured its position as the most useful method for constructing six-membered ring systems with a high degree of stereocontrol, and its importance is further manifested by accessibility and low cost of reactants and simple reaction conditions. The 1,3-dipolar reaction[4-6] is considered to be the most useful method for the synthesis of five-membered heterocyclic ring systems containing one or more heteroatoms. Accessibility and low cost of reactants combined with simple reaction conditions also apply in certain cases to this reaction. Conceivably, we have here another carbon-carbon coupling reaction of potential value comprising a large variety of reactants.

However, heterocycles have not yet received full recognition as useful relay compounds in synthetic schemes, e.g., at instances when functionalities and fragments have to be temporarily masked or protected to be brought into action at the desired moment. The work on the chemistry of heterocycles dealt primarily with their synthesis and properties and was usually not extended to investigations of their potential utilization as building blocks in synthesis.

There are, indeed, several exceptions to this statement, as is evident from a monograph on the subject.[7] Classical examples of the use of heterocycles in synthesis are furans and thiophenes serving as templets for constructing a variety of functionalized alkanes. Readily available furans are first functionalized by electrophilic substitution and then subjected to oxidative ring opening, eq. (1.1).[8] Suitable 1,4-diketones formed could, e.g., be cyclized to prostanoids.[9,10] Functionalization of thiophene followed by reductive cleavage with Raney-Ni leads conveniently to a multitude of hydrocarbons, eq. (1.2).[11,12]

(1.1)

(E^+ = electrophile)

(1.2)

An example of direct relevance to the discussions in the following chapters is given by the isoxazole **1**, which contains a masked 1,3-dicarbonyl function. It is successfully used in the annelation of ketones, whereby first the heterocycle is unmasked by reduction and then used in the construction of condensed systems, eq. (1.3).[13]

As we shall see later, there is a difference in strategy between the usage of the three first-mentioned heterocycles and the usage of the isoxazole, isoxazoline, and isoxazolidine groups discussed in the following sections. In the furan case, e.g., the substituents are first attached to the prebuilt nucleus, which then is unmasked by oxidative cleavage and eventually subjected to further reactions. The isoxazole, isoxazoline, or isoxazolidine heterocycles are, on the other hand, synthesized during the process by cycloaddition, then possibly functionalized before cleavage and the further transformations. At the same time it is possible to exert certain stereochemical control. In the cycloaddition step two fragments of the final molecule are coupled (Scheme 1.1).

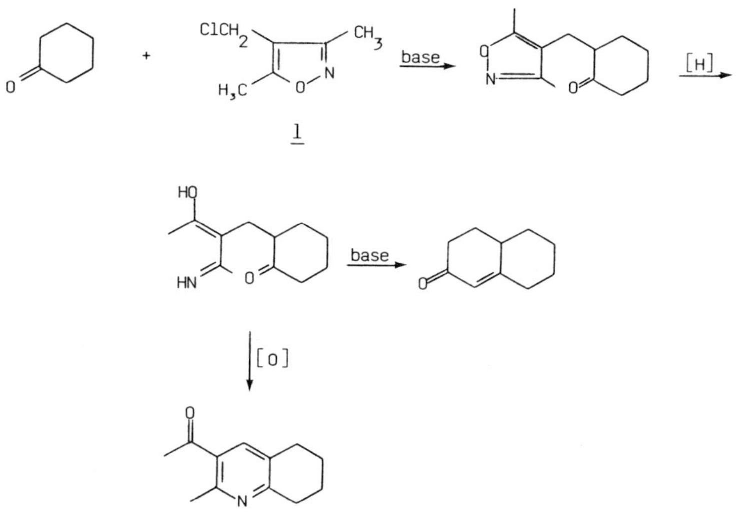

Scheme 1.1. *Routes to 2-isoxazolines and isoxazolidines: stereochemical control.*

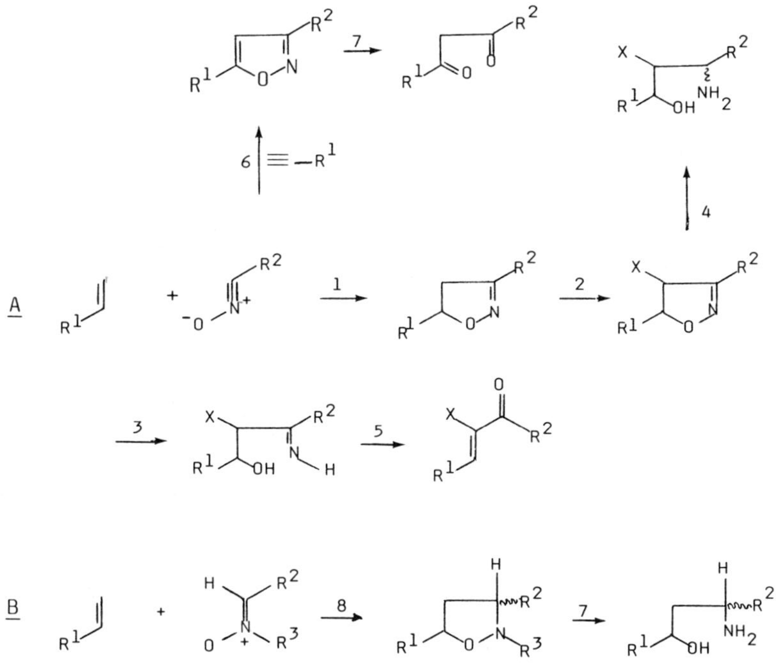

A. 1. Cycloaddition of a nitrile oxide to an alkene. Formation of 2-isoxazolines
2. Functionalization of the isoxazoline by introduction of a substituent X
3,4. Selective reductive cleavage
5. Final transformation, elimination of water and ammonia
6. Cycloaddition of a nitrile oxide to an acetylene. Formation of isoxazoles
7. Reductive cleavage
B. 8. Cycloaddition of a nitrone (R^3 = alkyl, aryl) or nitronate [R^3 = OSi(CH$_3$)$_3$, OCH$_3$, OC$_2$H$_5$] to an alkene. Formation of isoxazolidines

Stereochemical control can be exerted in steps 2, 4, and 8.

For a heterocycle to serve as an efficient tool in organic synthesis, it is desirable for some of the following requirements to be met. The heterocycle should

1. Be easy to prepare from inexpensive, simple starting material
2. Contain masked functionalities and structural units of potential synthetic interest
3. Be easy to functionalize
4. Have a reasonable stability towards reagents
5. Be cleaved or unmasked by selective reagents to produce the desired structural entity
6. Allow stereoselective operations

No compound is known to fulfill all these requirements. It is somewhat schizophrenic to demand stability and reactivity at the same time.

Reference has already been made to the 1,3-dipolar reactions as a suitable high-yielding procedure for carbon-carbon coupling involving a large variety of reactants. The question arises as to which are the most ideal heterocycles fulfilling at least some of requirements 1–6. It appears from a survey of the literature that several dipoles are disqualified by

Structural limitations to a few substituents, occasionally rather exotic ones
Laborious preparative procedures
Formation of acid, base, or heat-labile heterocycles.
Formation of heterocycles giving unsuitable cleavage products

This can be exemplified, e.g., by the nitrile imines **2**, which until now were limited mostly to aromatic derivatives, and by the isomeric diazoalkanes **3**

$R^1 - \overset{+}{C}=N-\overset{-}{N}-R^2$ $\qquad\qquad$ $\underset{R^2}{\overset{R^1}{>}}\overset{+}{C}-N=\overset{-}{N}$

$\qquad\qquad$ **2** $\qquad\qquad\qquad\qquad\qquad$ **3**

R^1 or R^2 = Ar $\qquad\qquad$ R^1, R^2 = H, Alk, Ar belonging to a small class of compounds that are comparatively difficult to synthesize. The pyrazolines obtained by the reaction of compound **2** with olefins, eqs. (1.4),(1.5), do possess masked 1,3-diamino functions, but they are difficult to cleave. Expulsion of nitrogen from the pyrazoline in eq. (1.5) gives cyclopropanes.

(1.4)

(1.5)

Of all the 1,3-dipoles, a set of three stands out, which happens to meet several of requirements 1–6: the nitrile oxides, the nitrones, and the silyl nitronates. They react with olefins and acetylenes to form isoxazoles, isoxazolines, and isoxazolidines (as exemplified in Scheme 1.1) in high yields, which, in the light of

Isoxazole   2-Isoxazoline   Isoxazolidine

recent developments, turn out to be extraordinary, valuable intermediates in organic synthesis. The chemistry of the isoxazole derivatives is discussed in the next sections with emphasis on their synthetic applications in relation to the requirements mentioned previously. In Chapters 2–4 the chemistry of the nitrile oxide, nitrone, and nitronate ester dipoles is briefly surveyed, mainly as a support to Chapter 5, which surveys their synthetic applications. That interest in the isoxazole, isoxazoline, and isoxazolidine heterocycles is of very recent vintage appears from the fact that the monograph (1974) on the use of heterocycles in organic synthesis contains few references to these particular compounds. Quite recently two comprehensive monographs appeared, which in part surveyed the role of heterocycles in general organic synthesis.[5,6]

1.3 STARTING MATERIAL. REACTION CONDITIONS

To satisfy the requirement of a useful reaction, one must consider the expense. Reactions of Claisen or aldol types belong to the most useful reactions for joining molecules together. This depends on the accessibility of carbonyl compounds, satisfactory yields, and simple reaction conditions, i.e., an ideal reaction, which is considered in every synthetic scheme. In the following discussion the dipolar route is compared with the Claisen procedure because, as we will see, it leads eventually to the same kind of products.

Aldehydes, eq. (1.6), or primary nitro compounds, eq. (1.7), the Mukaiyama-Hoshino procedure, serve as starting material for the nitrile oxides, and olefins or acetylenes are dipolarophiles in the cycloaddition reaction that produces 2-isoxazolines and isoxazoles in good yield.[14] The reaction conditions are simple. The critical point of the reaction sequence starting from aldehydes is the chlorination, which sometimes is difficult to control and which does not allow other sensitive functions to be present. This has been handled by chlorination with N-chlorosuccinimide, which leaves most other functional groups untouched.[15]

$$RCHO \xrightarrow[\text{2. Cl}_2]{\text{1. NH}_2\text{OH}} RCCl=NOH \xrightarrow{\text{base}} RC{\equiv}N^+{-}O^- \qquad (1.6)$$

$$RCH_2NO_2 \xrightarrow[\text{Et}_3\text{N}]{\text{C}_6\text{H}_5\text{N=C=O}} RC{\equiv}N^+{-}O^- + (C_6H_5NH)_2CO \qquad (1.7)$$

In case the two reacting functions are located in the same molecule, intramolecular cyclization occurs,[16-19] leading to polycyclic systems.

Aldehydes or ketones are the starting material of choice for the preparation of nitrones by condensation with hydroxylamines, eq. (1.8).[20] They form isoxazolidines and 4-isoxazolines with olefins and acetylenes, respectively. Several useful intramolecular cyclizations have been carried out with bifunctional molecules, leading to polycyclic structures.[21-23]

$$RR^1CO \xrightarrow{R^2NHOH} RR^1C=N^+\!\!\begin{array}{c}R^2\\O^-\end{array} \qquad (1.8)$$

$R^1, R^2 =$ H, alkyl, aryl

The silyl nitronates are conveniently prepared by reacting primary and secondary nitro compounds with chlorotrimethylsilane and triethylamine in acetonitrile at room temperature or with chlorotrimethylsilane and lithium diisopropylamide in tetrahydrofuran at $-70°$C or with silylated amides.[24-27,159] Silyl nitronates cycloadd to olefins and form N-silyloxyisoxazolidines, eq. (1.9).

$$RR^1CHNO_2 \xrightarrow[Et_3N]{ClSi(CH_3)_3} RR^1C=\overset{+}{N}\underset{O^-}{\overset{OSi(CH_3)_3}{\diagup}}$$

$$\downarrow \parallel_X^{R^1}$$

(1.9)

[isoxazoline structure with R^2, X, OSi(CH$_3$)$_3$]

R, R^1 = H, alkyl, aryl

Conclusion: The starting material and reagents are inexpensive, easily accessible compounds, which can undergo facil 1,3-dipolar cycloaddition to form isoxazole derivatives; consequently, requirement 1 (Section 1.2) is fulfilled.

1.4 ISOXAZOLES.[28,29,58] REACTIONS AND PHYSICOCHEMICAL PROPERTIES. THE MASKED FUNCTIONALITY

The isoxazoles are weakly basic compounds possessing aromatic properties similar to those of pyridine. They are resistant to strong acids and oxidizing conditions, which is evident from the nitration of 3,5-dimethylisoxazole with mixed nitric and sulfuric acids at 100°C. 4-Nitro-3,5-dimethylisoxazole is obtained in 86% yield, eq. (1.10). All electrophilic reactions studied so far occur at the 4-position.[28,29] In phenyl-substituted isoxazoles nitration occurs almost exclusively in the *p*-position of the phenyl nucleus, but if the aryl ring is deactivated by a nitro group, nitration occurs predominantly at the 4-position of the isoxazole nucleus, eq. (1.11).[30-32] Sulfonation of 5-alkylisoxazoles gives 5-alkylisoxazole-4-sulfonic acid.[33] The position of the sulfonyl group was proved by treatment with sodium hydroxide, which gave the disodium salt of 1-cyano-2-hydroxypropensulfonic acid, eq. (1.12). This cleavage, which is characteristic for isoxazoles with a hydrogen at C^3, was discovered several years earlier and constitutes a general synthesis of α-cyanoketones.[34] Sulfonation of 5-phenylisoxazole takes place primarily in the phenyl ring and leads to a 2:1 mixture of *m*- and *p*-sulfonic acids, eq. (1.13).[35]

Mercuration,[30,38] eq. (1.14), X=HgOAc, and halogenation[30,36-39] with molecular halogen or with *N*-bromosuccinimide occur likewise at C^4 of the isoxazole ring. For R = C$_6$H$_5$ and R^1 = H, eq. (1.14), bromination takes place predominantly at the *p*-position of the phenyl nucleus.[32]

$$H_3C\underset{O}{\overset{CH_3}{\diagdown\diagup}}N \xrightarrow[100 \ °C]{HNO_3/H_2SO_4} H_3C\underset{O}{\overset{O_2N\ \ \ CH_3}{\diagdown\diagup}}N$$

86%

(1.10)

Of special interest is the chloroalkylation, which leads to 4-chloroalkylisoxazoles.[40–42] The 3,5-dimethyl derivative, eq. (1.15), is used as a masked vinyl ketone equivalent[13] (Section 5.4). Also, introducing a formyl group at C[4] according to the method of Vilsmeier and Haak produces isoxazole derivatives of synthetic interest.[43,44]

As weak bases, the isoxazoles are quaternized by alkylsulfonates or trialkyloxonium salts, eq. (1.16).[45–47] In contrast, formaldehyde and dimethylformamide attack isoxazoles at C^4. The different substitution patterns are explained by the supposition that alkylation as well as acylation actually first occur at the nitrogen atom, but chloroalkylation, hydroxyalkylation, and formylation are reversible, eq. (1.17). The acyl cations are eventually irreversibly and competitively trapped at C^4 according to eq. (1.15). N-Methylation or N-ethylation, eq. (1.16), are consequently not reversible to any measurable extent. Isoxazolium salts are efficient reagents in peptide synthesis (Section 5.3).[35,48,49]

$$\text{3,5-dimethylisoxazole} \xrightarrow{(C_2H_5)_3O^+ \text{ or } (C_2H_5)_2SO_4} \text{N-ethyl-3,5-dimethylisoxazolium} \quad (1.16)$$

$$\text{3,5-dimethylisoxazole} + CH_2O + HCl \rightleftharpoons \text{N-chloromethyl-3,5-dimethylisoxazolium} + H_2O \quad (1.17)$$

Whereas isoxazoles are comparatively stable against oxidants, such as nitric acid, halogens, chromic acid,[41,42,50,52,53] acid permanganate,[51] peracids,[54,55] manganese dioxide,[56] N-bromosuccinimide,[106-108] and dichlorodicyanoquinone,[113,114] the reaction with alkaline potassium permanganate destroys the heterocycle. Aryl-substituted isoxazoles give with this oxidant the corresponding benzoic acids, a degradation used for determining the orientation of electrophilic substitution in the phenyl ring.[30,39] With chromic acid it is possible to oxidize a side-chain hydroxyl group to the corresponding carbonyl group.[52]

The reaction of isoxazoles with nucleophiles often disrupts the heterocycle, and nucleophilic substitution is therefore of less synthetic importance. The chlorine at the C^5 position is reactive and undergoes nucleophilic substitution with alkoxy, thioalkoxy, and amino groups.[28,29,57] Isoxazoles unsubstituted at C^3 are smoothly cleaved by base to α-cyanoketones, eq. (1.12), a reaction studied by several workers.[28,29,34,58] Decarboxylation of 3-carboxyisoxazoles causes the same type of fragmentation, eq. (1.18).[59,60] When C^3 is substituted but C^5 unsubstituted, the fragmentation takes a different course, eq. (1.19), and both the N—O and C^3—C^4 bonds are cleaved.[28,29,58,61-63]

$$\text{3-carboxyisoxazole} \xrightarrow{\Delta} \text{intermediate} \xrightarrow{H^+} \text{α-cyanoketone} \quad (1.18)$$

The isoxazole ring activates α-protons to undergo aldol or Michael-type reactions,[28,29,58,64-71] which enables further functionalization of the isoxazole ring, especially at C^5, where the α-protons are most acidic, eqs. (1.20),(1.21). The remarkable ketone **4** formed in eq. (1.21) is actually a masked 1,3,5,7,9-

pentaketone.[68,69] Equation (1.22) depicts a simple synthesis of curcumin.[64,70] The isoxazolium nucleus activates all three methyl groups. These reactions at the α-carbons of the isoxazole ring can be used as alternatives to the usual dianion methodology applied to 1,3-dicarbonyl systems.

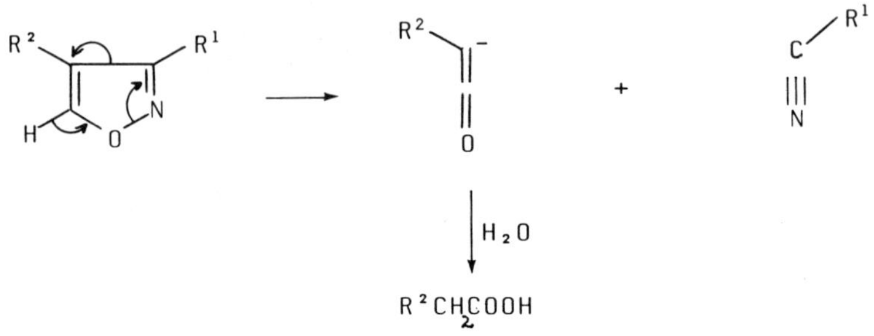

(1.19)

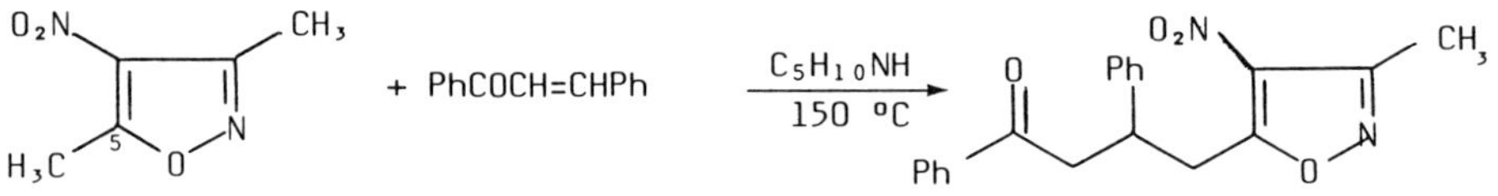

(1.20)

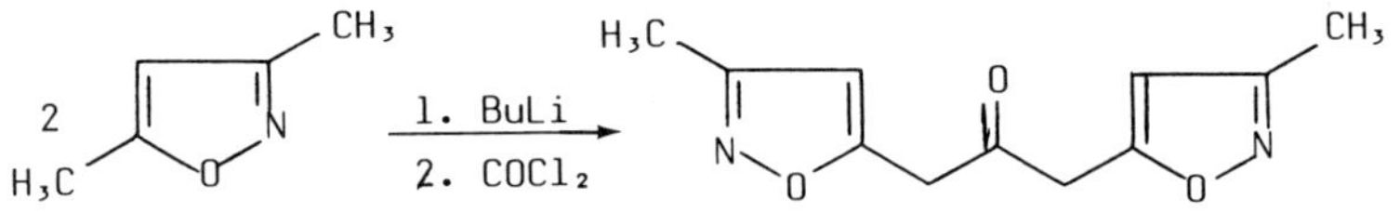

(1.21)

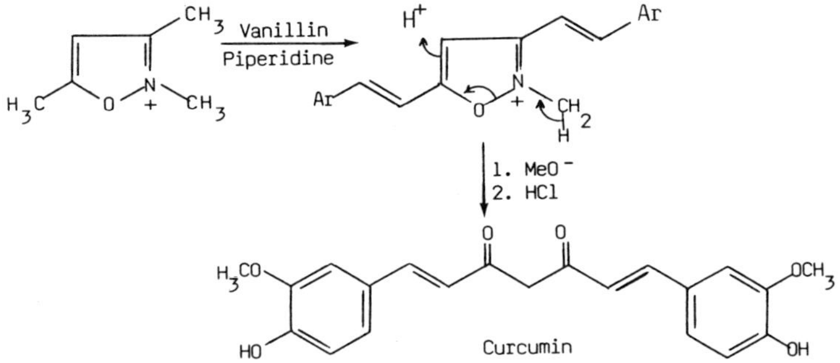

(1.22)

Noteworthy is the selective reductive cleavage of the N—O bond in the isoxazoles, because this reaction leads to unmasking of the versatile 1,3-dicarbonyl or the 1,3-iminocarbonyl systems, eq. (1.23), which subsequently can be integrated into molecular structures in various ways. Equation (1.25) shows how the reduced isoxazole unit is built into a novel, condensed heterocyclic system. The reduction can be effected by sodium in alcohol,[59,72] sodium amalgam,[73] Grignard reagents,[74,75] $Fe(CO)_5$ and $Mo(CO)_6$,[76a] dihydrolipoamide-iron II,[76b] electrochemistry[77,78] and now, most frequently, by catalytic reduction over Pd, Pt, or Raney-Ni.[13,79-88] Zinc in ethanol[52] does not affect the isoxazole ring, nor does treatment with lithium aluminum hydride,[89] which can be used for selective reduction of ring substituents, e.g., acyl groups to the corresponding carbinols (cf. Section 1.6). However, lithium aluminum hydride can selectively reduce the 4,5 double bond in isoxazoles functionalized at C^4 with electron attracting groups, eq. (1.24).[90]

A number of UV[80,91-94] and IR spectra[95,96] of isoxazoles have been recorded and analyzed. The NMR spectra of the parent isoxazole show the following data: ^{1}H NMR (CCl$_4$): δ 6.32 (H^4, t, J = 1.6 Hz), 8.19 (H^3, d, J = 1.6 Hz), 8.44 (H^5, d, J = 1.6 Hz).[97] ^{13}C NMR (CDCl$_3$): δ 103.7 (C^4), 149.1 (C^3), 157.9 (C^5).[98]

The prototropic tautomerism of hydroxy- and amino-substituted isoxazoles have been studied by these spectral methods.[99] The following general conclusions can be drawn:

1. Amino derivatives irrespective of position occur in the amino form.

14   Introduction and General Considerations

2. 3-Hydroxy derivatives occur in the hydroxy form.
3. 5-Hydroxy derivatives occur predominantly in the oxo form as a mixture of **A** and **B**, eq. (1.26).

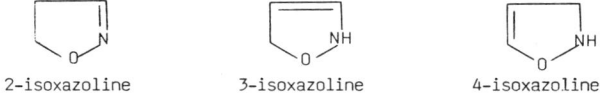

$$\underset{\underline{A}}{} \rightleftharpoons \underset{\underline{B}}{} \rightleftharpoons \underset{\underline{C}}{} \qquad (1.26)$$

Earlier contributions to the physicochemical properties of isoxazoles and derivatives have been reviewed (Ref. 29).

Conclusion: The isoxazoles contain a protected 1,3-dicarbonyl function with a certain lability associated with the N—O bond. They can be functionalized, readily unmasked, and elaborated for further structural operations; i.e., they fulfill requirements 2–5 (Section 1.2) of a synthetic tool with wide applications.

1.5 ISOXAZOLINES. REACTIONS AND PHYSICOCHEMICAL PROPERTIES. THE MASKED FUNCTIONALITY AND THE ALDOL CONCEPT

The isoxazolines occur in three isomeric forms: 2-, 3-, and 4-isoxazolines, of which the 2-isoxazolines are the most common, most stable, and, as we shall see also synthetically, the most versatile. Loss of aromaticity makes the isoxazolines

more susceptible to various reagents. 2-Isoxazolines are readily available by 1,3-dipolar addition of nitrile oxides or silyl nitronates to olefins. A promising route to 2-isoxazolines, still unexploited, is nitrosation of cyclopropanes accompanied by ring cleavage and recylization,[100,101] eq. (1.27). They are easily dehydrogenated by chromic acid,[102–105] *N*-bromosuccinimide,[106–108] manganese dioxide,[109,110] potassium permanganate,[111,112] and dichlorodicyanobenzoquinone.[113,114] The aromatization of 2-isoxazolines to isoxazoles is important because 1,3-dipolar addition is easier with olefins than with the corresponding acetylenes, and olefins are generally more available. Isoxazolines are stable towards peracids. It is thus possible to epoxidize a vinyl substituent without oxidzing the heterocycle, eq. (1.28).[114] The C^4 and C^3 αH atoms of 2-isoxazolines are activated by the iminoxy function, and treatment with lithium isopropylamide at $-78°C$ can generate an anion at C^4 or $C^{3\alpha}$ which is prone to react with electrophiles, eq. (1.29).[115-118,120] The attack occurs predominantly from the less hindered side—cis to C^5—H. At higher temperatures the 2-isoxazoline ring is cleaved and α,β-unsaturated oximes are formed. It is thus possible to functionalize 2-isoxazolines under certain conditions. Isoxazolines with hydrogen at C^3 are cleaved by bases and α,β-cyanohydroxy derivatives are formed, eq. (1.30).[25,114,161]

Thermal decarboxylation of 3-carboxy-2-isoxazolines causes a similar fragmentation.[126] Treatment of 5-acylated 2-isoxazolines with base[121,122] or acid[134] also causes cleavage of the ring, according to eq. (1.31). Related to this cleavage is the conversion of 5-trimethylsilyl-2-isoxazolines into silylenol ethers by flash-vapor pyrolysis.[141]

2-Isoxazolines are quite stable at ordinary temperatures, but cycloreversions have been observed under flash-vacuum conditions at 600–650°C[123] and by irradiation.[124,125] 2-Isoxazolines are weak bases and are *N*-alkylated by dimethyl sulfate, eq. (1.32)[25,127–129], or triethyloxonium tetrafluoroborate, eq. (1.33).[119] The ammonium salts can be rearranged into nitrones[25] or 3-isoxazolines.[119,127,128]

$$R^1CN + R^2\text{-CO-} \quad (1.31)$$

$$(1.32)$$

$$(1.33)$$

The reductive cleavage of the N—O bond can be accomplished readily and selectively by various methods. Catalytic reduction over Pd, Pt, or Raney-Ni (most frequently used) cleaves the N—O bond without reducing the imino function, eq. (1.34)[24,130–140]; cf. catalytic reduction of isoxazoles in Section 1.4. Catalytic reduction of 2-isoxazolines with a carboxy group at the 3-position is difficult to stop at the imino level. In this case a mixture of stereoisomeric amino acids is obtained, eq. (1.35).[142,143] Titanous salts are mild but slow reductants for selective N—O bond cleavage,[25,114,133,134,144,145] and this is a useful procedure, especially when the 2-isoxazolines contain olefinic functions, which are reduced catalytically. The N—O bond is also cleaved with zinc powder and acid.[25,133,160] Under certain conditions[133] simultaneous cleavage of the C—O bond also occurs, eq. (1.36). Electrolytic reduction under controlled potential to the aldol stage has also been achieved.[78,245] In contrast to isoxazoles, 2-isoxazolines are readily reduced to 1,3-aminoalcohols by lithium aluminum hydride,[114,146–158] borane complexes,[153,159] sodium amalgam, and sodium in ethanol, eq. (1.37).[153] Selective reduction of the imine bond without cleavage of the N—O bond is difficult to accomplish. It has been reported that sodium borohydride reduces 5-phenyl-2-isoxazoline-3-carboxylic acid to the corresponding isoxazolidine, eq. (1.38).[160]

$$(1.34)$$

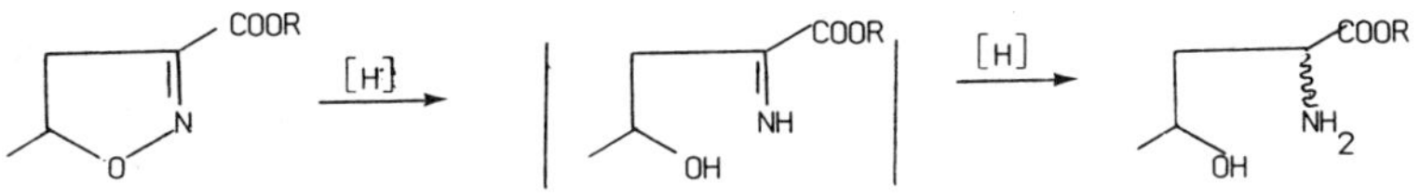

The Masked Functionality and the Aldol Concept

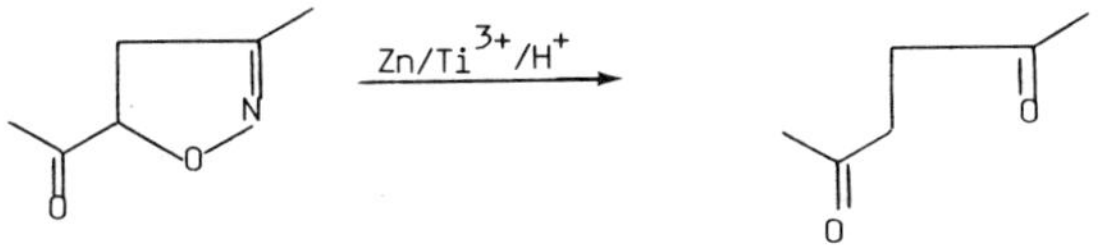

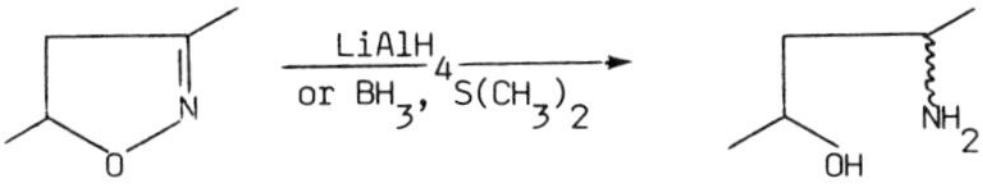

$$\text{isoxazoline-COOR} \xrightarrow{[H]} \text{HO-CH-CH}_2\text{-C(=NH)-COOR} \xrightarrow{[H]} \text{HO-CH-CH}_2\text{-CH(NH}_2\text{)-COOR} \quad (1.35)$$

$$\text{acyl-isoxazoline} \xrightarrow{\text{Zn/Ti}^{3+}/\text{H}^+} \text{1,3-diketone} \quad (1.36)$$

$$\text{isoxazoline} \xrightarrow[\text{or BH}_3,\ \text{S(CH}_3)_2]{\text{LiAlH}_4} \text{1,3-aminoalcohol} \quad (1.37)$$

$$\text{Ph-isoxazoline-COOH} \xrightarrow{\text{NaBH}_4} \text{Ph-isoxazolidine-COOH} \quad (1.38)$$

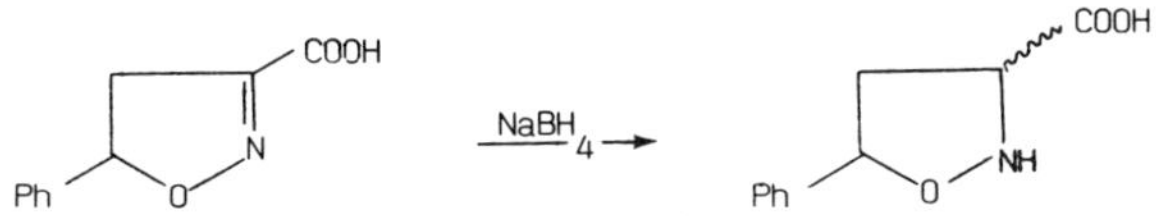

Reduction equation (1.34) represents the unmasking of the protected function and results in the familiar aldol moiety, which normally is the product of the classical Claisen reaction. Thus, a novel entrance[24,25,114,115,135,137] is opened to the synthetically important aldols, and quite different starting material can be used for the purpose: inexpensive and readily available olefins, aldehydes, and primary nitro compounds (Scheme 1.2). The routes are complementary, and the choice a matter of synthetic planning. Complete reduction of 2-isoxazolines gives 1,3-aminoalcohols (route C).

Scheme 1.2. *Routes to aldols and the α,β-unsaturated carbonyl system. A: Claisen route. B: Isoxazoline route.*

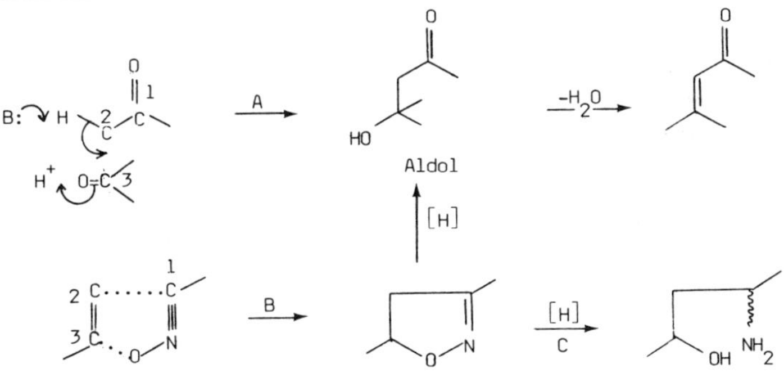

The IR spectra of 2-isoxazolines show an absorption in the 1570–1630 cm^{-1} region (w–m) characteristic of the C=N stretching.[107,117,161] A great number of ^{1}H NMR data have been reported; see, e.g., Refs. 15, 24, 25, 107, 114, 117, 134, 152, 153, 161. The ^{13}C NMR data are as yet scarce.[119,132,162]

Few papers have been devoted to the isomeric 3-isoxazolines. It is reported that some 5-unsubstituted isoxazolium salts react with Grignard reagents to form 5-substituted 3-isoxazolines, eq. (1.39).[163] They are also formed by N-alkylation of 2-isoxazolines with dimethyl sulfate or triethyloxonium tetrafluoroborate and subsequent removal of a C^4-proton by trimethylamine or sodium methoxide, eq. (1.33).[119,127,128] The compounds can be distilled in vacuo and appear to be stable up to ca. 100°C. In contrast to the enamines, which are alkylated at the β-position, the 3-isoxazolines are alkylated on nitrogen to give enammonium salts, eq. (1.40).

Although the 3-isoxazolines are unimportant in the context of synthetic intermediates, the 4-isoxazolines[164] attract some interest. The lability of the heterocycle is partly associated with the weak N—O bond and partly with the enol moiety. Characteristic of the 4-isoxazolines is their tendency for rather capricious and unpredictable rearrangements into other heterocycles such as 4-oxazolines, aziridines, pyrroles, etc., eq (1.41).[165-168] This is astonishing in view of the simple structure of the 4-isoxazolines, and early workers in the field were certainly not prepared for the unexpected lability of the system. 4-Isoxazolines are best prepared by 1,3-dipolar addition of nitrones to acetylenes, but the reaction is sluggish with acetylenes that do not contain electron withdrawing groups. There is less regioselectivity in the 1,3-dipolar addition with acetylenes than with alkenes. Reaction of Grignard reagents and sodium borohydride with some isoxazolium salts gives 4-isoxazolines.[167,168]

The reductive cleavage of the 4-isoxazoline ring and subsequent elimination of amine is expected to give α,β-unsaturated carbonyl compounds, eq. (1.42). This has been achieved, but the carbonyl function can induce retroaldol reactions, eq. (1.43).[164,169] Lithium aluminum hydride reduction, which is supposed to give 1,3-aminoalcohols, also gives unpredictable products, i.e., aziridines, eq. (1.43). Some simple 4-isoxazolines are reported to be inert towards lithium aluminum hydride, and 4-isoxazolines are actually the products from lithium aluminum hydride reductions of N-methylisoxazolium salts, eq. (1.44).[170] If an electron withdrawing group is attached to C^4, a mixture of 3-, 4-isoxazolines and isoxazolidine is obtained.

The IR spectra of 4-isoxazolines show a characteristic absorption (w-m) in the 1615–1680 cm^{-1} region originating from the enol group. In the ^{1}H NMR spectra the C^4H and C^5H are located in the regions of δ 5.5 and δ 7.0, respectively.[164]

Conclusion: 2-Isoxazolines contain a masked aldol moiety. This class of compounds is formed from inexpensive accessible starting material by simple procedures. They are reasonably stable, can be functionalized, allow steric operations, and, consequently, are suitable as versatile structural entities in organic synthesis in concordance with the requirements presented in Section 1.2. 4-Isoxazolines are less suitable as relay compounds in organic synthesis, partly as a result of inaccessibility, and partly as a result of their unpredictable reactions. 3-Isoxazolines are not suitable as synthetic intermediates.

1.6 ISOXAZOLIDINES.[171] REACTIONS AND PHYSICOCHEMICAL PROPERTIES. CYCLOREVERSIONS AND THE MASKED FUNCTIONALITY

1,3-Dipolar addition of nitrones to olefins is the most useful method for synthesizing isoxazolidines. The addition proceeds smoothly at room temperature in high yields and within hours with reactive nitrones (e.g., C-acylnitrones) and olefins activated with electron withdrawing groups, e.g. CN, COOCH$_3$, RCO, etc.[172-175] Heating to ca. 100°C for 1–2 days is necessary for combining less reactive compounds. Cyclization to the saturated isoxazolidine ring forms isomers. Each regioselective route, eq. (1.45a,b), gives one pair of stereoisomers, i.e., four products. Degeneracy (e.g., $R^1 = R^2$), eq. (1.45), or strong regio- and stereoselectivity, which often occurs, decreases the number of products (Section 1.7).

(1.45)

The inversion barrier of nitrogen is usually low, and it is not possible to observe it on the time scale of the NMR spectrometer. Under certain conditions the inversion barrier is so high that invertomers can be isolated. This occurs because ring strain combines with electron repulsion in the inversion transition state in N-chloroaziridine (5)[176,177] and in N-methoxyaziridine (6),[178] or solely because of electron repulsion in the inversion transition state in N-alkoxyisoxazolidines.[179-181] For the rapidly inverting N-methylisoxazolidine (7) ($R^1, R^2, R^3, R^4 = H$), the activation energy is $E_a = 15.6 \pm 0.5$ kcal mol^{-1},[182] and for (7) ($R^1, R^2 = NO_2, R^3, R^4 = CH_3$) $E_a = 14.6 \pm 0.5$ kcal mol^{-1},[183] according to ^{1}H NMR measurements. For the more stable N-methoxy-3,3-dimethoxycarbonyl-5-cyanoisoxazolidine, eq. (1.46, **8a** ⇌ **8b**) and N-methoxy-3,5-

dicyanoisoxazolidine (**9**), the activation energies are 29.7 ± 0.8 kcal mol^{-1},[179] and 28.7 ± 0.2 kcal mol^{-1},[181] respectively. The half-life of **8** at 29°C is 47 days. The conformation with the methoxy group in axial position, depicted in eq. (1.46), is the most likely one based on the anomeric effect. *N*-Trimethylsilyloxyisoxazolidines behave in the same way. Two invertomers of **10** were observed spectroscopically. Both give the same product **11**, 5-phenyl-2-isoxazoline, by acid-catalyzed elimination of trimethylsilanol, eq. (1.47).[24] The elimination of alcohol is a general and facile reaction characteristic for *N*-alkoxy and *N*-silyloxyisoxazolidines.[24,25,114,134,145,184-192] Cleavage of the cyclic N—O bond also occurs on acid treatment. Thus, the isoxazolidine **13** gives a mixture of 2-isoxazoline **14** and iminoxy alcohol **15**, eq. (1.48).[185] Treating *N*-methoxyisoxazolidines with sodium or potassium hydroxide has been reported to give 2-isoxazolidines[189] by eliminating methanol. The *N*-trimethylsilyloxyisoxazolidine **10** undergoes ring cleavage in basic methanol, eq. (1.47).[24,186]

Compound **12** is ultimatively obtained by hydrolysis of the intermediate oxime. Ring opening of the isoxazolidinium salt under basic conditions gives **16** and **17**, eq. (1.49).[193–196] Another mode of cleavage has been observed with lithium aluminum hydride,[196] eq. (1.50).

$$\text{13} \xrightarrow{H^+} \text{14} + \text{15} \quad (1.48)$$

$$\text{(structure)} \xrightarrow{OH^-} RCOCH_2-CH\overset{Ph}{\underset{N(CH_3)_2}{}} + RCOCH=CHPh \quad (1.49)$$

$$\text{16} \qquad \text{17}$$

$$\text{(structure)} \xrightarrow{LiAlH_4} \text{(structure)} \quad (1.50)$$

Isoxazolidines are thermally unstable because they cleave in a retro-1,3-dipolar cycloaddition.[173,197–212,326] On distillation in vacuo above 100–150°C, isoxazolines slowly reform nitrone and olefin. That means that a kinetically controlled addition mixture [see eq. (1.51)] will slowly change its composition in favor of the most stable derivatives on prolonged heating. It was observed that the trans-fused isoxazolidine **18** on thermal equilibration gave a mixture of two cis-fused isoxazolidines **19** and **20**, eq. (1.51).[202,203] An attempted distillation of isoxazolidine **21** caused practically complete fragmentation into the starting material, cyclohexanone oxime and methyl acrylate, eq. (1.52).[206] The initial intramolecular cycloaddition product **22** obtained in refluxing toluene rearranges thermally at 195°C into the spiroisomer **23**, eq. (1.53).[209]

$$\text{18} \overset{\Delta}{\rightleftharpoons} \text{19} + \text{20} \quad (1.51)$$

$$\text{21} \overset{\Delta}{\rightleftharpoons} \text{(cyclohexanone oxime)} + \text{COOCH}_3 \quad (1.52)$$

$$(1.53)$$

Retrocycloaddition is especially facile in tetrasubstituted isoxazolidines which does not permit isolation of adducts.[205] The retrocycloaddition has been applied as a step in the synthesis of cocaine for temporary protection of a nitrone function, eq. (1.54).[213] 1,3-Dipolar retrocycloadditions were recently reviewed.[214]

$$(1.54)$$

Few systematic investigations of the action of oxidants on isoxazolidines have been carried out. Demethylation of N-methylisoxazolidines by peracids gives oxazines via ring cleavage and formation of an intermediate nitrone, eq. (1.55).[193,213,215–217] Oxidation of the parent isoxazolidine with lead tetraacetate gives 2-isoxazoline, eq. (1.56).[218] The isoxazolidine ring is decomposed by potassium permanganate[171] and chromic acid[219] but is resistant to oxidation with dimethylsulfoxide; oxidations in the side chain can be achieved with this reagent.[219]

$$(1.55)$$

$$\text{(structure 24)} \xrightarrow{Pb(OAc)_4} \text{(product)} \qquad (1.56)$$

24

The reductive cleavage, which unmasks the 1,3-aminoalcohol, eq. (1.57), can be easily achieved by catalytic hydrogenation over Raney-Ni,[24,173,206,220-222] Pd,[173,219,223-234] Pt,[173,227,235,236] and Rh,[237,238] by aluminum amalgam[173] sodium in ammonia or alcohol,[173] by zinc and acetic acid,[173,213,236,239-242] by lithium aluminum hydride,[173,235,242,243,244] by Fe^{2+}-dihydrolipoamide,[76b] and by electrolytic reduction.[78,245] Under certain conditions lithium aluminum hydride is able to reduce selectivity an ester function in the isoxazolidine without cleavage of the N—O bond, eq. (1.58).[173,233,246]

$$\text{isoxazolidine-NR} \xrightarrow{[H]} \text{NHR, OH} \xrightarrow{[H]} \text{NH}_2, \text{OH} \qquad (1.57)$$

$(R = CH_2C_6H_5)$

$$\text{(ester isoxazolidine)} \xrightarrow{LiAlH_4} \text{(hydroxymethyl isoxazolidine)} \qquad (1.58)^{173}$$

To obtain the primary amines from reaction (1.57), the cycloaddition is carried out with N-benzyl nitrones. The benzyl group is reductively eliminated under concomitant opening of the ring. Alternatively, the N-alkyl group can be eliminated oxidatively, eq. (1.55). Hydrolysis of the oxazine and reduction of the hydroxylamine give the primary amine.[213]

Isoxazolidines are stronger bases than isoxazoles and isoxazolines. The pK_a of the parent isoxazolidine hydrochloride 24 is 5.05.[247]

Infrared and ultraviolet spectra are of little diagnostic value. ^{1}H NMR shifts and couplings for two monocyclic isoxazolidines follow:

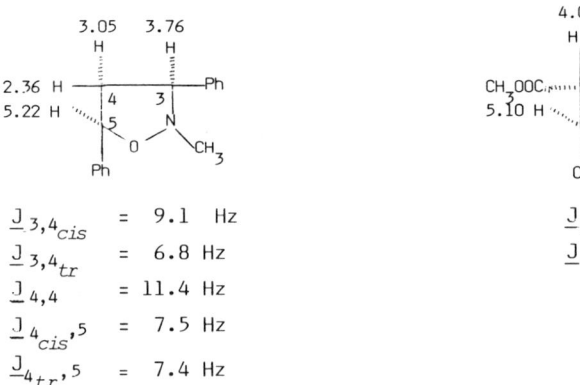

$J_{3,4_{cis}}$ = 9.1 Hz
$J_{3,4_{tr}}$ = 6.8 Hz
$J_{4,4}$ = 11.4 Hz
$J_{4_{cis},5}$ = 7.5 Hz
$J_{4_{tr},5}$ = 7.4 Hz

$J_{3,4}$ = 6.4 Hz
$J_{4,5}$ = 4.8 Hz

The stereostructure of the isoxazolidines can often be deduced from the coupling constant,[173] $J_{cis} > J_{tr}$.

Conclusion: The isoxazolidines are accessible from simple starting material: olefins, aldehydes, ketones, nitro compounds, and hydroxylamines. They are reasonably stable, undergo thermal retroaddition, allow considerable steric control during the cycloaddition, and can be functionalized to a limited extent. The unmasking of the important 1,3-aminoalcohol moiety is achieved by reductive cleavage. The intramolecular cycloaddition of isoxazolidines is an important method for construction of carbocyclic rings in the synthesis of natural products. It is therefore expected that the isoxazolidines will attract considerable interest as relay compounds in organic synthesis.

1.7 MECHANISMS OF THE 1,3-DIPOLAR CYCLOADDITION. STEREOCHEMICAL AND REGIOCHEMICAL CONSIDERATIONS. ASYMMETRIC INDUCTION

There are certain characteristic features of the 1,3-dipolar cycloadditions irrespective of the structures of the reactants.[172] The reactants are oriented in a two-plane complex and interact via their π-orbitals in a π^4s-π^2s process, eq. (1.59). Much attention has been given to the problem about the timing of c–d versus a–c bond formation. Is it synchronized or are the bonds formed successively? Calculations on transition state (TS) geometry give ambiguous results. They depend on the method chosen.[358] Ab initio calculation favors a symmetrical or close to symmetrical (and early) transition state with synchronized bond formation, whereas parameterized MINDO calculations result in a highly unsymmetrical and late transition state of zwitterionic or biradical character. Apparently calculations are not refined enough to allow reliable quantitative predictions.

(1.59)

A wealth of existing experimental data permits several conclusions to be drawn. Second-order rate constants for the cycloaddition of benzonitrile oxide to various acetylenes are throughout smaller than those for the corresponding ethylenes. Consequently, the reactions with acetylenes, which give aromatic isoxazoles, do not profit from gain in aromaticity, and the transition state is therefore located early on the reaction path,[259] eq. (1.60):

$$\text{(1.60)}$$

aromatic   nonaromatic

The effects of solvent polarity on the reaction rate are small, which indicates the absence of zwitterionic intermediates in the TS. The solvation energy of the reactants is thus similar to that of the TS. The large negative entropy of activation,[174,353-357] ΔS^* ca. -30 eu, reveals a highly ordered TS resembling that of the Diels-Alder reaction. The activation enthalpies are modest: ΔH^* ca. 15 kcal mol^{-1}. All these data favor an early TS and synchronized formation of bonds.

Preserving the steric structure of the reactants in the product formed is a criterion for synchronized a–e and c–d bond formation, eq. (1.59). If stepwise formation of bonds occurs, i.e., formation of zwitterionic or biradical intermediates, there is also a possibility for bond rotation and loss of stereospecificity. *Cis*- and *trans*-isomeric dipolarophiles or dipoles should presumably to some extent give rise to isomerized cycloadducts, eq. (1.61), unless the rotation is slow—rotation barrier high—in comparison to the rate of cyclization. Careful analysis of the reaction product by gas chromatography demonstrates the very high stereospecificity of diazomethane addition to methyl angelate and methyl tiglate,[359a] eq. (1.62).

$$\text{(1.61)}$$

zwitterion   biradical   cis   trans
                                   mixture of isomers

The calculated rotation barriers for the hypothetical biradical intermediates, (ca. 4–6 kcal mol^{-1}) are considerably higher than anticipated for an alkyl radical. Barriers to rotation of simple primary, secondary, and tertiary alkyl radicals are in the range of 0–1.2 kcal mol^{-1}. Experiments with cycloadditions of *p*-nitrobenzonitrile oxide with *cis*- and *trans*-1,2-dideuteroethylenes establish a stereospecificity of at least 98%, which means that the rotation barrier would have to be at least 2.3 kcal mol^{-1} higher than the cyclization barrier.[359b] Since the rotation barrier for the primary radical [eq. (1.61), R = H] is only 0.4 kcal mol^{-1} or less, there can actually be no cyclization barrier. These data

[Structures showing two stereospecific cycloaddition reactions with stereospecificity >99.94% and >99.97%]

(1.62)

make the biradical structure unlikely. They offer strong evidence for a concerted reaction, i.e., for a symmetrical or at least a nearly symmetrical TS. In summary: The investigations of stereospecificity of the 1,3-dipolar cycloaddition support synchronized bond formation and so do the frontier molecular orbital (FMO) calculations,[2,172,259,269–272,358] which satisfactorily explain reactivities (Section 1.8) and the regiochemical and stereochemical phenomena of the cycloaddition.

In Tables 1.1–1.4 some representative experimental data on the regio-and stereoselectivity of the cycloaddition are collected. In addition to the investigations cited there a great number of structural studies have been carried out. See Refs. 175, 181, 208, 255, 257, 263, 266, 267, 276–294, 330–351, 360–367, 373–377. The following general conclusions can be drawn from the extensive experimental work:

1. Monosubstituted olefins and acetylenes show high regioselectivity and give 5-substituted derivatives for both electron donating (e.g., —OR) and electron withdrawing groups (e.g., —COOR). Very strong electron withdrawing groups, e.g., —SO$_2$R, give predominantly 4-substituted derivatives.
2. 1,1-Disubstituted olefins show high regioselectivity and give 5,5-disubstituted products. Strong electron withdrawing groups give 4,4-disubstituted products.
3. 1,2-Disubstituted olefins and acetylenes give mixtures of regioisomers.
4. Both electron withdrawing and electron donating groups and strain in the dipolarophiles increase the reactivity of the dipolarophiles.
5. The addition is a concerted cis-addition (suprafacial).
6. The regioselectivity of acetylenes are less pronounced than that of olefins.
7. The stereoisomeric ratio of the nitrone additions varies widely and unpredictably with the substituents.

Several of these points need comment. It was argued from polarizability considerations that a dipolarophile like vinyl ethyl ether with an electron

Table 1.1. Orientation of Monosubstituted Ethylenes and Acetylenes in the 1,3-Dipolar Addition with Nitrile Oxides, Eqs. (1.63), (1.64)

Olefin, acetylene	Nitrile oxide, RCNO	Ratio 5:4-substitution	Ref.
=/COOCH$_3$	CH$_3$	95:5	161
=/COOCH$_3$	C$_6$H$_5$	95:5	161
=/	C$_6$H$_5$	~100:0	249
=/CH$_3$	C$_6$H$_5$	~100:0	250
=/OBu	C$_6$H$_5$	~100:0	250
=/Ph	C$_6$H$_5$	99.5:0.5	251
=/SO$_2$Ph	3,5-Cl$_2$-2,4,6-(CH$_3$)$_3$C$_6$	91:9	252
≡—Bu	C$_6$H$_5$	~100:0	250
≡—COOCH$_3$	CH$_3$	69:31	161
≡—COOCH$_3$	H	84:16	161

$$\diagdown_X + RCNO \longrightarrow \underset{X}{\overset{R}{\boxed{}}}_{5\;N}^{\;\;O} + \underset{X}{\overset{R}{\boxed{}}}_{4\;N}^{\;\;O} \qquad (1.63)$$

$$\equiv -X + RCNO \longrightarrow \underset{X}{\overset{R}{\boxed{}}}_{5\;N}^{\;\;O} + \underset{X}{\overset{R}{\boxed{}}}_{4\;N}^{\;\;O} \qquad (1.64)$$

donating substituent and a dipolarophile like methyl acrylate with an electron withdrawing substituent should align themselves in the opposite direction to the 1,3-dipole. This is not the case, and only 5-substituted isoxazolines are obtained from monosubstituted ethylenes with nitrile oxides (Table 1.1). However, it was observed that with very strong electron withdrawing groups, such as the sulfono group, the 4:5-substitution ratio increased. This effect is more evident for the reactions with nitrones—e.g., the reactions between methyl acrylate, acrylonitrile, phenyl vinyl sulfone, nitroethylene, and the cyclopropyl nitrones (Table 1.2), and the reactions with sulfide, sulfoxide, and sulfone (Table 1.3); the reactions with *t*-butylnitrone seem not to follow the general trend. Furthermore, it was difficult to understand why both electron withdrawing and electron donating groups on ethylene should increase the reactivity of the dipolarophile.[259] This is a case where traditional organic reaction mechanism is incapable of giving a satisfactory explanation.

Table 1.2. Orientation of Monosubstituted Olefins and Acetylenes in the 1,3-Dipolar Addition with Nitrones and Nitronates, Eqs. (1.65), (1.66)

Olefin acetylene	Nitrone or nitronate $R^1R^2C=N(O)R^3$	Ratio 5:4-substitution	Ref.
=/Ph	$R^1 = R^3 = Ph, R^2 = H$	~100:0[a]	173
=/Ph	$R^1 = Ph, R^2 = H, R^3 = CH_3$	~100:0[b]	173
=/OBu	$R^1 = Ph, R^2 = H, R^3 = CH_3$	~100:0	173
=/CH_2OH	$R^1 = R^3 = Ph, R^2 = H$	~100:0	173
=/CN	$R^1 = R^3 = Ph, R^2 = H$	~100:0[c]	173
=/$COOCH_3$	$R^1 = R^3 = Ph, R^2 = H$	~100:0[d]	173
=/Ph	$R^1 = $ cyclopropyl, $R^2 = H, R^3 = CH_3$	~100:0[e]	253
=/$COOCH_3$	$R^1 = $ cyclopropyl, $R^2 = H, R^3 = CH_3$	80:20[e]	253
=/CN	$R^1 = $ cyclopropyl, $R^2 = H, R^3 = CH_3$	67:33[e]	253
=/SO_2Ph	$R^1 = $ cyclopropyl, $R^2 = H, R^3 = CH_3$	38:62[e]	253
=/$COOCH_3$	$R^1 = R^2 = $ cyclopropyl, $R^3 = CH_3$	50:50	253
=/CN	$R^1 = R^2 = $ cyclopropyl, $R^3 = CH_3$	25:75	253
=/SO_2Ph	$R^1 = R^2 = $ cyclopropyl, $R^3 = CH_3$	~0:100	253
=/NO_2	$R^1 = Ph, R^2 = H, R^3 = CH_3$	~0:100[f]	254
=/NO_2	$R^1 = R^2 = H, R^3 = t$-Bu	~100:0	254
=/OEt	$R^1 = Ph, R^2 = H, R^3 = CH_3$	~100:0[g]	256

Table 1.2. (continued)

Olefin acetylene	Nitrone or nitronate $\begin{matrix}R^1\\R^2\end{matrix}{=}N{<}\begin{matrix}O\\R^3\end{matrix}$	Ratio 5:4-substitution	Ref.
=/OAc	$R^1 = R^3$ = Ph, R^2 = H	~100:0[h]	256
=/SO$_2$Ph	$R^1 = R^2$ = H, R^3 = t-Bu	70:30	254
=/SO$_2$Ph	R^1 = p-OCH$_3$-Ph, R^2 = H, R^3 = Ph	~0:100[i]	262
=/OEt	R^1 = COOEt, R^2 = H, R^3 = CH$_3$	~100:0[i]	256
=/OEt	R^1 = H, R^2 = COOEt, R^3 = CH$_3$	~100:0[i]	256
=/SO$_2$Ph	R^1 = Ph, R^2 = H, R^3 = CH$_3$	32:68	254
=/COOCH$_3$	$R^1 = R^2$ = CH$_3$, R^3 = CH$_2$CH$_2$COOCH$_3$	80:20	206
≡—COOCH$_3$	R^1 = Ph, R^2 = H, R^3 = CH$_3$	42:52	257
≡—COOC$_2$H$_5$	$R^1 = R^2$ = H, R^3 = t-Bu	70:30	254
≡—CN	R^1 = Ph, R^2 = H, R^3 = CH$_3$	~100:0	254
≡—CN	$R^1 = R^2$ = H, R^3 = t-Bu	50:50	254
=/CN	$R^1 = R^2$ = H, R^3 = t-Bu	~100:0	254
=/COOCH$_3$	R^1 = CH$_3$, R^2 = H, R^3 = OSi(CH$_3$)$_3$	~100:0	24
=/	$R^1 = R^2$ = H, R^3 = OSi(CH$_3$)$_3$	~100:0	114
=/COOH	$R^1 = R^2$ = NO$_2$, R^3 = OCH$_3$	~100:0	192
=/CH$_3$	⟨⟩N—O	~100:0[j]	274
=/CH$_2$OH	⟨⟩N—O	~95:5[k]	274
=/Ph	⟨⟩N—O	~100:0[l]	274

Table 1.2. (continued)

Olefin acetylene	Nitrone or nitronate $R^1R^2C{=}N(O)R^3$	Ratio 5:4-substitution	Ref.
CH$_2$=CH–COOCH$_3$	(piperidine N-oxide)	~0:100^m	274

^{a}C^3-Ph:C^5-Ph/cis:tr (exo:endo), 90:10.
^{b}C^3-Ph:C^5-Ph, cis:tr, 67:33.
^{c}C^3-Ph:C^5-CN, cis:tr, ~100:0.
^{d}Two stereoisomers, ~57:43.
^{e}Each regioisomer is a mixture of cis:tr stereoisomers.
^{f}cis:tr, 2:1.
^{g}cis:tr, 1:1.
^{h}One stereoisomer, cis.
^{i}One isomer, tr.
^{j}cis:tr (endo:exo), ~0:100.
^{k}cis:tr < 5:95.
^{l}cis:tr = 22:78.
^{m}cis:tr = 18:82.

$$\text{CH}_2{=}\text{CHX} \;+\; R^1R^2C{=}N(O)R^3 \;\longrightarrow\; \text{(5-X isoxazolidine)} \;+\; \text{(4-X isoxazolidine)} \qquad (1.65)$$

$$\text{HC}{\equiv}\text{CX} \;+\; R^1R^2C{=}N(O)R^3 \;\longrightarrow\; \text{(5-X isoxazoline)} \;+\; \text{(4-X isoxazoline)} \qquad (1.66)$$

Table 1.3. Orientation of Disubstituted Olefins and Acetylenes in the 1,3-Dipolar Addition with Nitrile Oxides, Eq. (1.67)

Olefin, acetylene (y)CH=CH(x)	Nitrile oxide, RCNO	Product, ratio, A:B	Ref.
(y)Ph–CH=CH–COOCH$_3$ (x)	Ph	30:70	161
Ph–CH=CH–COOCH$_3$	CN	15:85	161
CH$_3$–CH=CH–COOCH$_3$	CH$_3$	36:64	161
CH$_3$–CH=CH–COOCH$_3$	H	62:38	161

Table 1.3. (continued)

Olefin, acetylene	Nitrile oxide, RCNO	Product, ratio, A:B	Ref.
H₃C-C(=O)-CH=CH-Ph	CH_3	55:45	275
Ph-C(=O)-CH=CH-CH₃	Ph	68:32	275
cyclopentanone with =O exocyclic	Ph	9:91	275
cyclohexanone with =O exocyclic	Ph	25:75	275
CH₃-CH=CH-SCH₃	Ph	96:4	258
2,3-dihydrofuran	Ph	99.4:0.6	258
2,3-dihydrothiophene	Ph	88:12	258
Ph-CH=CH-SCH₃	Ph	97:3	251
Ph-CH=CH-SOCH₃	Ph	45:55	251
Ph-CH=CH-SO₂CH₃	Ph	29:71	251
Ph-CH=CH-SCH₃ (cis)	Ph	~100:0	251
Ph-CH=CH-SOCH₃ (cis)	Ph	90:10	251
Ph-CH=CH-SO₂CH₃ (cis)	Ph	22:78	251
CH₃-C(CH₃)=CH-CH₃ (with CH₃)	Ph	37:63	138

Table 1.3. (Continued)

Olefin, acetylene	Nitrile oxide, RCNO	Product, ratio, A:B	Ref.
H₃C-C(CH₃)=CH-CH₃ (with CH₃)	Ph	0:100	138
Ph-C(CH₃)=CH-CH₃	Ph	51:49	259
Ph-C(pyrrolidinyl)=CH₂	Ph	~100:0	259
Ph-CH=CH-NO₂	Ph	33:67	250
CH₃-≡-COOCH₃	Ph	1.3:98.7	259
Ph-≡-COOCH₃	Ph	1.2:98.8	259
CH₂=C(COOCH₃)(CH₃)	Ph	~100(5,5):0	259
CH₃OOC-CH=C(CH₃)₂	Ph	~100(5,5-di-Me):0	259
CH₂=C(Cl)(Cl)	Ph	~100:0ᵃ	260

ᵃThe product is 3-phenyl-5-chloroisoxazole.

$$\begin{array}{c} \underset{y}{\overset{x}{\diagup}}= \end{array} + RC\equiv NO \longrightarrow \underset{A}{\begin{array}{c} y \\ \diagdown 4 \\ 5 \diagup \\ x \diagdown O \diagup N \end{array}}^{R} + \underset{B}{\begin{array}{c} x \\ \diagdown 4 \\ 5 \diagup \\ y \diagdown O \diagup N \end{array}}^{R} \quad (1.67)$$

Fukui's frontier orbital concept[268] provides us with a rationalization of the experimental results.[2,172,269-272,358] The expression for the energy change in molecules undergoing cycloaddition is a sum of three terms: a closed-shell repulsion term, a Coulombic term, and an interorbital overlapping term.[273] As a first approximation, the two first terms can be disregarded for a qualitative discussion. The third term is an expression of the generalized statement that

Table 1.4. Orientation of Disubstituted Olefins in the 1,3-Dipolar Addition of Nitrones, Eqs. (1.68), (1.69)

Olefin $\overset{x}{\underset{y}{\diagup}}\!=\!\diagup$ $\overset{x}{\diagup}\!=\!\underset{y}{\diagdown}^{x}$	Nitrone $\begin{array}{c}R^1\diagdown\quad\diagup O\\ \quad=N\\ R^2\diagup\quad\diagdown R^3\end{array}$	Product ratio A:B	Ref.
(y) H₃C−CH=CH−COOCH₃ (x)	$R^1 = R^3 = Ph,\ R^2 = H$	$\sim 0:100^a$	173, 248
H₃C−CH=CH−COOCH₃	$R^1 = i\text{-Pr},\ R^2 = H,\ R^3 = i\text{-Pr-CHCN}$	$\sim 0:100^b$	261
H₃C−CH=CH−CN	$R^1 = i\text{-Pr},\ R^2 = H,\ R^3 = i\text{-Pr-CHCN}$	$\sim 0:100^c$	261
Ph−CH=CH−NO₂	$R^1 = PhCO,\ R^2 = H,\ R^3 = Ph$	$\sim 0:100^b$	248
O=C(Ph)−CH=CH−SO₂Ph	$R^1 = R^3 = Ph,\ R^2 = H$	$\sim 0:100^b$	262
maleic anhydride	$R^1 = R^3 = Ph,\ R^2 = H$	d	248
maleic anhydride	$R^1 = PhCO,\ R^2 = H,\ R^3 = Ph$	e	248
H₃C−CH=CH−COOCH₂Ph	$R^1 = CH_2\text{-}\underset{\overset{|}{O\diagdown\!\diagup\!O}}{C}\text{-}CO_2Et,\ R^2 = H,\ R^3 = PhCH_2$	$\sim 0:100^f$	263
CH₂=C(Ph)(Ph)	$R^1 = Ph,\ R^2 = H,\ R^3 = CH_3$	$\sim 100(5,5):0$	173
CH₂=C(Ph)(CH₃)	$R^1 = Ph,\ R^2 = H,\ R^3 = CH_3$	$\sim 100(5,5):0^g$	173
CH₂=C(NHCOCH₃)(COOEt)	$R^1 = R^3 = Ph,\ R^2 = H$	$\sim 100(5,5):0^h$	264
CH₂=C(COOCH₃)(CH₃)	$R^1 = R^3 = Ph,\ R^2 = H$	$\sim 100(5,5):0^i$	173

Table 1.4. (Continued)

Olefin	Nitrone $R^1\!\!\searrow\!\!\nearrow\!\!O$ $R^2\!\!\nearrow\!\!=\!\!N\!\!\searrow\!\!R^3$	Product ratio A:B	Ref.
(CH=C(COOCH$_3$)$_2$)	$R^1 = R^3 = $ Ph, $R^2 = $ H	$\sim 0:100(4,4)$	265

^{a}Two stereoisomers, 15:85, major, H^3, H^4, H^5; tr, tr.
^{b}One stereoisomer, H^3, H^4, H^5; tr, tr.
^{c}Two stereoisomers, 2:1, major, tr, tr.
^{d}Two stereoisomers, 55:45, minor, H^3, H^4, H^5; tr, cis.
^{e}One stereoisomer, H^3, H^4, H^5; tr, cis.
^{f}Two stereoisomers, 82:18, major, tr, tr.
^{g}Two stereoisomers, 55:45.
^{h}Two stereoisomers, 5:2.
^{i}Two stereoisomers, 96:4.

$$\text{olefin} + \text{nitrone} \longrightarrow A + B \qquad (1.68)$$

$$\text{olefin} + \text{nitrone} \longrightarrow \text{product} \qquad (1.69)$$

energy gain in bond formation is highest when those orbitals interact that are closest in energy and have the best overlap, eq. (1.70):

$$\Delta E = \frac{(C_1^1 \cdot C_a \cdot \gamma_{1a} + C_3^1 \cdot C_b \cdot \gamma_{3b})^2}{E_1} + \frac{(C_1 \cdot C_a^1 \cdot \gamma_{1a} + C_3 \cdot C_b^1 \cdot \gamma_{3b})^2}{E_2} \qquad (1.70)$$

where E_1 = energy gap between the lowest unoccupied molecular orbital (LUMO) (dipole) and the highest occupied molecular orbital (HOMO) (dipolarophile); E_2 = energy gap between LUMO (dipolarophile) and HOMO (dipole). C and C^1 are the orbital coefficients of the molecular orbitals of the HOMOs and LUMOs, respectively, and γ is the resonance integral.

HOMO energies (and to a certain extent also LUMO energies) are related to the ionization potentials of the molecule; a high ionization potential of a reactant lowers its HOMO energy, and vice versa. LUMO energies can be approximately calculated or estimated from reduction potentials (electron affinities). A schematic representation of the orbital stabilization in the transition state due to different coefficient magnitudes is shown in Figure 1.1.

FIGURE 1.1. Schematic representation of the orbital interaction between the LUMO of a 1,3-dipole, 1-3, and the HOMO dipolarophile, a-b. The orbital size represents the magnitude of the coefficients. The stabilization is more effective in A than in B because this alignment makes the numerator in eq. (1.70) larger.

The size and energies of frontier orbitals of alkenes are effected by substituents in the following way:[2,272]

1. Electron withdrawing groups decrease the energies of the HOMO and LUMO, the latter more than the former. In both HOMO and LUMO the unsubstituted end has the larger orbital coefficient.
2. Electron donating groups increase the energy of the HOMO and LUMO, the former more than the latter. The unsubstituted HOMO is larger than the substituted HOMO, Figure 1.1, but the magnitudes of the coefficients are reversed in the LUMO.
3. In conjugated alkenes the HOMO energy is raised and the LUMO energy is lowered. The coefficient at the unsubstituted end is larger in both HOMO and LUMO.

Depending on the relative position of energies of the interacting frontier orbitals, the 1,3-dipolar reaction can either be controlled by the dipolar HOMO and dipolarophile LUMO (HO controlled, Case I, Figure 1.2) or by the dipolar LUMO and dipolarophile HOMO (LU controlled, Case II). In the former case the energy gap between the dipolar HOMO and dipolarophile LUMO is the smallest; i.e., the denominator in eq. (1.70) becomes smaller, and consequently ΔE becomes larger. In the latter case the energy gap between the dipolar LUMO and dipolarophile HOMO is the smallest.

A reaction with a dipolarophile carrying an electron donating group will thus preferably be LU controlled (Case II). Conversely, when the dipolarophile carries electron withdrawing groups, the HO controlled reaction becomes faster, Case I, Figure 1.2. This explains why both electron donating and electron withdrawing substituents can accelerate a 1,3-dipolar reaction. This simultaneously offers a qualitative explanation for the regioselectivity of the reaction with monosubstituted alkenes and acetylenes. Figure 1.3 explains the outcome of the reactions between benzonitrile oxide and the electron-rich vinyl ethyl ether and the reaction

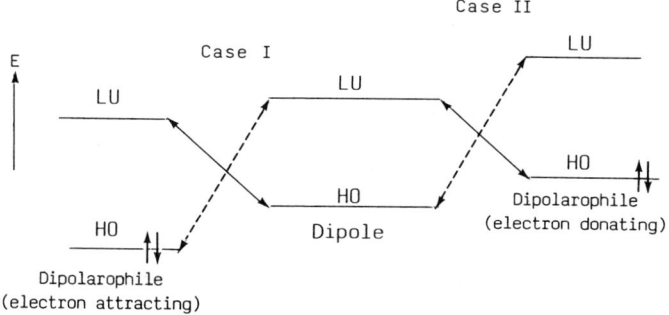

FIGURE 1.2. Case I, dipole-HO controlled reaction. Case II, dipole-LU controlled reaction.

between benzonitrile oxide and the electron-deficient methyl acrylate, both giving the same substitution pattern in the product. Both additions are LU controlled, and the magnitudes of the orbital coefficients align the alkene so that the carbon atom carrying the substituent approaches the oxygen atom of the nitrile oxide. The energy separation for the HO controlled reaction is larger, but, provided that it contributes, the vinyl ether will still give the 5-substituted derivative because the LUMO coefficient at the unsubstituted end is smaller. For methyl acrylate the situation is the opposite in the HO controlled reaction, which will give the 4-substituted derivative. In fact, experimentally ca. 5% of the 4-methoxycarbonyl

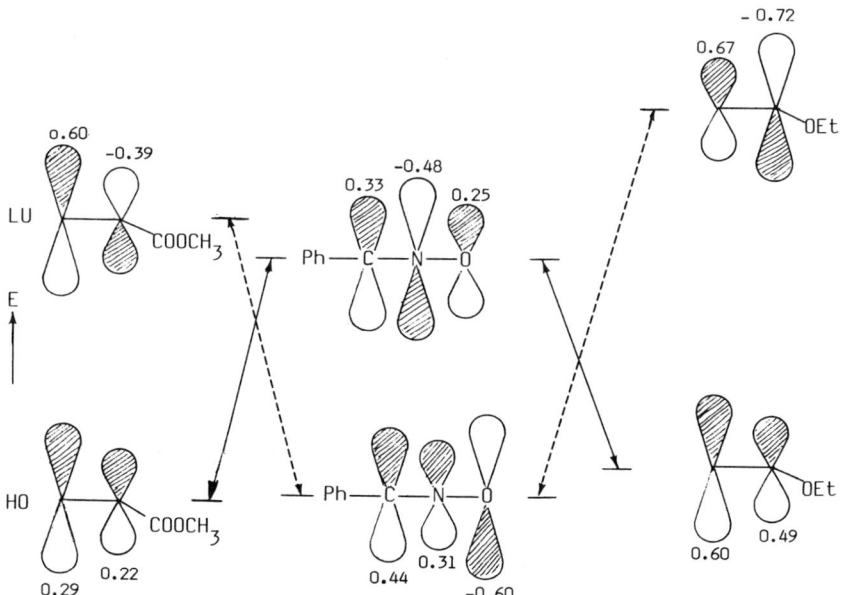

FIGURE 1.3. Regioselective LU alignment of vinyl ethyl ether and methyl acrylate to benzonitrile oxide.

derivative is obtained with nitrile oxide (Table 1.1), whereas vinyl ethyl ether gives the five-substituted derivative only. It is anticipated that electron donating groups on the nitrile oxide or nitrone and very strong electron withdrawing groups on the dipolarophile should increase the 4:5-substitution ratio. This prediction is borne out in practice, as is evident from the reaction of electron donating C-cyclopropyl nitrones with electron-deficient alkenes (Table 1.2). The reactions of diethyl methylenemalonate[265] and trimethyl ethylenetricarboxylate[266] with C,N-diphenylnitrone give exclusively the 4,4-disubstituted isoxazolidine (Table 1.4), and the effect is also noticeable in the reaction of a C-alkoxynitrone with methyl propriolate.[267]

The corresponding acetylenic derivatives have generally higher ionization potentials than the alkenes, which is reflected in a higher 4:5-substitution ratio for the dipolar addition.

When steric requirements hamper the formation of a 5-substituted cyclization product, it is possible to bypass the orbital control,[295-299] e.g., eq. (2.28).

The factors controlling the exo-endo orientation of the reactants are more difficult to evaluate. Steric effects on the orientation of the reactants have been noted. One striking case is found for the reaction of 1,2-disubstituted cis- and trans-alkenes with a nitrone[138] (Table 1.3). The cis-substituted alkene gives only one regioisomer, whereas the trans-substituted alkene gives a mixture of regioisomers, 37:63. The orbital coefficients for the two alkenes are most probably identical, and the route to the transition state appears to be delicately balanced by steric effects.

α-Substituents in cyclic olefins have a remarkable effect on the mode of approach of benzonitrile oxide (Table 1.5), eq. (1.71). Benzonitrile oxide adds with equal ease to both sides of cis-3,4-dichlorocyclobutene (25), $x = Cl$, $y,$—.

$$\text{25} + \varphi\text{CNO} \rightarrow \text{26} + \text{27} \quad (1.71)$$

The cis-diacetoxy derivative, $x = OCOH_3$, $y,$—, shows a stronger preference for syn addition, which still is more evident for the dihydroxy derivative, probably as a result of hydrogen bonding in the transition state.[300,301] But the cis-dimethoxycarbonyl derivative 25, $x = COOCH_3$, $y,$—, directs the cycloaddition toward the anti side. These results are in contrast to the reaction with cis-3,5-disubstituted cyclopentenes, (25), $x = Br$, OCH_3, $OCOCH_3$, $OCOPh$, OH; $y = CH_2$ and cis-2,5-disubstituted-2,5-dihydrofuranes (25), $x = OCH_3$, $OCOCH_3$; $y = O$, which all are attacked by benzonitrile oxide from the less-hindered side.[302] The ratio of **26:27** is in the range 0–20:100–80.

Table 1.5. syn and anti Additions of Benzonitrile Oxide to α-Substituted Cyclic Olefins

Olefin 25	α-Substituent x	Ratio **26:27**
y,— Cyclobutenes	COOCH$_3$ Cl OCOCH$_3$ OH	5:95 48:52 90:10 100:0
y = CH$_2$ Cyclopentenes	Br OCOPh OCH$_3$ OCOCH$_3$ OH	0:100 4:96 7:93 9:91 20:80
y = O 2,5-Dihydrofurans	OCH$_3$ OCOCH$_3$	0:100 0:100

3-Monosubstituted cyclopentenes also show a strong preference for addition to the anti-side.[303-305] The unexpected results in the cyclobutene series are interpreted in terms of an intrinsic pyramidalization of the sp_2-carbons towards a transition state favoring compound **26**, y,—, thus compensating for the repulsion between the reactants. This out-of-plane bending is negligible in the cyclopentene series, but it is again noticeable (ca. 3°) in the bridged norbornene system **28**.[306,307] The distortion, which can be as high as ca. 20° in certain fused norbornyl systems (cf. Ref. 308 for further references), favors the exo attack of benzonitrile oxide;[309] exo:endo = 90:10. Dewar benzene (**29**) gives exclusively the exo product.[310,311] The out-of-plane bending increases the reactivity of the olefin. Norbornene (**28**) is ca. 75 times more reactive than cyclopentene.[259] Ring strain in combination with pyramidalization accounts for this increased reactivity.

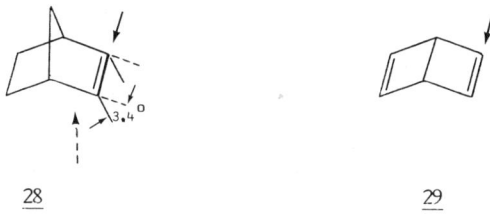

28   29

The dipolar cycloaddition of nitrile oxides or nitrones to allylic ethers or alcohols is of interest in connection with the stereocontrolled synthesis of functionalized acyclic derivatives. It was found that nitrile oxides add to **30**, giving an erythro:threo, **31:32**, ratio of ca. 3–4:1,[312] eq. (1.72). The mechanistically related osmium tetroxide catalyzed *cis*-hydroxylation of compound **30** gives predominantly the *erythro*-isomer in approximately the same erythro:threo ratio[132,368,369] (Section 5.21). Adding nitrile oxides to compounds of the general structure **33** (eq. (1.73), Table 1.6) takes the same course.[312-318] Computations indicate that conformation **34** leading to erythro derivatives is the favored one.[313] It has the ether group in the "inside" position, and cycloaddition takes place from the sterically less-hindered direction, anti to the large R' group. In this conformation the repulsion between the ether and nitrile oxide oxygens is minimized. For $x = $ OH or NH_3^+ the threo structure **35** is favored due to the effect of hydrogen bondings (entries 1 and 11 in Table 1.6). There is low steric induction for the alkyl and halogen derivatives (entries 9 and 10 in Table 1.6). This is also the case for chiral nitrile oxides, eq. (1.74).[15,319-322] The diastereomeric ratio in the product **36** is close to 1. For 1,2-disubstituted

Table 1.6. Diastereomeric Ratio of Nitrile Oxide Cycloadditions to Chiral Vinyl Compounds

R	33		34:35	Ref.
	R'	x		
1 Ph	CH_3	OH	40:60	313
2 Ph	CH_3	OCH_2Ph	64:36	313
3 Ph	CH_3	$OSi(CH_3)_3$	71:29	313
4 Ph	—$CH_2OC(CH_3)_2O$—		85:15	313
5 COOEt	—$CH_2OC(CH_3)_2O$—		80:20	315
6 Ph	—CH_2O—		69:31	313
7 p-NO_2Ph	Ph	OCH_3	67:33	313
8 p-NO_2Ph	t-Bu	OCH_3	>95:5	313
9 p-NO_2Ph	CH_3	Cl	50:50	313, 317
10 p-NO_2Ph	CH_3	Ph	55:45	313
11 Br	COO^-	NH_3^+	25:75	316
12 C_6H_{13}	—$CH_2C(C_6H_{13})$=N—O—		77:23	312
13 p-NO_2Ph			83:17	318

olefins, where the new asymmetric center is closer to the inducing center, the ratio is ca. 1:2. The related nitrone **37** gives with vinyl ether practically complete steric induction in compound **38**, eq. (1.75). Here the asymmetric centers are located on adjacent carbon atoms.[323] High asymmetric induction (ee > 80–90%) is observed in the cycloaddition of (R)-(+)-p-tolyl vinyl sulfoxide (**39**), with acyclic nitrones, eq. (1.76).[324] Cycloadditions with nitrones **40** containing the menthyl moiety give a small asymmetric induction (ee = 6%, 8%), eq. (1.77). When the menthyl moiety is located on the dipolarophile **41**, the induction is ee = 12%, 20%, eq. (1.78),[325] cf. Addendum.

Enantiomerically pure isoxazolines are obtained by *exo*-alkylation of 3-methyl isoxazolines with (−)-menthyl *p*-toluenesulfinate (ee ∼ 8–20%) and subsequent chromatographic separation of the diastereomers.[370,371] Desulfuration is achieved with NaHg in methanol or catalytically with Raney-Ni, eq. (1.79). This reaction allows an entry into chiral β-hydroxy ketones **42, 43**, β,β-dihydroxy ketones **44**, and 3-amino alcohols. The addition of magnesium and lithium reagents to 3-acylisoxazolines proved to be highly stereoselective (80:20–99:1), but surprisingly the reagents attack the 3-acyl group from opposite faces. It is suggested that the acyl group occupies different conformations (**45, 46**) in the two reactions. The direction of attack is controlled by the R^2 substituent on C-4.[372]

$$\tag{1.79}$$

The cycloaddition of *N*-chiral nitrones with prochiral olefins gives fair to good asymmetric inductions (ee 20–100%).[326–329] Nitrone **47** reacts with styrene to give the diastereomeric cis-trans mixture **48–51** in a relative amount of 76:11:8:5, i.e., an ee of 75% and 23%, respectively, eq. (1.80).[326]

Deprotonation of 3-phenyl-5-methylisoxazoline at −85°C in tetrahydrofuran (THF) with lithium diisopropylamide and subsequent deuteration of the C^4 anion gives a cis:trans ratio of 22:78. Alkylation by methyl and ethyl iodide proceeds with still better stereoselectivity, cis:trans = 7:93,[116,117] eq. (1.29). Reduction of 3,5-disubstituted 2-isoxazolines with lithium aluminum hydride proceeds with considerable stereoselectivity.[146] The hydride is added from the sterically least-hindered side, trans to the C^5 substituent, and the enantiomeric excess is in the range of ca. 90%.[153–156]

Mechanisms of the 1,3-Dipolar Cycloaddition

$$\text{(1.80)}$$

47 + nitrone → 48 (76% 3(R),5(S)) + 49 (11% 3(S),5(R))

$R^* = C(Ph)(CH_3)(H)$

+ 50 (8% 3(R),5(R)) + 51 (5% 3(S),5(S))

Monosubstituted olefins react with nitrones to give rise to a diastereomeric ratio of cis-trans, C^3,C^5-disubstituted isoxazolidines. The exo-approach, eq. (1.81), gives the cis-product, provided that the N—R^1 and C—R^2 substituents are in a trans-relationship; the endo-approach gives the trans-product, eq. (1.82). Inspection of the available data tends to show that the exo-approach is favored, eq. (1.81), in most cases when secondary orbital interactions are negligible [curved arrow in eq. (1.82), x = alkyl, alkoxy]. When x = $COOCH_3$, CN, or Ph such secondary interactions could stabilize the transition state in eq. (1.82) and give predominantly the trans-adduct. These secondary interactions are analogous to those directing the endo-exo approach in the Diels-Alder reaction. However, there are data available that are inconsistent with this interpretation; see, e.g., Refs. 173, 256, 288. Predictions are therefore unreliable, and it is advisable to thoroughly determine structures for every reaction. It has been suggested that *cis-trans*-isomerization of the starting nitrone prior to cycloaddition could account for inconsistent results even though the isomerizations at a measurable rate normally occur at higher temperatures than those at which

$$\text{(1.81)}$$ cis

$$\text{(1.82)}$$ trans

$$\text{(1.83)}$$

the cycloadditions are carried out. The unpredictable behavior of nitrones is demonstrated by the reactions of N,C_α-diphenylnitrone and N-methylnitrone with diethyl methylenemalonate, which in the former case gives only the 4,4-disubstituted regioisomer **52**, and in the latter case the 5,5-disubstituted regioisomer **53**.[265]

<u>52</u>   <u>53</u>

In most cases aldonitrone additions to 1,2-disubstituted olefins favor formation of 3,4-*trans*-substituted isoxazolidines, eq. (1.83).

1.8 RELATIVE REACTIVITIES OF REACTANTS

The FMO calculations discussed in Section 1.7, eq. (1.70), offer simultaneously an explanation for the relative reactivities of dipoles and dipolarophiles. Substituents in the reactants affect the HOMO-LUMO interaction in the transition state, i.e., the activation energies, and consequently the reaction rates.

In Table 1.7 relative rates are collected for the addition of benzonitrile oxide and N-methyl-C-phenylnitrone to various olefins, acetylenes, and heterodipolarophiles. Some characteristic features appear on inspection of the data. The rates range over four to five powers of 10. In monosubstituted alkenes n- and π-conjugation increase the reactivity, whereas halogens and alkyl groups decrease the reactivity. Disubstitution decreases the reactivity and 1,2-disubstitution more so than 1,1-disubstitution. Acetylene derivatives are less reactive than the corresponding olefins. This contradicts predictions, since a gain in aromaticity is expected to decrease the energy of the transition state. Hence, the transition state appears early on the reaction path and is reactantlike.

The relative reactivities for benzonitrile oxide and N-methyl-C-phenylnitrone in the two series are roughly parallel. Since most nitrile oxides are very unstable and have to be generated in situ, it is not possible to study relative reactivities of this group of compounds, but the reaction rates of several nitrones with ethyl crotonate have been studied[174] (Table 1.8). A slight increase in rate occurs on changing C_α- and N-alkyl groups with aryl groups (entries 1–7). The C-benzoyl nitrone (entry 8) shows a remarkable enhanced reactivity and so do the E-nitrones (entries 9, 10), as compared to 1–7 and the trisubstituted cyclic nitrone (entry 11), presumably due to steric factors. The aromatic nitrone (entry 12) reacts very slowly as a result of disruption of the aromaticity, which indicates a more productlike transition state.

Reaction rates and activation parameters have been determined for several

Table 1.7. Relative Rates for the Addition of Benzonitriles and N-Methyl-C-Phenylnitrone to Various Dipolarphiles.

Dipolarophiles	Benzonitrile K_{rel} in ether at 5°C	N-methyl-C-phenylnitrone K_{rel} in toluene at 120°C
A. *Olefins*		
Ethylene	1[a]	
Propylene	0.32	
1-Hexene	0.31	1[b]
Methyl acrylate	8.3	145
Acrylonitrile		106
Butyl vinyl ether	2.1	
Methyl vinyl ether		2.8
Styrene	1.15	4.4
4-Nitrostyrene	2.3	19
Vinyl chloride	0.081	
Maleic anhydride		2,523
Dimethyl fumarate	6.1	229
Diethyl maleate		86
Methyl crotonate	0.082	12
Methyl cinnamate	0.071	
Methyl 3,3-dimethylacrylate	0.0062	1
Methyl 2-methylacrylate	3.6	49
trans-Stilbene	0.023	
α-Methylstyrene		0.40
β-Isopropyl styrene	0.014	
1,1-Diphenylethylene	0.40	0.37
trans-Dibenzoylethylene		902
Cyclopentene	0.21	0.31
Cyclopentadiene	0.44	
Norbornene	15.3	3.3
Cyclohexene	0.0025	
B. *Acetylenes*		
Acetylene	0.40	
1-Hexyne	0.066	
Methyl propiolate	1.24	576
Dimethyl acetylenedicarboxylate	3.1	12,200
Phenylacetylene	0.112	
C. *Heterodipolarophiles*		
Phenylisocyanate		648
Diethyl mesoxalate	11.2	
N-Benzylidenemethylamine	4.5	
Benzaldehyde	0.0024	
Benzonitrile	0.0023	
$\rangle C=S \gg \rangle=O^{352}$		

[a] Relative to ethylene.
[b] Relative to 1-hexene.

Table 1.8. Relative and Absolute Rates for the Addition of Various Nitrones, with Ethyl Crotonate in Toluene at 100°C

Nitrone	K_{rel}	$10^5 \cdot K_2$ $1\ \text{mol}^{-1}\ \text{s}^{-1}$
1 Ph–CH=N(O)–CH$_3$	1.1	10.7
2 Ph–CH=N(O)–Ph	5.7	55.4
3 p-NO$_2$Ph–CH=N(O)–Ph	7.8	75
4 (alkyl)CH=N(O)–Ph	0.71	6.8
5 (alkyl)CH=N(O)–cyclohexyl	1	9.6
6 (alkyl)CH=N(O)–CH$_2$Ph	4.5	43
7 Ph$_2$C=N(O)–Ph	1.1	10.5
8 Ph–CO–CH=N(O)–Ph	642	6,200
9 (pyrroline N-oxide)	166	1,600
10 (3,4-dihydroisoquinoline N-oxide)	248	2,400
11 (2-Ph-pyrroline N-oxide)	2.1	20
12 (isoquinoline N-oxide)	0.007	0.067

other nitrones and olefins.[353-357] A kinetic study of the addition of phenylacetylene to a series of benzonitrile oxides demonstrates competitive concerted 1,3-cycloaddition leading to isoxazoles and two-step 1,3-addition leading to acetylenic oximes[357]; cf. also the reactions of CF_3CNO.[362]

REFERENCES

1. R. B. Woodward; R. Hoffmann, "The Conservation of Orbital Symmetry", Verlag Chemie, Weinheim, 1970.
2. K. N. Houk; J. Sims; R. E. Duke, Jr.; R. W. Strozier; J. K. George, *J. Am. Chem. Soc. 95* (1973), 7287.
3. K. N. Houk; J. Sims; C. R. Watts; L. J. Luskus, *J. Am. Chem. Soc. 95* (1973), 7301.
4. R. Huisgen, *Angew. Chem. 75* (1963), 604.
5. A. Padwa, "1,3-Dipolar Cycloaddition Chemistry", Wiley, New York, 1984.
6. A. Katritzky and C. Rees, Eds., "Comprehensive Heterocyclic Chemistry", Pergamon, 1984.
7. A. I. Meyers, "Heterocycles in Organic Synthesis", Wiley, New York, 1974.
8. N. Elming In "Advances in Organic Chemistry, Methods and Results" (R. A. Raphael, E. C. Taylor, and H. Wynberg, Eds.), Vol. 2, Interscience, New York, 1960, p. 67.
9. N. Finch; J. J. Fitt; I. H. C. Hsu, *J. Org. Chem. 36* (1971), 3191.
10. G. Piancatelli; A. Scettri, *Tetrahedron Lett. 1977*, 1131.
11. S. Gronowitz In "Advances in Heterocyclic Chemistry" (A. R. Katritzky, Ed.), Vol. 1, Academic Press, New York, 1963, p. 1.
12. G. R. Pettit; E. E. van Tamelen, *Organic Reactions 12* (1962), 356. J. Wiley & Sons, New York.
13. G. Stork; S. Danishefsky; M. Ohashi, *J. Am. Chem. Soc. 89* (1967), 5459; M. Ohashi, H. Kamachi; H. Kakisawa; G. Stork, *J. Am. Chem. Soc. 89* (1967), 5460; G. Stork; J. E. McMurry, *J. Am. Chem. Soc. 89* (1967), 5461, 5463, 5464.
14. C. Grundmann and P. Grünanger, "The Nitrile Oxides", Springer-Verlag, Berlin, 1971.
15. K. E. Larsen; K. B. G. Torssell, *Tetrahedron 40* (1984), 2985.
16. R. Fusco; L. Garanti; G. Zecchi, *Chim. Ind. (Milan) 57* (1975), 16; L. Garanti and G. Zecchi, *J. Heterocycl. Chem. 17* (1980), 609; L. Garanti; A. Sala; G. Zecchi, *J. Org. Chem. 40* (1975), 2403.
17. O. Isuge; K. Ueno; S. Kanemasa, *Chem. Letters 1984*, 285.
18. A. P. Kozikowski; P. D. Stein, *J. Org. Chem. 49* (1984), 2301.
19. P. Caramella; P. Grünanger In "1,3-Dipolar Cycloaddition Chemistry" (A. Padwa, Ed.), Vol. 1, Wiley, New York, 1984, p. 291.
20. W. Rundel In Houben-Weyl: "Methoden der Organischen Chemie" *10:4* (E. Müller, Ed.), G. Thieme Verlag, Stuttgart, 1968, p. 311.
21. N. A. LeBel; J. J. Whang, *J. Am. Chem. Soc. 81* (1959), 6334.
22. D. St. C. Black; R. F. Crozier; V. C. Davis, *Synthesis 1975*, 205.
23. J. J. Tufariello In "1,3-Dipolar Cycloaddition Chemistry" (A. Padwa, Ed.), Vol. 2, Wiley, New York, 1984, p. 83.
24. K. B. G. Torssell; O. Zeuthen, *Acta Chem. Scand. B32* (1978), 118.
25. S. H. Andersen; N. B. Das; R. D. Jørgensen; G. Kjeldsen; J. S. Knudsen; S. C. Sharma; K. B. G. Torssell, *Acta Chem. Scand. B36* (1982), 1.
26. E. W. Colvin; A. K. Beck; B. Bastani; D. Seebach; Y. Kai; J. D. Dunitz, *Helv. Chim. Acta 63* (1980), 697.
27. S. L. Joffe; M. V. Kashutina; V. M. Shitkin; A. Z. Yankelevich; A. A. Levin; V. A. Tartakovskii, *Izv. Akad. Nauk SSSR, Ser. Khim Nauk 1972*, 1341.
28. N. K. Kochetkov; S. D. Sokolov In "Advances in Heterocyclic Chemistry", Vol. 2, 1963, p. 365.
29. A. Quilico In "The Chemistry of Heterocyclic Compounds" (A. Weissberger, Ed.), Vol. 17, 1962, p. 1.
30. N. K. Kochetkov; E. D. Khomutova, *Zhur. Obshchei Khim. 28* (1958), 359; *29* (1959), 535.
31. S. D. Sokolov; I. M. Yudintseva; P. V. Petrovskii; V. G. Kalyuzhnaya, *Zh. Org. Khim. 7* (1971), 1979.
32. V. Bertini; A. De Munno; V. Dell'Amico; P. Pino, *Gazz. Chim. Ital. 97* (1967), 1604.
33. A. Quilico; R. Justoni, *Gazz. Chim. Ital. 70* (1940), 1, 11.
34. L. Claisen; R. Stock, *Chem. Ber. 24* (1891), 130.

35. R. B. Woodward; R. A. Olofson; H. Mayer, *J. Am. Chem. Soc. 83* (1961), 1010.
36. A. Quilico; R. Justoni, *Rend. Ist. Lombardo Sci. 69* (1936), 587.
37. P. Pino; F. Piacenti; G. Fatti, *Gazz. Chim. Ital. 90* (1960), 356.
38. N. K. Kochetkov; E. D. Khomutova, *Zh. Obshchei Khim. 30* (1960), 1269.
39. N. K. Kochetkov; S. D. Sokolov; N. M. Vagurtova, *Zh. Obshchei Khim. 31* (1961), 2326; *32* (1962), 325.
40. N. K. Kochetkov, E. D. Khomutova, M. Ya. Karpeysky; R. M. Khomotov, *Zh. Obshchei Khim. 27* (1957), 3210.
41. N. K. Kochetkov; E. D. Khomutova; M. V. Bazilevsky, *Zh. Obshchei Khim. 28* (1958), 2376.
42. N. K. Kochetkov; S. D. Sokolov; V. E. Zhvirblis, *Zh. Obshchei Khim. 30* (1960), 3675.
43. R. K. M. R. Kallury; P. S. U. Devi, *Tetrahedron Lett. 1977*, 3655.
44. Y. Yamanaka; T. Sakamoto; A. Shiozawa, *Heterocycles 7* (1977), 51.
45. O. Mumm; G. Münchmeyer, *Chem. Ber. 43* (1910), 3335.
46. R. B. Woodward; R. Olofson, *J. Am. Chem. Soc. 83* (1961), 1007.
47. D. J. Woodman, *J. Org. Chem. 33* (1968), 2397.
48. R. B. Woodward; R. A. Olofson; H. Mayer, *Tetrahedron Suppl. 8* (1966), 321.
49. R. B. Woodward; R. A. Olofson, *Tetrahedron Suppl. 7* (1966), 415.
50. A. Quilico; G. Speroni; E. Galeffi, *Gazz. Chim. Ital. 69* (1939), 508.
51. C. Musante, *Gazz. Chim. Ital. 71* (1941), 172.
52. P. Grünanger; S. Mangiapan, *Gazz. Chim. Ital. 88* (1958), 149.
53. N. R. Natale; D. A. Quincy, *Synth. Commun. 13* (1983), 817.
54. M. M. Botwinnik; N. I. Gawrilow, *J. Prakt. Chem. 148* (1937), 170.
55. C. Musante, *Gazz. Chim. Ital. 71* (1941), 553.
56. A. Barco; S. Benetti; G. P. Pollini; P. G. Baraldi, *Synthesis 1977*, 837.
57. R. G. Micetich; C. G. Chin, *Can. J. Chem. 48* (1970), 1371.
58. B. J. Wakefield; D. J. Wright In "Advances in Heterocyclic Chemistry", Vol. 25, 1979, p. 147.
59. L. Claisen, *Chem. Ber. 24* (1891), 3900.
60. S. Cusmano; T. Tiberio, *Gazz. Chim. Ital. 78* (1948), 896.
61. H. Wieland, *Liebigs Ann. 328* (1903), 154.
62. L. Claisen, *Chem. Ber. 36* (1903), 3664.
63. U. Schöllkopf; I. Hoppe, *Angew. Chem. 87* (1975), 814.
64. W. Lampe; J. Smolinska, *Roczniki Chem. 28* (1954), 163; *29* (1955), 934; *Bull. Akad. Polon Sci. 6* (1958), 481.
65. G. Renzi; V. Dal Piaz; S. Pinzauti, *Gazz. Chim. Ital. 99* (1969), 753.
66. C. Kashima; Y. Yamamoto; Y. Tsuda, *Heterocycles 6* (1977), 805.
67. N. K. Kochetkov; S. D. Sokolov; V. M. Luboshnikova, *Zh. Obshchei Khim. 32* (1962), 1778.
68. S. Auriccio; R. Colle; S. Morrochi; A. Ricca, *Gazz. Chim. Ital. 106* (1976), 823.
69. T. Tanaka; M. Miyazaki; I. Iijima, *J. Chem. Soc., Chem. Commun. 1973*, 233.
70. C. Kashima; N. Mukai; Y. Yamamoto; Y. Tsuda; Y. Omote, *Heterocycles 7* (1977), 241.
71. (a) A. Alberola; A. M. Gonzales Nogal; F. J. Pulido, *Heterocycles 20* (1983), 1035.
    (b) D. J. Brunelle, *Tetrahedron Lett. 1981*, 3699.
    (c) N. R. Natale; C.-S. Niou, *Tetrahedron Lett. 1984*, 3943; N. R. Natale; J. I. McKenna; C.-S. Niou; M. Borth; H. Hope, *J. Org. Chem. 50* (1985) 5660.
72. G. Büchi; J. C. Vederas, *J. Am. Chem. Soc. 94* (1972), 9128.
73. J. Thiele; H. Landers, *Liebigs Ann. 369* (1909), 300.
74. L. Claisen, *Chem. Ber. 59* (1926), 144.
75. E. P. Kohler; N. K. Richtmyer, *J. Am. Chem. Soc. 50* (1928), 3093.
76. (a) M. Nitta; T. Kobayashi, *Tetrahedron Lett. 1982*, 3925; *J. Chem. Soc., Chem. Commun. 1982*, 877; *Chem. Lett. 1983*, 51. M. Nitta; A. Yi; T. Kobayashi, *Bull. Soc. Chim. Japan 58* (1985), 991; P. G. Baraldi; A. Barco; S. Benetti; M. Guarneri; S. Manfredini; G. P. Pollini; D. Simoni, *Tetrahedron Lett. 1985* 5319.
    (b) M. Kijima; Y. Nambu; T. Endo, *J. Org. Chem. 50* (1985), 1140.
77. J. G. Markova; M. K. Polievktov; S. D. Sokolov, *Zh. Obshch. Khim. 46* (1976), 398.
78. H. Lund; J. Tabaković In "Advances in Heterocyclic Chemistry", Vol. 36, 1984, p. 235; I. Surov; H. Lund, *Acta. Chem. Scand. B40* (1986) 831.
79. P. Pino; F. Piacenti; G. Fatti, *Gazz. Chim. Ital. 90* (1960), 356.
80. P. Bravo; G. Gaudiano; A. Quilico; A. Ricca, *Gazz. Chim. Ital. 91* (1961), 47.
81. G. Stagno d'Alcontres, *Gazz. Chim. Ital. 80* (1950), 441.
82. G. Shaw; G. Sugowdz, *J. Chem. Soc. 1954*, 665.
83. S. Auricchio; S. Morrocchi; A. Ricca, *Tetrahedron Lett.* (1974), 2793.

References

84. D. N. McGregor; U. Corbin; J. E. Swigor; L. C. Cheney, *Tetrahedron 25* (1969), 389.
85. R. V. Stevens, *Tetrahedron 32* (1976), 1599.
86. A. A. Akhrem; F. A. Lakhvish; U. A. Khripack; I. B. Klebanovich, *Tetrahedron Lett.* (1976), 3983.
87. S. Plescia; S. Petruso; V. Sprio, *J. Heterocycl. Chem. 11* (1974), 623.
88. C. Kashima, *J. Org. Chem. 40* (1975), 526.
89. P. N. Craig; M. P. Olmsted, *J. Org. Chem. 22* (1957), 559.
90. A. Alberola; A. M. Gonzales; M. A. Laguna; F. J. Pulido, *Synthesis 1983*, 413.
91. P. Pino; G. Speroni; V. Fuga, *Gazz. Chim. Ital. 84* (1954), 759.
92. G. Gaudiano; A. Quilico; A. Ricca, *Tetrahedron 7* (1959), 24.
93. P. Grünanger; M. R. Langella, *Gazz. Chim. Ital. 89* (1959), 1784.
94. A. J. Boulton; A. R. Katritzky, *Tetrahedron 12* (1960), 41, 51.
95. S. Califano; F. Piacenti; G. Speroni, *Spectrochim. Acta 15* (1959), 86.
96. A. R. Katritzky; A. J. Boulton, *Spectrochim. Acta 17* (1961), 238.
97. S. D. Solokov; I. M. Yudintseva; P. V. Petrovskii, *Zh. Org. Khim. 6* (1970), 2584.
98. J. Gainer; G. A. Haworth; W. Hoyle; S. M. Roberts, *Org. Magn. Res. 8* (1976), 226.
99. J. Elguero; C. Marzin; A. R. Katritzky; P. Linda In "Advances in Heterocyclic Chemistry," Suppl. 1, 1976.
100. Yu. S. Shabarov; L. G. Saginova; R. A. Gazzaeva, *Khim. Geterosikl. Soed. 19* (1983), 738.
101. R. A. Gazzaeva; Yu. S. Shabarov; L. Saginova, *Khim. Geterosikl. Soed. 1984*, 309.
102. A. H. Blatt, *J. Am. Chem. Soc. 53* (1931), 1133.
103. P. Grünanger; L. Grasso, *Gazz. Chim. Ital. 85* (1955), 1271.
104. P. Grünanger; P. Vita Finzi, *Gazz. Chim. Ital. 89* (1959), 1771.
105. R. P. Barnes; F. E. Chigbo, *J. Org. Chem. 23* (1958), 1777.
106. G. Bianchi; P. Grünanger, *Tetrahedron 21* (1965), 817.
107. M. Christl; R. Huisgen; R. Sustmann, *Chem. Ber. 106* (1973), 3275.
108. K. P. Park; C.-Y. Shiue; L. B. Clapp, *J. Org. Chem. 35* (1970), 2065.
109. A. Barco; S. Benetti; G. Pollini; P. G. Baraldi, *Synthesis 1977*, 837.
110. A. Barco; S. Benetti; G. Pollini; P. G. Baraldi; M. Guarneri; C. B. Vicentini, *J. Org. Chem. 44* (1979), 105.
111. W. R. Vaughan; J. L. Spenser, *J. Org. Chem. 25* (1960), 1160.
112. P. Venturella; A. Bellino; S. Cusmano, *Ann. Chim. Roma 51* (1961), 1074.
113. G. Bianchi; M. De Amici, *J. Chem. Research (S), 1979*, 311.
114. N. B. Das; K. Torssell, *Tetrahedron 39* (1983), 2247.
115. V. Jäger; H. Grund, *Angew. Chem. 88* (1976), 27.
116. V. Jäger; W. Schwab, *Tetrahedron Lett. 1978*, 3129.
117. V. Jäger; H. Grund, *Liebigs Ann. 1980*, 80.
118. W. Schwab; V. Jäger, *Angew. Chem. 93* (1981), 578.
119. S. Shatzmiller; E. Shalom; R. Lidor; E. Tartkovski, *Liebigs Ann. 1983*, 906.
120. M. J. Fray; E. J. Thomas, *Tetrahedron 40* (1984), 673; A. P. Kozikowski; A. K. Ghosh, *J. Org. Chem. 49* (1984), 2762.
121. G. Bianchi; R. Gandolfi; P. Grünanger, *J. Heterocyclic Chem. 5* (1968), 49.
122. G. Bianchi; A. Gamba Invernizzi; R. Gandolfi, *J. Chem. Soc., Perkin I 1974*, 1757.
123. D. N. Reinhoudt; C. G. Kouwenhoven, *Rec. Trav. Chim. Pays-Bas 95* (1976), 67.
124. A. Osawa; H. Arai; H. Igeta; T. Akimoto; A. Tsuji; Y. Iitaka *Tetrahedron 35* (1979), 1267.
125. Y. Ito; T. Matsuura, *Tetrahedron 31* (1975), 1373.
126. J. Kalvoda; H. Kaufmann, *J. Chem. Soc., Chem. Commun. 1976*, 209; A. P. Kozikowski; M. Adamczyk, *J. Org. Chem. 48* (1983), 366.
127. A. Belly; F. Petrus; J. Verducci, *Bull. Soc. Chim. France 1973*, 1390.
128. A. Belly; C. Petrus; F. Petrus, *Bull. Soc. Chim. France 1974*, 1025.
129. A. H. Blatt; N. Gross, *J. Am. Chem. Soc. 77* (1955), 5424.
130. L. Panizzi, *Gazz. Chim. Ital. 76* (1946), 44.
131. A. R. Katritzky; S. Øksne, *Proc. Chem. Soc. 1961*, 387.
132. K. B. G. Torssell; A. C. Hazell; R. G. Hazell, *Tetrahedron 41* (1985), 5569.
133. S. H. Andersen; K. K. Sharma; K. B. G. Torssell, *Tetrahedron 39* (1983), 2241.
134. S. K. Mukerji; K. K. Sharma; K. B. G. Torssell, *Tetrahedron 39* (1983), 2235.
135. D. P. Curran, *J. Am. Chem. Soc. 104* (1982), 4024; *105* (1983), 5826.
136. A. P. Kozikowski; P. D. Stein, *J. Am. Chem. Soc. 104* (1982), 4023.
137. A. P. Kozikowski; M. Adamczyk, *Tetrahedron Lett. 1982*, 3123.
138. S. F. Martin; B. Dupré, *Tetrahedron Lett. 1983*, 1337.

139. M. Asaoka; T. Mukuta; H. Takei, *Tetrahedron Lett. 1981*, 735; *Chem. Lett. 1982*, 215.
140. T. Kametani; H. Furuyama; T. Honda, *Heterocycles 9* (1982), 357.
141. R. F. Cunico, *J. Organomet. Chem. 212* C 51 *1981*.
142. G. Drehfahl; H.-H. Hörhold, *Chem. Ber. 97* (1964), 159.
143. T. Kusumi; H. Kakisawa; S. Suzuki; K. Harada; C. Kashima, *Bull. Chem. Soc. Japan 51* (1978), 1261.
144. G. H. Timms; E. Wildschmidt, *Tetrahedron Lett.* (1971), 195.
145. N. B. Das; K. B. G. Torssell, *Tetrahedron 39* (1983), 2227.
146. G. W. Perold; F. V. K. von Reiche, *J. Am. Chem. Soc. 79* (1957), 465.
147. E. E. van Tamelen; J. E. Brenner, *J. Am. Chem. Soc. 79* (1957), 3839.
148. N. Barbulesco; A. Quilico, *Gazz. Chim. Ital. 91* (1961), 326.
149. P. Grünanger; M. R. Langella, *Gazz. Chim. Ital. 91* (1961), 1112.
150. R. Paul; S. Tchelitcheff, *Bull. Soc. Chim. France 1962*, 2215.
151. P. Grünanger; L. Grasso, *Gazz. Chim. Ital. 85* (1955), 1271.
152. G. Bianchi; A. Cogoli; R. Gandolfi, *Gazz. Chim. Ital. 98* (1968), 74.
153. V. Jäger; V. Buss, *Liebigs Ann. 1980*, 101.
154. V. Jäger; V. Buss; W. Schwab, *Liebigs Ann. 1980*, 122.
155. P. N. Confalone; E. D. Lollar; G. Pizzolato; M. R. Uskoković, *J. Am. Chem. Soc. 100* (1978), 6291.
156. K. F. Burri; R. A. Cardone; W. Y. Chen; P. Rosen, *J. Am. Chem. Soc. 100* (1978), 7069.
157. V. Jäger; H. Grund: V. Buss; W. Schwab; I. Müller; R. Schohe; R. Franz; R. Ehrler, *Bull. Soc. Chim. Belg. 92* (1983), 1039, and references cited therein.
158. J. Hoenicke; H. Steudle; H. Stamm, *Arch. Pharm. 317* (1984), 474.
159. M. V. Kashutina; S. L. Joffe; V. A. Tartakovskii, *Dokl. Akad. Nauk SSSR Ser. Khim. 218* (1974), 109.
160. G. S. King; P. D. Magnus; H. S. Rzepa, *J. Chem. Soc., Perkin I* (1972), 437.
161. R. Huisgen; M. Christl, *Chem. Ber. 106* (1973), 3291, 3345.
162. V. Jäger; R. Schohe, *Tetrahedron 40* (1984), 2199.
163. I. Adachi; H. Kano, *Chem. Pharm. Bull. 17* (1969), 2201.
164. J. P. Freeman, *Chem. Rev. 83* (1983), 241.
165. R. Grigg, *J. Chem. Soc. Chem. Commun. 1966*, 607.
166. J. E. Baldwin; R. G. Pudussery; A. K. Qureshi; B. Sklarz, *J. Am. Chem. Soc. 90* (1968), 5325.
167. I. Adachi; K. Harada; R. Miyazaki; H. Kano, *Chem. Pharm. Bull. 22* (1974), 61.
168. I. Adachi; R. Miyazaki; H. Kano, *Chem. Pharm. Bull. 22* (1974), 70.
169. K. Niklas, Dissertation, Univ. of München, 1975.
170. A. Alberola; A. M. Gonzales; M. A. Laguna; F. J. Pulido, *Synthesis 1982*, 1067; *1983*, 413; *1984*, 510.
171. Y. Takeuchi; F. Furusaki In "Advances in Heterocyclic Chemistry", Vol. 21, 1977, p. 209.
172. R. Huisgen, *Angew. Chem. 75* (1963), 604; In "1,3-Dipolar Cycloaddition Chemistry" (A. Padwa, Ed.), Vol. 1, Wiley, New York, 1984, p. 1.
173. R. Huisgen; R. Grashey; H. Hauck; H. Seidl, *Chem. Ber. 101* (1968), 2043, 2548, 2559, 2568; *102* (1969), 736.
174. R. Huisgen; H. Seidl; I. Brüning, *Chem. Ber. 102* (1969), 1102; K. Bast; M. Christl; R. Huisgen; W. Mack, *Chem. Ber. 106* (1973), 3312.
175. R. Huisgen; H. Hauck; H. Seidl; M. Burger, *Chem. Ber. 102* (1969), 1117.
176. S. J. Brois, *J. Am. Chem. Soc. 90* (1968), 506, 508.
177. D. Felix; A. Eschenmoser, *Angew. Chem. 80* (1968), 197.
178. R. Grée; R. Carrié, *J. Am. Chem. Soc. 99* (1977), 6667.
179. K. Müller; A. Eschenmoser, *Helv. Chim. Acta 52* (1969), 1823.
180. V. A. Tartakovskii; I. A. Savostyanova; S. S. Novikov, *Zh. Org. Khim. 4* (1968), 240.
181. R. Grée; R. Carrié, *Tetrahedron 32* (1976), 675, 683.
182. F. G. Riddel; J. M. Lehn; J. Wagner, *Chem. Commun. 1968*, 1403.
183. G. V. Lagodzinskaya, *Zh. Strukt. Khim. 1970*, 27.
184. V. A. Tartakovskii; I. E. Chlenov; N. S. Morozova; S. S. Novikov, *Izv. Akad. Nauk SSSR 1966*, 370.
185. V. A. Tartakovskii; Z. Ya. Lapshina; I. A. Savostyanova; S. S. Novikov, *Zh. Org. Khim. 4* (1968), 236.
186. M. U. Kashutina; S. L. Joffe; V. A. Tartakovskii, *Dokl. Akad. Nauk SSSR 218* (1974), 109.
187. V. A. Tartakovskii; O. A. Lukyanov; N. I. Shlykova; S. S. Novikov, *Zh. Org. Khim. 3* (1967), 980.
188. V. A. Tartakovskii; I. A. Savostyanova; S. S. Novikov, *Zh. Org. Khim. 4* (1968), 240.

189. V. A. Tartakovskii; I. E. Chlenov; S. L. Joffe; G. V. Lagodzinskaya; S. S. Novikov, *Zh. Org. Khim. 2* (1966), 1593.
190. S. C. Sharma; K. Torssell, *Acta Chem. Scand. B33* (1979), 379.
191. V. A. Tartakovskii; S. S. Smagin; I. E. Chlenov; S. S. Novikov, *Izv. Akad. Nauk SSSR 1965*, 552.
192. V. A. Tartakovskii; I. E. Chlenov; G. V. Lagodzinskaya; S. S. Novikov, *Dokl. Akad. Nauk SSSR 161* (1965), 136.
193. N. A. LeBel, *Trans. N.Y. Acad. Sci. 27* (1965), 858.
194. J. Meinwald; O. L. Chapman, *J. Am. Chem. Soc. 81* (1959), 5800.
195. J. J. Tufariello; Sk. A. Ali, *J. Am. Chem. Soc. 101* (1979), 7114.
196. A. Liguori; G. Sindona; N. Uccella, *Tetrahedron 39* (1983), 683; *J. Heterocycl. Chem. 20* (1983), 1207.
197. C. W. Brown; K. Marsden; M. A. T. Rogers; C. M. B. Tylor; R. Wright, *Proc. Chem. Soc. 1960*, 254.
198. G. R. Delpierre; M. Lamchen, *Proc. Chem. Soc. 1960*, 386.
199. G. R. Delpierre; M. Lamchen, *J. Chem. Soc. 1963*, 4693.
200. M. Lamchen; T. W. Mittag, *J. Chem. Soc. C 1968*, 1917.
201. N. A. LeBel; G. M. J. Sluzarczuk; L. A. Spurlock, *J. Am. Chem. Soc. 84* (1962), 4360.
202. N. A. LeBel; T. A. Lajiness, *Tetrahedron Lett. 1966*, 2173.
203. N. A. LeBel; E. G. Bannucci, *J. Org. Chem. 36* (1971), 2440.
204. M. Joucla; J. Hamelin; R. Carrié, *Bull. Soc. Chim. France 1973*, 3116.
205. M. Masui; K. Suda; M. Yamauchi; C. Yijima, *Chem. Pharm. Bull. 21* (1973), 1605.
206. N. K. A. Dalgård; K. E. Larsen; R. B. G. Torssell, *Acta Chem. Scand. B38* (1984), 423.
207. G. Bianchi; C. DeMicheli; R. Gandolfi, *Angew. Chem. 91* (1979), 781.
208. D. St. C. Black; R. F. Crozier; J. D. Rae, *Aust. J. Chem. 31* (1978), 2239.
209. E. Gössinger; R. Imhof; A. Wehrli, *Helv. Chim. Acta 58* (1975), 96.
210. F. De Sarlo; A. Guarna; A. Brandi, *J. Heterocycl. Chem. 20* (1983), 1505.
211. R. A. Reamer; M. Sletzinger; I. Shinkai, *Tetrahedron Lett. 1980*, 3447.
212. L. Fisera; J. Kovác; J. Poliacikova, *Heterocycles 12* (1979), 1005.
213. J. J. Tufariello; G. B. Mullen; J. J. Tegeler; E. J. Trubulski; S. C. Wong; Sk. A. Ali, *J. Am. Chem. Soc. 101* (1979), 2435.
214. G. Bianchi; R. Gandolfi In "1,3-Dipolar Cycloaddition Chemistry" (A. Padwa, Ed.), Vol. 2, Wiley, New York, 1984, p. 451.
215. N. A. LeBel; L. A. Spurlock, *J. Org. Chem. 29* (1964), 1337.
216. N. A. LeBel; M. E. Post; D. Hwang, *J. Org. Chem. 44* (1979), 1819.
217. J. J. Tufariello; G. B. Mullen, *J. Am. Chem. Soc. 100* (1978), 3638.
218. R. O. C. Norman; R. Purchase; C. B. Thomas, *J. Chem. Soc., Perkin I 1972*, 1701.
219. C. M. Tice; B. Ganem, *J. Org. Chem. 48* (1983), 5048.
220. A. Vasella, *Helv. Chim. Acta 60* (1977), 426.
221. A. Vasella; R. Voeffray, *Helv. Chim. Acta 65* (1982), 1134.
222. T. Koizumi; H. Hirai; E. Yoshii, *J. Org. Chem. 47* (1982), 4004.
223. R. Grashey; R. Huisgen; H. Leiterman, *Tetrahedron Lett. 12* (1960), 9.
224. A. Lablache-Combier; M. L. Villaume, *Tetrahedron 24* (1968), 6951.
225. A. Belly; R. Jacquier; F. Petrus; J. Verducci, *Bull. Soc. Chim. France* (1972), 330.
226. V. N. Chistokletov; A. A. Petrov, *Zh. Obshch. Khim. 32* (1962), 2385.
227. V. N. Chistokletov; L. Kvagina; A. A. Petrov, *Zh. Org. Khim. 1* (1965), 369.
228. P. Burns; W. A. Waters, *J. Chem. Soc. C 1969*, 27.
229. J. J. Tufariello; J. P. Tette, *Chem. Commun. 1971*, 469.
230. R. Brambilla; R. Friary; A. Ganguly; M. S. Puar; B. R. Sunday; J. J. Wright; K. D. Onan; A. T. McPhail, *Tetrahedron 37* (1981), 3615.
231. P. DeShong; J. M. Leginus, *J. Am. Chem. Soc. 105* (1983), 1686.
232. R. L. Funk; L. H. M. Horcher, II; J. U. Daggett; M. M. Hansen, *J. Org. Chem. 48* (1983), 2632.
233. W. Oppolzer; K. Keller, *Tetrahedron Lett. 1970*, 1117, 4313.
234. W. C. Lumma, *J. Am. Chem. Soc. 91* (1969), 2820.
235. J. B. Bapat; D. St. C. Black; R. F. C. Brown; C. Ichlov, *Aust. J. Chem. 25* (1972), 2445.
236. S. Ito; S. Narita; K. Endo, *Bull. Chem. Soc. Japan 46* (1973), 3517.
237. A. Vasella, *Helv. Chim. Acta 60* (1977), 1273.
238. A. Vasella; R. Voeffray, *J. Chem. Soc. Chem. Commun. 1981*, 97.
239. R. R. Fraser; Y. S. Lin, *Can. J. Chem. 46* (1968), 81.
240. N. A. LeBel; N. D. Ojha; J. R. Menke; R. J. Newland, *J. Org. Chem. 37* (1972), 2896.
241. M. Ishidate; Y. Sakurai; M. Torigoe, *Chem. Pharm. Bull. 9* (1961), 485.

242. J. J. Tufariello; E. J. Trubulski, *Chem. Commun. 1973*, 720.
243. W. Oppolzer; M. Petrzilka, *J. Am. Chem. Soc. 98* (1976), 6722.
244. J. J. Tufariello, *Accts. Chem. Res. 12* (1979), 396.
245. V. N. Leibzon, S. G. Mairanovskii; E. I. Konnik, *Izv. Akad. Nauk SSSR 1971*, 1429.
246. I. Panfil; C. Belzecki, *Pol. J. Chem. 55* (1981), 977.
247. H. J. Brass; J. O. Edwards; N. J. Fina, *J. Chem. Soc., Perkin 2* (1972), 726.
248. M. Joucla; D. Grée; J. Hamelin, *Tetrahedron 29* (1973), 2315.
249. A. Quilico; P. Grünanger; R. Massini, *Gazz. Chim. Ital. 82* (1952), 349.
250. K. Bast; M. Christl; R. Huisgen; W. Mack; R. Sustmann, *Chem. Ber. 106* (1973), 3258.
251. P. Caramella; E. Albini; T. Bandiera; A. C. Coda; P. Grünanger; F. M. Albini, *Tetrahedron 39* (1983), 689.
252. M. Barzaghi; P. L. Beltrame; P. D. Croce; P. D. Buttero; E. Licandro; S. Maiorana; G. Zecchi, *J. Org. Chem. 48* (1983), 3807.
253. A. Z. Bimanand; K. N. Houk, *Tetrahedron Lett. 24* (1983), 435.
254. J. Sims; K. N. Houk, *J. Am. Chem. Soc. 95* (1973), 5798.
255. A. Padwa; L. Fisera; K. F. Koehler; A. Rodriguez; G. S. K. Wong, *J. Org. Chem. 49* (1984), 276.
256. P. DeShong; C. M. Dicken; R. R. Staib; A. J. Freyer; S. M. Weinreb, *J. Org. Chem. 47* (1982), 4397.
257. H. Seidl; R. Huisgen; R. Knorr, *Chem. Ber. 102* (1969), 904.
258. P. Caramella; T. Bandiera; P. Grünanger; F. M. Albini, *Tetrahedron, 40* (1984), 441.
259. K. Bast; M. Christl; R. Huisgen; W. Mack, *Chem. Ber. 106* (1973), 3312.
260. R. G. Micetich, *Org. Prep. Proc. 2* (1970), 225.
261. M. Masui; K. Suda; M. Yamaguchi; C. Yijima, *Chem. Pharm. Bull. 21* (1973), 1605.
262. P. D. Croce; C. La Rosa; R. Stradi; M. Ballabio, *J. Heterocycl. Chem. 20* (1983), 519.
263. R. V. Stevens; K. Albizati, *J. Chem. Soc., Chem. Commun. 1982*, 104.
264. H. Harikawa; T. Nishitani; T. Iwasaki; I. Inoue, *Tetrahedron Lett. 24* (1983), 2193.
265. Sk. A. Ali; P. A. Senaratne; C. R. Illig; H. Meckler; J. J. Tufariello; *Tetrahedron Lett. 1979*, 4167.
266. M. Joucla; J. Hamelin, *J. Chem. Res. (S) 1978*, 276.
267. J. B. Hendrikson; D. A. Pearson, *Tetrahedron Lett. 1983*, 4657.
268. K. Fukui In "Molecular Orbitals in Organic Chemistry" (P. O. Löwdin and B. Pullman, Eds.), Academic Press, New York, 1964, p. 513.
269. R. Sustmann, *Tetrahedron Lett. 1971*, 2717.
270. J. Bastide; N. ElGhandour; O. Henri-Rousseau, *Tetrahedron Lett. 1972*, 4225; *Bull. Soc. Chim. 1973*, 2290.
271. J. Bastide; O. Henri-Rousseau, *Bull. Soc. Chim. 1973*, 2294.
272. K. N. Houk, *Accts. Chem. Res. 8* (1975), 361.
273. L. Salem, *J. Am. Chem. Soc. 90* (1968), 543, 553.
274. J. J. Tufariello; Sk. A. Ali, *Tetrahedron Lett. 1978*, 4647.
275. G. Bianchi; C. De Micheli; R. Gandolfi; P. Grünanger; P. Vita Finzi; O. V. De Pava, *J. Chem. Soc., Perkin I 1973*, 1148.
276. M. Nitta; A. Omata; S. Hirayama; Y. Yajima, *Bull. Chem. Soc. Japan 56* (1983), 514.
277. F. De Sarlo; A. Guarna; A. Brandi, *J. Heterocycl. Chem. 20* (1983), 1505.
278. G. Bianchi; R. Gandolfi; P. Grünanger; A. Perotti, *J. Chem. Soc. C 1967*, 1598.
279. G. Bailo; P. Caramella; G. Cellerino; A. Gamba Invernizzi; P. Grünanger, *Gazz. Chim. Ital. 103* (1973), 47.
280. B. Bianchi; C. De Micheli; R. Gandolfi, *J. Chem. Soc., Perkin I 1976*, 1518.
281. D. N. Nicolaides; A. G. Catsasoumis, *Chim. Chronica 7* (1978), 189.
282. M. Joucla; J. Hamelin; D. Grée, *J. Chem. Res (S) 1978*, 276.
283. Y. Iwakura; K. Uno; Y. Kihara; M. Setsu; Y. Yamamoto, *Nippon Kagaku Kaishi* (1972), 1448.
284. P. Caramella; G. Cellerino; P. Grünanger; F. M. Albini; M. R. Cellerino, *Tetrahedron 34* (1978), 3545.
285. Y. Inouye; Y. Watanabe; S. Takahashi; H. Kakisawa, *Bull. Chim. Soc. Japan 52* (1979), 3763.
286. J. J. Tufariello; G. E. Lee; P. A. Senaratne; M. Al-Nuri, *Tetrahedron Lett. 1979*, 4359.
287. T. Kametani; S.-P. Huang; S. Yokohama; Y. Suzuki; M. Ihara, *J. Am. Chem. Soc. 102* (1980), 2060.
288. (a) A. Padwa; K. F. Koehler; A. Rodriguez, *J. Am. Chem. Soc. 103* (1981), 4974; *J. Org. Chem. 49* (1984), 282.
    (b) M. J. Fray; R. H. Jones; E. J. Thomas, *J. Chem. Soc., Perkin Trans. I, 1985*, 2753.
289. F. M. Albini; P. Ceva; A. Mascherpa; E. Albini; P. Caramella, *Tetrahedron 38* (1982), 3629.
290. M. Joucla; F. Tonnard; D. Grée; J. Hamelin, *J. Chem. Res. (S) 1978*, 240.

291. A. Padwa; J. G. MacDonald, *Tetrahedron Lett. 1982*, 3219; *J. Org. Chem. 48* (1983), 3189.
292. P. G. De Benedetti; S. Quartieri; A. Rastelli; M. De Amici; C. De Micheli; R. Gandolfi; P. Gariboldi, *J. Chem. Soc., Perkin II 1982*, 95.
293. A. Bened; R. Durand; D. Pioch; P. Geneste; J.-P. Declercq; G. Germain; R. Rambaud; R. Roques; C. Guimon; G. P. Guillovzo, *J. Org. Chem. 47* (1982), 2461.
294. P. Caramella; E. Cereda, *Synthesis 1971*, 433.
295. R. Fusco; L. Garanti; G. Zecchi, *Chim. Ind. (Milan) 57* (1975), 16.
296. T. Kusumi; S. Takahashi; Y. Sato; H. Kakisawa, *Heterocycles 10* (1978), 257.
297. R. L. Funk; L. M. H. Horcher, II; J. U. Dagget; M. M. Hansen, *J. Org. Chem. 43* (1983), 2632.
298. R. J. Ferrier; R. H. Furneaux; P. Prasit; P. J. Tyler; K. L. Brown; G. J. Gainsford; J. N. Diehl, *J. Chem. Soc., Perkin I 1983*, 1621.
299. D. St. C. Black; R. F. Crozier; J. D. Rae, *Aust. J. Chem. 31* (1978), 2013.
300. G. Bianchi; C. De Micheli; A. Gamba Invernizzi; R. Gandolfi, *J. Chem. Soc., Perkin I 1974*, 137.
301. C. De Micheli; A. Gamba Invernizzi; R. Gandolfi; L. Scevola, *J. Chem. Soc., Chem. Commun. 1976*, 246.
302. P. Caramella; F. M. Albini; D. Vitali; N. G. Rondan; Y.-D. Wu; T. R. Schwartz; K. N. Houk, *Tetrahedron Lett. 25* (1984), 1875.
303. P. Caramella; G. Cellerino, *Tetrahedron Lett. 1974*, 229.
304. E. J. McAlduff; P. Caramella; K. N. Houk, *J. Am. Chem. Soc. 100* (1978), 105.
305. P. Beltrame; P. L. Beltrame; P. Caramella; G. Cellerino; R. Fantechi, *Tetrahedron Lett. 1975*, 3543.
306. N. G. Rondan; M. N. Paddon-Row; P. Caramella; K. N. Houk, *J. Am. Chem. Soc. 103* (1981), 2436.
307. P. Caramella; N. G. Rondan; M. N. Paddon-Row; K. N. Houk, *J. Am. Chem. Soc. 103* (1981), 2438.
308. C. See; L. A. Paquette; G. De Lucca; K. Ohkala; J. C. Galucci, *J. Am. Chem. Soc. 107* (1985), 1015, and references cited therein.
309. R. Lazar; F. G. Cocu; N. Barbulescu, *Rev. Chim. (Bucharest) 20* (1969), 3. *C.A. 70*, 114341n (1969).
310. G. Brüntrup; M. Christl, *Tetrahedron Lett. 1973*, 3369.
311. M. G. Barlow; R. N. Hazeldine; W. D. Morton; D. R. Woodward, *J. Chem. Soc., Perkin I 1973*, 1798.
312. N. B. Das; K. B. G. Torssell, *Tetrahedron 39* (1983), 2247.
313. K. N. Houk; S. R. Moses; Y.-D. Wu; N. G. Rondan; V. Jäger; R. Schohe; F. R. Fronczet, *J. Am. Chem. Soc. 106* (1984), 3880.
314. V. Jäger; R. Schohe; E. F. Paulus, *Tetrahedron Lett. 1983*, 5501.
315. A. P. Kozikowski; A. K. Ghosh, *J. Am. Chem. Soc. 104* (1982), 5788; *J. Org. Chem. 49* (1984), 2762.
316. A. A. Hagedorn, III; B. J. Miller; O. J. Nagy, *Tetrahedron Lett. 1980*, 229.
317. T. G. Burrows; W. R. Jackson; S. Faulks; I. Sharp, *Aust. J. Chem. 30* (1977), 1855.
318. J. M. J. Tronchet; S. Thorndahl-Jaccard; L. Faivre; R. Massard, *Helv. Chim. Acta 56* (1973), 1103.
319. J. M. J. Tronchet; A. Jotterand; N. Le Hong; F. Perret; S. Thorndahl-Jaccard; J. Tronchet; J. M. Chalet; L. Faivre; C. Hausser; C. Sébastian, *Helv. Chim. Acta 53* (1970), 1484.
320. J. M. J. Tronchet; N. LeHong, *Carbohydrate Res. 29* (1973), 311.
321. R. H. Jones; G. C. Robinson; E. J. Thomas, *Tetrahedron 40* (1984), 177.
322. A. P. Kozikowski; Y. Kitagawa; J. P. Springer, *J. Chem. Soc., Chem. Commun. 1983*, 1460.
323. P. De Shong; M. Leginus, *J. Am. Chem. Soc. 105* (1983), 1636.
324. T. Koizumi; H. Hirai; E. Yoshii, *J. Org. Chem. 47* (1982), 4004.
325. I. Panfil; C. Belzecki, *Pol. J. Chem. 55* (1981), 977.
326. I. Panfil; C. Belzecki, *J. Chem. Soc., Chem. Commun. 1977*, 303; *J. Org. Chem. 44* (1978), 1212.
327. P. M. Wovkulich; M. R. Uskoković, *J. Am. Chem. Soc. 103* (1981), 3956.
328. A. Vasella, *Helv. Chim. Acta 60* (1977), 426, 1273.
329. A. Vasella; R. Voeffrey, *J. Chem. Soc., Chem. Commun. 1981*, 97; *Helv. Chim. Acta 65* (1982), 1134; *66* (1983), 1241; B. Bernet; E. Krawczyk; A. Vasella, *Helv. Chim. Acta 68* (1985), 2299; R. Huber; A. Knierzinger; J. P. Obrecht; A. Vasella, *Helv. Chim. Acta 68* (1985), 1730.
330. I. Müller, *Dissertation*, Würzburg 1984.
331. O. Tsuge; M. Tashiro; Y. Nishihara, *Nippon Kagaku Zasshi 92* (1971), 72. *C.A. 77* (1972), 5392 d.
332. L. Fisera; P. Oravec, *Collect. Czech. Chem. Commun. 52* (1987), 1315.

333. Ya. D. Samuilov; S. E. Soloveva; A. I. Konovalov, *Zh. Org. Khim. 16* (1980), 1228. *C.A. 93* (1980), 185323 u.
334. T. Hayakawa; K. Araki; S. Shiraishi, *Bull. Soc. Chim. Japan 57* (1984), 1643.
335. S. Shiraishi; B. S. Holla; K. Imamura, *Bull. Soc. Chim. Japan 56* (1983), 3457.
336. C. M. Dicken; P. De Shong, *J. Org. Chem. 47* (1982), 2047.
337. A. Baranski, *Pol. J. Chem. 56* (1982), 1585.
338. M. Nitta; A. Omata; S. Hirayama; Y. Yajima, *Bull. Soc. Chem. Japan 56* (1983), 514.
339. S. K. Dubey; E. E. Knaus, *J. Org. Chem. 49* (1984), 123.
340. T. Hayakawa; K. Araki; S. Shiraishi, *Bull. Soc. Chim. Japan 57* (1984), 1643.
341. L. I. Vasileva; G. S. Akimova; V. N. Chistokletov, *Zh. Org. Khim. 20* (1984), 148.
342. A. Corsaro; U. Chiacchio; G. Perrini; P. Caramella; G. Purello, *J. Chem. Res.* (S) *1984*, 402.
343. F. M. Albini; E. Albini; T. Bandiera; P. Caramella, *J. Chem. Rec.* (S) *1984*, 36.
344. N. G. Argyropoulos; E. Coutouli-Argyropoulou; P. Pistikopoulos, *J. Chem. Res.* (S) *1984*, 362.
345. N. G. Argyropoulos; E. Coutouli-Argyropoulou, *J. Heterocycl. Chem. 21* (1984), 1397.
346. N. N. Magdesieva; T. A. Sergeeva, *Zh. Org. Khim. 20* (1984), 1590.
347. L. R. Hepp; J. Bordner; T. A. Bryson, *Tetrahedron Lett. 1985*, 595.
348. C. Parini; S. Colombi; A. Ius; C. Renato; R. Longhi; G. Vecchio, *Gazz. Chim. Ital. 107* (1977), 559.
349. C. De Micheli; R. Gandolfi; P. Grünanger, *Tetrahedron 30* (1974), 3765.
350. G. L'abbe; G. Mathys, *J. Org. Chem. 39* (1974), 1221.
351. S. Rajappa; B. G. Advani; R. Sreenivasan, *Synthesis 1974*, 656.
352. R. Huisgen; W. Mack, *Chem. Ber. 105* (1972), 2815.
353. N. A. Akmanova; Yu. M. Shaul'skii; Ya. V. Svetkin, *Zh. Org. Khim. 12* (1976), 88.
354. Ya. D. Samuilov; S. E. Solov'eva; T. F. Girutskaya; A. I. Konovalov, *Zh. Org. Khim. 14* (1978), 1693.
355. Ya. D. Samuilov; S. E. Solov'eva; A. I. Konovalov; T. G. Mannafov, *Zh. Org. Khim. 15* (1979), 279.
356. Ya. D. Samuilov; S. E. Solov'eva; A. I. Konovalov; *Zh. Obshch. Khim. 49* (1979), 637; *Dokl. Akad. Nauk SSR 255* (1980), 606.
357. A. Dondoni; G. Barbaro, *J. Chem. Soc., Perkin II, 1974*, 1591.
358. K. N. Houk; K. Yamaguchi In "1,3-Dipole Cycloaddition Chemistry" (A. Padwa, Ed.,), Vol. 2, Wiley, New York, 1984, p. 407.
359. (a) W. Bihlmeyer; J. Geittner; R. Huisgen; N.-U. Reissig, *Heterocycles 10* (1978), 147.
(b) K. N. Houk; R. A. Firestone; L. L. Munchausen; P. H. Mueller; B. H. Arison; L. A. Garcia, *J. Am. Chem. Soc. 107* (1985), 7227.
360. P. Battioni; L. Vo Quang; Y. Vo Quang, *Bull. Soc. Chim. France 1978*, 415.
361. E. Arlandini; M. Ballabio; L. Da Prada; M. L. Rossi; P. Trimarco, *J. Chem. Res.* (S) *1983*, 170.
362. K. Tanaka; H. Masuda; K. Mitsuhashi, *Bull. Chem. Soc. Japan, 57* (1984), 2184; *58* (1985), 2061.
363. I. Panfil; M. Chmielewski, *Tetrahedron 41* (1985), 4713.
364. T. Minami; T. Hanamoto; I. Hirao, *J. Org. Chem. 50* (1985), 1278.
365. S. Niwayama; S. Dan; Y. Inouye; H. Kakisawa, *Chem. Lett. 1985*, 957.
366. A. N. Frolkov; Yu. I. Kheruze; V. N. Chistokletov; A. A. Petrov, *Zh. Org. Khim. 21* (1985), 730.
367. Y. Shen; J. Zheng; Y. Huang, *Synthesis 1985*, 970.
368. R. Annunziata; M. Cinquini; F. Cozzi; L. Raimondi, *J. Chem. Soc., Chem. Commun. 1985*, 403.
369. R. Annunziata; M. Cinquini; F. Cozzi; L. Raimondi; A. Restelli, *Helv. Chim. Acta 68* (1985), 1217.
370. M. Cinquini; F. Cozzi; A. Gilardi, *J. Chem. Soc., Chem. Commun. 1984*, 551.
371. R. Annunziata; M. Cinquini; F. Cozzi; A. Restelli, *J. Chem Soc., Chem. Commun. 1984*, 1253; *J. Chem Soc., Perkin I 1985*, 2293; R. Annunziata; M. Cinquini; F. Cozzi; A. Gilardi; A. Restelli, *J. Chem. Soc., Perkin I 1985*, 2289.
372. P. A. Wade; D. T. Price; J. P. McCauley; P. J. Carroll, *J. Org. Chem. 50* (1985), 2804.
373. H. Gnichtel; L. Autenrieth; P. Luger; K. Vangehr, *Liebigs Ann. 1982*, 1091; H. Gnichtel; L. Autenrieth-Ansorge, *Liebigs Ann. 1985*, 2217.
374. W. Adam; N. Carballiera; E. Crämer; U. Lucchini; E.-M. Peters; K. Peters, H. G. von Schnering, *Chem. Ber. 120* (1987) 695.
375. K. Tanaka; M. Kishida; S. Maeno; K. Mitsuhashi, *Bull. Chem. Soc. Japan 59* (1986) 2631.
376. L. Fisera; N. D. Kosina; L. A. Badovskaya; L. Stribanyi, *Chem. Pap. 40* (1986) 685.
377. A. Hassner; K. S. K. Murthy, *Tetrahedron Lett. 1987*, 4097.

Chapter 2

The Nitrile Oxides

2.1 REACTIONS AND PHYSICOCHEMICAL PROPERTIES

Most nitrile oxides (**1**) are short-lived, reactive species, structurally isomeric with isocyanates (**2**) and cyanates (**3**). The parent member of the series, fulminic acid (**1**: R = H) and oxalodinitrile oxide (**4**) are explosives and have to be handled with great care. Generally, compounds with more than one —C≡N—O group in the molecule are potentially dangerous compounds. Since it is unnecessary to isolate the nitrile oxides in pure condition for synthetic purposes, this instability is, in practice, no serious problem.

$$R—C≡N—O \qquad\qquad R—N=C=O$$
$$\mathbf{1} \qquad\qquad\qquad\qquad \mathbf{2}$$

$$R—O—C≡N \qquad\qquad O—N≡C—C≡N—O$$
$$\mathbf{3} \qquad\qquad\qquad\qquad \mathbf{4}$$

In the absence of suitable reactants such as olefins or acetylenes, the nitrile oxides dimerize to furoxans; 1,2,5-oxadiazole 2-oxides (**5**), rearrange into isocyanates **2** or polymerize, eq. (2.1).[1] Macrocyclic oligomers ($n = 6,7,8$) have been isolated in low yields in this process.[2] Trimethylamine catalyzed dimerization leads to the formation of 1,2,4-oxadiazole 4-oxides (**6**), and pyridine directs the dimerization to the 1,4,2,5-dioxadiazine **7**.[3,6] Triethylamine gives substantial amounts of 1,2,4-oxadiazole (**8**) and *N*,*N*-diethylvinylamine, which can be trapped as a cycloadduct.[4] Depending on the conditions trifluoroacetonitrile oxide gives either **5** or **7** (R = CF$_3$) on dimerization, eq. (2.1).[5] The true structure of the polymer is uncertain. Thermal depolymerization leads, in some cases, to isocyanate, and treatment with hydrogen chloride leads to hydroximic acid chloride.

The cycloaddition of two nitrile oxides to furoxan (**5**) is frontier orbital controlled and favors carbon-carbon and nitrogen-oxygen couplings.[7] Calculations show that the energy surface of the cyclodimerization passes an energy minimum corresponding to a 1,2-dinitroso structure, eq. (2.1).[8] The transition state is thus asymmetric, and the synchronous character of the 1,3-dipolar cycloadditions has also been questioned.[9] The furoxans exist as a tautomeric

x = nucleophile
y = electrophile

mixture. The equilibrium constant and free energy difference for unsymmetrical furoxanes have been determined: $E_a \approx 130\text{–}150$ kJ mol^{-1}.[10]

The half-life of aromatic nitrile oxides is approximately minutes to days, and that of aliphatic nitrile oxides approximately seconds to minutes.[1] Bulky substituents in the ortho-positions of the benzene ring or on the α-carbon of the aliphatic derivatives enhance the stability considerably, and some sterically hindered nitrile oxides are permanently stable (e.g., 2,6-dimethylbenzonitrile oxide and di-*tert*-butylacetonitrile oxide). Fulminic acid is exceptional in that it does not give a stable furoxan. The half-life of a 0.4 M aqueous fulminic acid solution is 70 min at 0°C and pH = 1.[11] The stability is pH dependent; stability decreases with increasing pH. The reactions of fulminic acid are complex and give rise to a number of trimers and tetramers or polymers.[12-15] This is a consequence of its acidic nature. Equation (2.2) depicts the formation of isocyanilic acid **10a** and metafulminuric acid **10b** from fulminic acid.

$$H-C\equiv N-O \rightleftharpoons H^{\oplus} + {}^{\ominus}C\equiv N-O \xrightarrow{HCNO}$$

$$O-N\equiv C-CH=N-OH \xrightarrow{Dimerization} HON=HC\underset{\underset{O}{N}}{\overset{}{-}}\underset{\underset{O}{N}}{\overset{}{-}}CH=NOH$$
9 → Acetylene, HCNO → **10a**　　　(2.2)

10b: isoxazole with =NOH and NOH substituents

11: isoxazole with –CH=NOH

The existence of the transient dimer **9** was proved by trapping with acetylene, to compound **11**.[16,17] The structure of fulminic acid has been debated for more than a century.[18] The correct tautomeric structures, eq. (2.3), were suggested by Nef[19] and Ley.[20] Infrared (IR) measurements of gaseous fulminic acid showed eventually that formonitrile oxide represents the correct structure.[21]

$$HO-N=C \underset{(\leftarrow)}{\xrightarrow{}} O-N\equiv CH \qquad (2.3)$$

A small group of compounds are actually known to contain the frozen tautomeric isonitrile structure, namely, E. Müller's "isodiazomethane" **12** and the N,N-dialkylated aminoisonitrile **13**,[22,23] They behave as isonitriles and show no similarity to the nitrile oxides. Whereas the dimerization of nitrile oxides to furoxans occurs spontaneously at or below room temperature, the rearrangement to isocyanates requires heating.[24,25]

$$H_2C=N=N \qquad C=N-NH_2 \qquad C=N-N{\overset{R}{\underset{R}{\diagdown}}}$$

Diazomethane　　　　　**12**　　　　　**13**

The mechanism of the thermal rearrangement of the nitrile oxide into isocyanate **2** is not definitely known. It does not proceed via separated ions or radicals because there is no sign of scrambling when a mixture of two differently labeled nitrile oxides is subjected to thermal rearrangements, eq. (2.4).[26,27a]

Attempts to trap an intermediary acylnitrene failed, eq. (2.5). The rearrangement occurs with complete retention of configuration, and there is no loss of optical activity, as demonstrated by the rearrangement of (−)-2-methyl-

$$R^1\text{-}^{13}C{\equiv}N\text{-}O + R^2\text{-}C{\equiv}N\text{-}O \xrightarrow{\Delta} R^1N^{13}CO + R^2NCO$$

$$\begin{cases} R^{1+} \\ R^{2+} \end{cases} + \begin{array}{c} ^{13}C^-{\equiv}N\text{-}O \\ C^-{\equiv}N\text{-}O \end{array} \quad \text{or/and} \quad \begin{cases} R^{1\cdot} \\ R^{2\cdot} \end{cases} + \begin{array}{c} ^{13}\dot{C}{\equiv}N\text{-}O \\ \dot{C}{\equiv}N\text{-}O \end{array} \tag{2.4}$$

$$R^1NCO + R^2N^{13}CO$$

$$R\text{-}C{\equiv}N\text{-}O \xrightarrow{\Delta} (R\text{-}\overset{O}{\underset{\|}{C}}\text{-}N) \longrightarrow R\text{-}N{=}C{=}O \tag{2.5}$$

$$\downarrow$$

$$R\text{-}\overset{O}{\underset{\|}{C}}\text{-}N\hspace{-0.5em}\bigcirc$$

2-phenylbutyronitrile oxide, camphene nitrile oxide, and norbornene nitrile oxide.[26,27a]

The intramolecular transition state **14** was suggested for the rearrangement, actually implying that no oxygen exchange takes place. Decomposition via dimerization to **7** is attractive but could be excluded, because **7** decomposes slowly on prolonged heating to give a complex mixture containing no isocyanate.[27a] Double isotopic (^{2}H, ^{18}O) labeling shows that the rearrangement occurs exlusively between nitrile oxide molecules, and the kinetic measurements suggest a pathway involving polymerization and cleavage as depicted in **14b**.[27b]

<u>7</u>   <u>14a</u>   <u>14b</u>
                    polymer

Hydrolysis of nitrile oxides leads to hydroxamic acids and ultimately to carboxylic acids and hydroxylamine, eq. (2.6). This is analogous to the hydrolysis of nitriles into ammonia and carboxylic acids. The reverse reaction, i.e., dehydration of hydroxamic acids to nitrile oxides, is difficult to achieve. Formation of isocyanates, Lossen's rearrangement, takes place. On the other hand, dry hydrochloric or hydrobromic acid add reversibly to nitrile oxides with formation of hydroximoyl halides, from which the nitrile oxide is regenerated by adding a base, eq. (2.6); cf. Sections 2.4 and 2.5.

$$RC\equiv N\text{-}O \underset{-H_2O}{\overset{H_2O}{\rightleftharpoons}} \underset{\overset{\|}{O}}{RC\text{-}NHOH} \longrightarrow R\text{-}COOH + NH_2OH$$

$$\underset{\text{base}}{-HCl}\Bigg\Updownarrow HCl \qquad\qquad \Bigg\downarrow -H_2O \tag{2.6}$$

$$RCCl=NOH \qquad\qquad RN=C=O$$

Nitrile oxides act as mild oxidants and liberate iodine from a solution of potassium iodide. Deoxygenation of nitrile oxides is effected by several reducing agents, e.g., zinc dust or tin and hydrochloric acid.[28,29]

The nitrile oxide function imposes the same solubility characteristics on the molecule as the cyano group does. It gives weak hydrogen bonds with protic solvents, as demonstrated by IR spectra.[30] 2,4,6-Trimethylbenzonitrile oxide (15) shows two characteristic strong absorptions at frequencies 2290 cm^{-1} (—C≡N) and 1334 cm^{-1} (≡N—O) in the IR spectrum (KBr).[31] The absorption at 2290 cm^{-1} is stronger and broader than the corresponding nitrile absorption, which is located at ca. 70 cm^{-1} lower frequency. The nitrile oxide group is linear, in analogy to the geometry of the diazo and azide groups. The ^{13}C NMR spectrum of 15 shows some interesting features.[32] The magnitude of $^1J_{^{13}C^{15}N}$ is 77.5 Hz, by far the largest one-bond C—N spin coupling constant yet recorded. The ^{13}C≡N—O is found at δ 35.7 ppm, i.e., a high field shift of 81.8 ppm in comparison with the ^{13}C shift of the corresponding nitrile 16, δ 117.5 ppm (tetramethylsilane). This implies that the carbon atom carries a considerable negative charge; i.e., structure 17a contributes to the ground state of the nitrile oxide, which agrees with an x-ray analysis of 4-methoxy-2,6-dimethylbenzonitrile oxide.[33] Relevant data are shown in structure 18. The N—O bond is remarkably short, indicative of the electronic distribution represented by

$$\underset{\underline{15}}{\text{2,4,6-(CH}_3)_3\text{C}_6\text{H}_2\text{-CNO} \;\; (35.7)} \qquad \underset{\underline{16}}{\text{2,4,6-(CH}_3)_3\text{C}_6\text{H}_2\text{-CN} \;\; (117.5)}$$

$$\underset{\underline{17a}}{R\text{-}\overset{-}{C}=\overset{+}{N}=O} \longleftrightarrow \underset{\underline{17b}}{R\text{-}\overset{+}{C}\equiv\overset{-}{N}\text{-}O}$$

Structure 18: 4-methoxy-2,6-dimethylphenyl-C≡N-O with angles 178.3° (at C1-C7), 173.8° (at C7≡N), bond lengths:

C^{1}C^7	1.435 Å
C^7≡N	1.147 Å
N–O	1.249 Å

2.2 SYNTHESIS OF FULMINIC ACID (FORMONITRILE OXIDE)

Since the synthesis and reactions of fulminic acid deviate from those of other nitrile oxides, they are treated separately. Unstable fulminic acid can be generated in several ways. The classical method is by acidification of its salts. **Warning: The fulminates are all very toxic and potentially explosive!** They are quite sensitive to shock and heat. Alfred Nobel actually used mercuric fulminate as an initial explosive for detonating nitroglycerin. Mercuric fulminate is the most practical salt to begin with, and reliable methods for its preparation have been published.[41-45] Sodium fulminate is prepared from mercuric fulminate by treating it with sodium amalgam.[44,45] The acid is normally liberated by adding sodium fulminate to aqueous sulfuric acid below 0°.[11,17] The reverse addition is not recommended because fulminic acid is more stable in acid solution. If an organic solvent is desired, the acid can be extracted by ether.[11] Since polymerization[12-15] of fulminic acid is a serious competitive side reaction affecting the yield of the 1,3-dipolar cycloaddition, the freshly prepared solution of sodium fulminate and fulminic acid should be used at once. Moreover, the dimerization and polymerization can be repressed by slow liberation of fulminic acid, and this technique has sometimes been used.[46-48] The use of fulminates is restricted to reactants soluble in water. However, use of a vigorously stirred two-phase system (i.e., water/ether), where the fulminic acid is generated slowly in the aqueous phase and extracted into the ether layer, where the cycloaddition takes place, does not seem to have been tested rigorously. Acetone is unsuitable as a solvent because fulminic acid adds to the carbonyl group and forms a new nitrile oxide, eq. (2.7),[49,50] a reaction reminiscent of the cyanohydrin reaction.

$$HCNO + (CH_3)_2CO \longrightarrow (CH_3)_2\underset{}{\overset{OH}{C}}-C\equiv N-O \qquad (2.7)$$

Generating fulminic acid from formohydroximoyl iodide[16,48,51] by treating with a base, eq. (2.8), in an organic solvent alleviates the solubility problem, and the cycloaddition proceeds with improved yields.

$$HCI=NOH \xrightarrow{Et_3N} HC\equiv NO + Et_3NHI \qquad (2.8)$$

Formohydroximoyl iodide is a solid, mp 65°C (dec),[52] that slowly decomposes at room temperature or even explodes after being stored for a long time. It

Synthesis of Fulminic Acid (Formonitrile Oxide)

is best kept in ether solution at $-20°C$. The compound is prepared either from mercuric fulminate by treating it with hydroiodic acid and potassium iodide, eq. (2.9),[48] or from sodium fulminate and hydroiodic acid.[49]

$$Hg(CNO)_2 + 4\ HI + 2\ KI \longrightarrow 2\ HC\begin{smallmatrix}NOH\\I\end{smallmatrix} + K_2HgI_4 \qquad (2.9)$$

Fulminotrimethylsilane, which can be prepared from chlorotrimethylsilane, or better yet from bromotrimethylsilane and mercuric fulminate, is a stable colorless liquid at room temperature[53,54] highly sensitive to moisture with the formation of fulminic acid, eq. (2.10).

$$Hg(CNO)_2 + 2(CH_3)_3SiBr \longrightarrow (CH_3)_3SiC\equiv N-O \xrightarrow{H_2O} HCNO + (CH_3)_3SiOH$$

(2.10)

The generation of fulminic acid under neutral conditions can be controlled by carrying out the reaction in moist tetrahydrofuran.[55] This procedure gives good yields of cycloaddition products.

Fulminic acid is formed by nitrosation of nitromethane to formonitrolic acid. Elimination of nitrous acid and precipitation as silver salt completes the process, eq. (2.11).[56] The only method that does not pass via the explosive fulminates, is the flash pyrolysis at $450°C$ and 10^{-4} mmHg of isoxazol-5(4H)-ones **19**, eq. (2.12).[57] The mixture of carbon dioxide, nitrile, and fulminic acid produced is condensed in the Dewar flask, where it is brought into contact with other reactants. The yield of fulminic acid is reported to be practically quantitative.

$$CH_3NO_2 \xrightarrow{NO^+} O_2NCH=NOH \xrightarrow[Ag^+]{-HNO_2} AgCNO \qquad (2.11)$$

$$RCOCH_2COOR' \xrightarrow[2.\ NO^+]{1.\ NH_2OH} \mathbf{19} \xrightarrow{450\ °C} RCN + CO_2 + HCNO \qquad (2.12)$$
$$R = CH_3,\ C_6H_5$$

The Mukaiyama-Hoshino procedure[58] (Section 2.4) fails for nitromethane. Free fulminic acid does not appear in this process (route 2.13a). α-Nitroacetanilide (**20**) is formed first and subsequently transformed to the nitrile oxide (**21**) (route 2.13b).[48,59] In the silyl nitronate procedure[60] (Chapter 4) the trimethylsilylester of *aci*-nitromethane (**22**) is trapped by the olefin (route 2.14a). The isoxazolidine **23** can be isolated in good yield. The presence of free fulminic acid has not been demonstrated in this reaction; route 2.14b is therefore questionable.

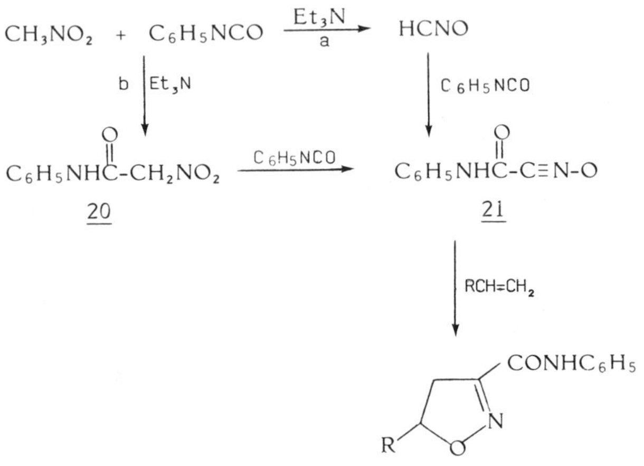

$$CH_3NO_2 + C_6H_5NCO \xrightarrow{Et_3N}_{a} HCNO \tag{2.13}$$

(Scheme 2.13: CH₃NO₂ + C₆H₅NCO route a with Et₃N gives HCNO; route b with Et₃N gives C₆H₅NHC(O)-CH₂NO₂ (**20**), which with C₆H₅NCO gives C₆H₅NHC(O)-C≡N-O (**21**); HCNO with C₆H₅NCO also gives **21**; **21** with RCH=CH₂ gives 5-R-4,5-dihydroisoxazole-3-CONHC₆H₅.)

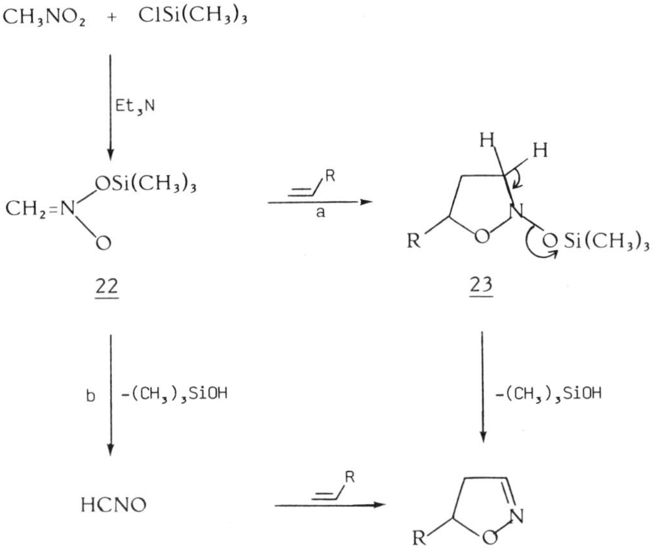

$$CH_3NO_2 + ClSi(CH_3)_3 \xrightarrow{Et_3N} CH_2=N(OSi(CH_3)_3)O \quad (\mathbf{22}) \tag{2.14}$$

(Scheme 2.14: **22** via route a with RCH=CH₂ gives isoxazolidine **23** bearing OSi(CH₃)₃; route b loses (CH₃)₃SiOH to give HCNO, which with alkene gives 5-R-4,5-dihydroisoxazole; **23** loses (CH₃)₃SiOH to give the same 4,5-dihydroisoxazole.)

2.3 SYNTHESIS OF HALOGEN AND SULFUR-SUBSTITUTED FULMINIC ACID

Metal fulminates (Na, Ag, Hg) are converted to dihaloformoximes **24** by halogens, eq. (2.15)[61–63] The unpleasant, toxic dichloro compound **24**, bp 47°C/18 mmHg, mp 39–40°C, and bromo compound, mp 70–71°C, are stable at room temperature, whereas the iodo compound decomposes. A preparative method for dichloroformoxime (**24**) (X = Cl), which avoids the fulminates, is reduction of chloropicrin with tin or iron powder in hydrochloric acid, eq. (2.16).[64] Treating the dichloroformoximes with a base gives the halonitrile oxides (**25**). In the absence of a reactive substrate dimerization of the halonitrile oxide to 3,4-dihalofuroxan **26** takes place, eq. (2.18). The yield of **25** is improved by using silver nitrate rather than a base.[65] Generating **25**, X = Cl, is of interest in the context of preparing[65-68] certain isoxazoline antibiotics **27**.[69] Benzenesulfonylnitrile oxide (**29**) is synthesized from benzenesulfonylnitromethane via phenylsulfonylbromoformoxime (**28**), eq. (2.19).[70-73a] It cycloadds to alkenes, forming 3-phenylsulfonylisoxazolines, which readily undergo a variety of substitution reactions. The reactions of sulfonyl substituted furoxans have also been investigated.[73b]

$$\text{MCNO} + X_2 \xrightarrow{H_2O} X_2C=NOH + MX \quad (2.15)$$

M = Na, Ag, Hg
X = Cl, Br, I
                                                     24

$$Cl_3CNO_2 \xrightarrow{[H]} Cl_3CNO \xrightarrow{[H]} Cl_2CHNO \longrightarrow Cl_2C=NOH \quad (2.16)$$

$$X_2C=NOH \xrightarrow[\text{or AgNO}_3]{\text{Base}} XCNO \quad (2.17)$$

24   **25**

$$2 \text{ XCNO} \longrightarrow \underset{\textbf{26}}{\text{(3,4-dihalofuroxan)}} \quad (2.18)$$

[Structure of compound **27**: chloro-isoxazoline with CH(NH₂)-COOH substituent]

$$PhSO_2CH_2NO_2 \xrightarrow[\text{2. CH}_2N_2, 0°C]{\text{1. Br}_2} PhSO_2CBr=N\overset{OCH_3}{\underset{O}{\diagdown}} \xrightarrow[-CH_2O]{40°C} \quad (2.19)$$

$$\underset{Br}{\overset{PhSO_2}{\diagdown}}C=N\overset{OH}{\diagup} \xrightarrow[\text{or AgNO}_3]{\text{Base}} PhSO_2CNO \xrightarrow{R^1} [\text{3-PhSO}_2\text{-isoxazoline}] \xrightarrow{Nu^-} [\text{3-Nu-isoxazoline}]$$

28   **29**

Nu⁻ = H⁻(BH₄⁻), R⁻, CN⁻, RO⁻

2.4 SYNTHESIS OF NITRILE OXIDES

Two methods of general applicability are available: (a) oxidation of aldoximes and (b) dehydration of nitro compounds. Methods using aldoximes as starting material are discussed first. The very first nitrile oxide prepared, benzonitrile oxide (31), was obtained in two steps: by chlorinating either E- or Z-benzaldoxime and subsequently dehydrohalogenating with a base, eq. (2.20);[74] this method is the most frequently used.

$$C_6H_5CH{=}NOH \xrightarrow{Cl_2} |C_6H_5CHClNO| \rightarrow \underset{Cl}{\overset{C_6H_5}{>}}{=}N{-}OH \xrightarrow[-HCl]{Base} C_6H_5CNO \quad (2.20)$$

$$\text{30a} \qquad \text{(Z) 30b} \qquad \text{31}$$

Ether and chloroform are preferred as solvent and chlorine is introduced at ca. $-20°C$ to $-40°C$ until the original blue color, indicative of a transient nitroso compound **30a** changes to light green.[75,97] Aqueous sodium hydroxide or sodium carbonate is used as a base, preferably at 0°C. The hydroximoyl chlorides are fairly stable, but the second step, the dehydrochlorination, has to be carried out in the presence of the substrate, since the lifetime of the reactive nitrile oxide is too short for isolation. The benzohydroximoyl chloride (**30b**) is formed in the Z-form.[76] The E-form can be prepared by photoisomerization of the acetate and subsequent hydrolysis.[77] The E-form loses HCl to give benzonitrile oxide 6×10^7 times slower than the Z-form. The dehydrochlorination procedure has been improved by the use of a one-phase system with a tertiary amine, such as triethylamine,[78-80] as base. The amine is added slowly to a mixture of hydroximoyl chloride and the olefin in order to keep the nitrile oxide concentration low, which is essential to prevent furoxan formation. Occasionally, side reactions occur with the use of tertiary amines, resulting in the formation of adducts with the hydroximoyl chlorides, eq. (2.21).[4,81-83]

$$R\text{-CClNOH} + Et_3N \rightarrow R\text{-}\underset{\overset{|}{{}^+NEt_3}}{C}{=}N{-}OH\ Cl^- \quad (2.21)$$

It is sometimes recommended to add equivalent amounts of tertiary amine and hydroximoyl chloride simultaneously in small portions to the dipolarophile.

The major disadvantage of this method is the chlorination step. It cannot be used for oximes containing other functions sensitive to chlorine, e.g., unsaturation,[84] ketones,[85] or certain aromatic rings.[84,86,87] Chlorination of tiglaldoxime (**32**) gives the chlorinated product **33**, eq. (2.22). Salicylaldoxime (**34**) and its 2-O-methyl derivative (**35**) are chlorinated on the benzene ring to give **36** and **37**, respectively, eq. (2.23).[84] Thiophenealdoxime (**38**) gives 5-chlorothiophene-2-hydroximoyl chloride (**39**), eq. (2.24).[86,87] Methyl groups activate the nucleus sufficiently to cause chlorination of the benzene ring.[88]

Synthesis of Nitrile Oxides

$$CH_3CH=C(CH_3)CH=NOH \xrightarrow{Cl_2} CH_3CHCl-CCl(CH_3)-CCl=NOH \quad (2.22)$$

<u>32</u>   <u>33</u>

(2.23)

<u>34</u> R = OH
<u>35</u> R = OCH$_3$

<u>36</u> R = OH, X = Cl
<u>37</u> R = OCH$_3$, X = H

(2.24)

<u>38</u>   <u>39</u>

Nitrosyl chloride is a milder reagent that has been used for selective chlorination in such cases.[86-88]

A selective dehydrogenating agent, which combines the two steps in eq. (2.20) and leaves sensitive functions, such as double bonds, untouched is alkaline aqueous sodium hypochlorite or hypobromite, eq. (2.25).[87,89-92] The reaction is best carried out in a two-phase system.[91] Apparently, the organic solvent protects the olefin from being attacked by the hypohalite ion. The nitrile oxide cycloadds to the olefin in the organic phase. Adding triethylamine has a favorable influence on the yield. The reaction fails when the reactants contain alkali labile functions or are otherwise sensitive to this particular oxidant.

$$R-CH=N-OH \xrightarrow[OH^-]{ClO^-} |RCCl=NOH| \longrightarrow R-CNO \quad (2.25)$$

N-Bromosuccinimide[93,94] and N-chlorosuccinimide (NCS)[95,96] have been used successfully in some cases as halogenating agents for aldoximes. This procedure is modified by using chloroform as the solvent for NCS chlorination, eq. (2.26).[97] NCS is only slightly soluble in chloroform, whereas succinimide (SI) is highly soluble. The end point of the chlorination is reached when the solid has dissolved and is therefore easy to observe. The reaction is catalyzed by pyridine and proceeds under neutral conditions at room temperature. Olefinic functions, thiophenes, furans, pyrroles, and methoxylated aromatic nuclei are not attacked. It is also possible to chlorinate salicylaldoxime selectively.[158]

$$R-CH=NOH + NCS \longrightarrow R-CCl=NOH + SI \quad (2.26)$$

Triethylamine is used as the base to liberate the nitrile oxide in the final step, and intramolecular cycloaddition of suitable unsaturated aldoximes can thus be achieved. The reaction is further simplified by using pyridine or a solid-phase base such as basic Al$_2$O$_3$ or Florisil, which are inert toward NCS.[98] Chlorination, dehydrochlorination, and addition can be performed as a one-step reaction simply by mixing all components in chloroform as solvent, eq. (2.27). Intramolecular cycloaddition has also been performed by oxidation with nitrogen dioxide, eq. (2.28).[99,100]

$$RCH=NOH + \underset{X}{\diagdown\!\!=\!\!\diagup} \xrightarrow{NCS, Al_2O_3} \text{(isoxazoline)} \qquad (2.27)$$

$$\text{(o-allyloxybenzaldoxime)} \xrightarrow{NO_2} \text{(fused isoxazoline)} \qquad (2.28)$$

Potassium ferricyanide[87] and lead tetraacetate[101-103] (at $-70°C$) have also been reported as dehydrogenating agents for aldoximes. The nitrile oxide and the hydrogen chloride are in equilibrium, eq. (2.29) with the corresponding hydroximoyl chloride. Thermolysis of hydroximoyl chlorides[104-109] in an inert solvent has therefore been used for generating a low concentration of the nitrile oxides that can be trapped by a suitable reagent.

$$R\,CCl=NOH \;\rightleftharpoons\; RCNO \;+\; HCl \qquad (2.29)$$

Nitrile oxides are formed in several reactions of less general applicability. Pyrolysis of furoxans[110-112] at ca. 500°C gives nitrile oxides, eq. (2.30). A number of cycloreversion studies of furoxans have been reported,[117-125] and it has been observed that dissociation into two moles of nitrile oxide also takes place in refluxing xylene. Isoxazolines are formed in good yields in the presence of 1-2 equivalent of olefins.

$$\text{(furoxan)} \underset{}{\overset{500\,°C}{\rightleftharpoons}} 2\;RCNO \qquad (2.30)$$

Furoxans react as nitrones with two moles of a dipolarophile according to eq. (2.31).[123,124] The chemistry of furoxans has been reviewed.[126] Nitration of α-ethynyl acetates with $NOF/NOBF_4$ or with NO_2BF_4 yields 3,4-bis[α-acetoxy acyl]furoxans.[113] The acetate group is essential for the outcome of the reaction, which is rationalized according to eq. (2.32). Treatment of α-bromophenylnitromethane with triphenylphosphine gives benzonitrile oxide,[114,115] as proved by trapping experiments. Hydroxamic acids undergo normally Lossen's rearrangement by treatment with acids. However,

Synthesis of Nitrile Oxides

the conversions of N-t-butylbenzhydroxamic acid into benzhydroximoyl chloride[116] (the precursor of benzonitrile oxide) with thionyl chloride, and trifluoroacetic anhydride into trifluoroacetonitrile oxide[127] with hydroxylamine hydrochloride and phosphorous pentachloride have been reported. Nitrile oxides are formed in the reaction of O-trimethylsilylhydroximoyl chloride with potassium fluoride at 25°C.[128a] The nitrile oxide dimer with the isomeric structure **7**, eq. (2.1), does not revert to nitrile oxide on flash pyrolysis but fragments into nitrile and a mixture of several other products.[110] α,β-Unsaturated nitro compounds are cleaved photolytically into carbonyl and nitrile oxide fragments that can be trapped by an olefin present, eq. (2.33).[128b]

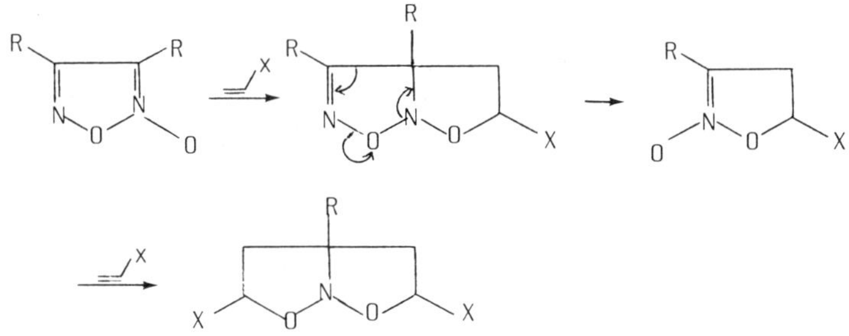

(2.31)

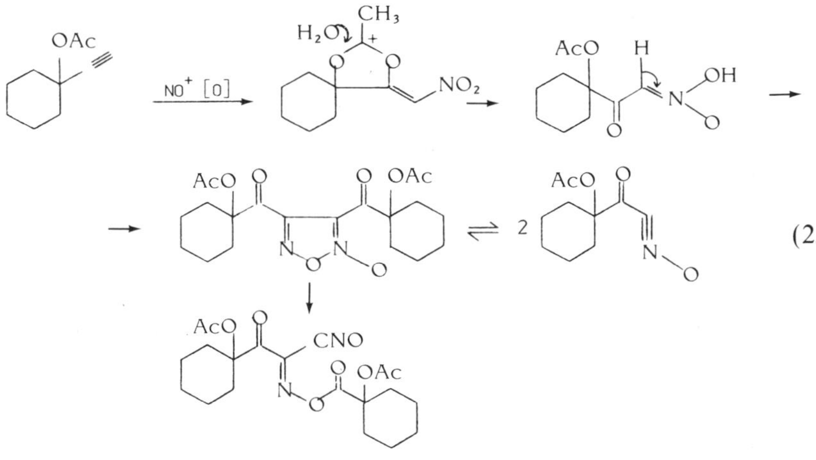

(2.32)

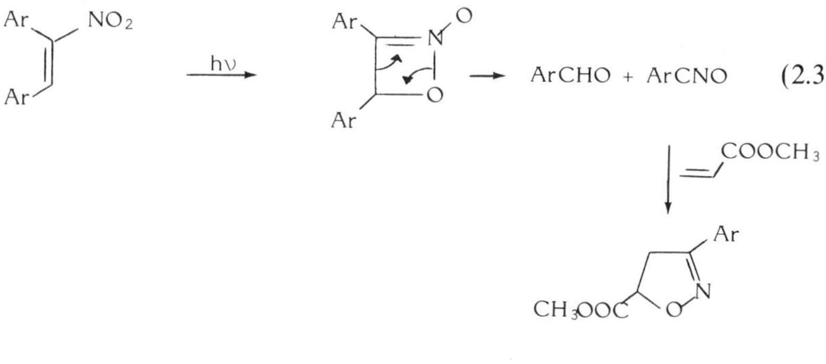

(2.33)

The second significant method is the dehydration of primary nitro compounds with phenylisocyanate, the Mukaiyama-Hoshino method.[58] Because of the reaction conditions, this method is not suitable for the isolation of nitrile oxides but only for in situ preparation and trapping of the intermediate nitrile oxide with a suitable reagent. In the absence of a scavenger the nitrile oxides dimerize to furoxans. The mechanism is depicted in eq. (2.34). Triethylamine is only needed in catalytic amounts and starts the reaction sequence by abstracting a proton from the primary nitro compound. The reaction has recently been applied to an increasing number of primary nitro compounds, and it proceeds with satisfactory yields (Chapter 5). It fails, however, for nitromethane, eq. (2.13). In this case phenylisocyanate first attacks the methyl carbon atom and forms α-nitroacetanilide, which in a second step is transferred into fulmidoformanilide. Phosphorous oxychloride, p-toluenesulfonic acid, acetic anhydride, ethyl chloroformate, benzenesulfonyl chloride, and acetyl chloride have also been used as dehydrating agents instead of phenylisocyanate, but they generally give inferior yields.[129-133]

The Mukaiyama-Hoshino procedure has the advantage over the aldoxime route in that the chlorinating step can be avoided. Aldehydes and derived oximes are, in general, the more readily available functional group, but recent developments in organonitro chemistry may compensate for that. The choice is ultimately a matter of synthetic planning.

Nitrolic acids are formed by nitrosation of primary nitro compounds, and they easily lose nitrous acid when heated, thus forming nitrile oxides, eq. (2.35). The reagent is thus regenerated, and in principle only a catalytic amount of nitrous acid should suffice to achieve dehydration. The scope of the reaction has not been fully investigated, but a few nitrile oxides have been prepared from nitrolic acids.[134-137]

The reaction between metal fulminates and organic halides is reported to produce isocyanates, eq. (2.36),[81,138-141] with one exception, namely the formation of the stable triphenylacetonitrile oxide (**40**), eq. (2.37).[142] It is conceivable that the nitrile oxides formed dimerized to furoxans and therefore were not observed in these early works. The reactions shown in eqs. (2.36),(2.37) are, however, of little practical importance.

$$AgCNO + RX \longrightarrow RNCO + AgX \qquad (2.36)$$

$$AgCNO + (C_6H_5)_3CCl \longrightarrow (C_6H_5)_3CCNO \qquad (2.37)$$
$$\underline{40}$$

2.5 NUCLEOPHILIC ADDITION TO NITRILE OXIDES

Nitrile oxides react with nucleophiles to form an array of hydroximic acid derivatives, as shown in eq. (2.38). Most of these additions have limited synthetic interest. The mechanism and kinetics for adding acetic acid to aromatic nitrile oxides have been studied.[143] On the other hand, the reverse reaction, e.g., elimination of hydrogen chloride from hydroximoyl chloride, is, as we have seen, the best choice for synthesizing nitrile oxides, since the hydroximoyl chlorides are readily accessible by chlorination of aldoximes.

$$RC{\equiv}N\text{-}O + Nu^- + H^+ \rightleftharpoons RC\overset{N\text{-}OH}{\underset{Nu}{\diagdown}} \qquad (2.38)$$

$Nu^- = Cl^-, Br^-, J^-, HO^-, RO^-, HS^-, RS^-, CN^-, R_3N, R^-MgX^+, RCOO^-, N_3^-, SCN^-$, etc.

Adding sulfides to nitrile oxides leads to derivatives of thiohydroxamic acid, eq. (2.39),[144-149] a group of compounds represented in nature by the glucosinolates (**41**).[150] They can be defined as the S-glucosides of thiohydroxamic acid with a sulfated iminoxy group. R in **41** derives from naturally occurring

$$\begin{array}{l} H^+ + R^1S^- + R^2CNO \\ Na^+ + R^1S^- + R^1CCl{=}NOH \end{array} \longrightarrow R^2C\overset{SR^1}{\underset{NOH}{\diagdown}} \qquad (2.39)$$

$$\underline{41}$$

amino acids, e.g., alanine, vinylglycine, phenylalanine, tyrosine, etc. On enzymatic hydrolysis the glucosinolates undergo a Lossen-type rearrangement, eq. (2.40), and form isothiocyanates 42 (mustard oils). Equation (2.39) has been applied in the chemical synthesis of these natural products.[144-147] Glucosinolates can also be prepared by nucleophilic addition of 1-thioglucose to a suitable silyl nitronate.[151] A comprehensive discussion of a number of nucleophilic additions is given in Ref. 1.

$$R^2-C\underset{NOSO_3H}{\overset{SR^1}{\diagup}} \longrightarrow R^2-N=C=S \qquad (2.40)$$
$$\underline{42}$$

2.6 THE 1,3-DIPOLAR CYCLOADDITION

The background for the development of the cycloaddition of nitrile oxides with olefins and acetylenes into an important and versatile synthetic reaction is given in Sections 1.1–1.5. The mechanistic and stereochemical aspects are discussed in Sections 1.7, 1.8. The synthetic applications are reviewed in Chapter 5 and organized according to classes of compounds and functionalities rather than reactions of the individual dipoles.

Since the reactions of nitrile oxides are focused on their use in synthesis, the reactions with other heterodipolarophiles have received peripheral attention. For more comprehensive discussions of the reactions with C=S, C=N, C=O, C≡N, C=P, N=N, N=B, N=P, N=S, and S=O bonds, see the reviews given in Section 1.1; cf. Section 5.16.

2.7 ALLERGENIC PROPERTIES OF NITRILE OXIDES

It is justified to call attention to the allergenic properties of ethyl chlorooximinoacetate (ethoxycarbonylchloroformoxime) 43, which when treated with a base gives 3,4-diethoxycarbonylfuroxan (45) via dimerization of the intermediate reactive ethoxycarbonylformonitriloxide (44), eq. (2.41). Two laboratory technicians contracted severe vesicant and itching blisters while working with compound 43. It had been observed several years ago that the furoxan 45 was a strong skin irritant.[152] Medical tests carried out with compound 43 showed that it is a highly active allergenic.[153] Benzohydroxamic acid did not show allergenic symptoms in our tests. 3-Ethoxycarbonyl-5-butylisoxazoline was also tested but found to be considerably less irritating than compound 45.

Little is known about the physiological effects of nitrile oxides, depending on their instability and capability to dimerize to furoxans.[1] These latter compounds

$$\underset{43}{\overset{EtOOC}{\underset{Cl}{>}}\!\!=\!\!NOH} \xrightarrow{\text{base}} \underset{44}{EtOOCCNO} \longrightarrow \underset{45}{\overset{EtOOC\quad COOEt}{\underset{O\diagdown N\diagup O}{\bigtriangleup}}} \qquad (2.41)$$

show a variety of antibiotic actions.[126] The toxicity of the parent fulminic acid is comparable to that of hydrocyanic acid.

Isoxazoles and their derivatives have found several applications as antibiotics, analgetics, anaesthetics, anabolics, anthelmintics, diuretics, gamma aminobutyric acid (GABA) agonists, anticonvulsants, muscle relaxants, herbicides, and they display antileprous, antitumor, mutagenic, plant growth controlling effects, etc.[154-156] The physiological effects of nitrones have been reviewed.[157]

REFERENCES

1. C. Grundmann; P. Grünanger, "The Nitrile Oxides", Springer-Verlag, Berlin, 1971.
2. F. De Sarlo; A. Guarna; A. Brandi; P. Mascagni, *Gazz. Chim. Ital. 110* (1980), 341; A. Brandi; F. De Sarlo; A. Guarna, *J. Chem. Soc., Perkin I, 1976*, 1827.
3. F. De Sarlo; A. Guarna, *J. Chem. Soc., Perkin I, 1976*, 1825.
4. P. Caramella; A. Corsaro; A. Compagnini; F. M. Albini, *Tetrahedron Lett. 1983*, 4377.
5. W. J. Middleton, *J. Org. Chem. 49* (1984), 919.
6. W. R. Mitchell; R. M. Paton, *J. Chem. Res. (S) 1984*, 58.
7. K. N. Houk; J. Sims; C. R. Watts; L. J. Luskus, *J. Am. Chem. Soc. 95*(1973), 7301.
8. V. E. Turs; N. M. Lyapkin; V. A. Shlyapochnikov, *Izv. Akad. Nauk SSSR. Ser. Khim. Nauk 1982*, 214.
9. M. Dewar, *J. Am. Chem. Soc. 106* (1984), 209.
10. A. Gasco; A. J. Boulton, *J. Chem. Soc., Perkin II, 1973*, 1613, and references therein.
11. L. Birkinbach; K. Sennewald, *Liebigs Ann. 512* (1934), 45.
12. H. Wieland; A. Baumann; C. Reisenegger; W. Scherer; J. Thiele; J. Will; H. Hausmann; W. Frank, *Liebigs Ann. 444* (1925), 7.
13. H. Wieland; Z. Kitasato; S. Utsino, *Liebigs Ann. 478* (1930), 43 and references therein.
14. C. Grundmann; R. K. Bansal; P. S. Osmanski, *Liebigs Ann. 1973*, 898.
15. C. Grundmann; G. W. Nickel; R. K. Bansal, *Liebigs Ann. 1975*, 1029.
16. A. Quilico; L. Panizzi, *Gazz. Chim. Ital. 72* (1942), 155.
17. A. Quilico; G. Stagno d'Alcontres, *Gazz. Chim. Ital. 79* (1949), 654, 703.
18. A historical account of all the shifting structure proposals is given in Ref. 1.
19. J. U. Nef, *Liebigs Ann. 280* (1894), 291.
20. H. Ley; M. Kissel, *Chem. Ber. 32* (1899), 1357.
21. W. Beck; E. Schuierer; K. Feldl, *Angew. Chem. 77* (1965), 722.
22. E. Müller; R. Beutler; B. Zeeh, *Liebigs Ann. 719* (1968), 72.
23. H. Bredereck; B. Föhlisch; K. Walz, *Liebigs Ann. 686* (1965), 92; *688* (1965), 93.
24. H. Wieland, *Chem. Ber. 42* (1909), 4207.
25. C. Grundmann; J. M. Dean, *Angew. Chem. 76* (1964), 682.
26. C. Grundmann; P. Kochs, *Angew. Chem. 82* (1970), 637.
27. (a) C. Grundmann; P. Kochs; J. R. Boal, *Liebigs Ann. 761* (1972), 162.
    (b) G. A. Taylor, *J. Chem. Soc., Perkin I, 1985*, 1181.
28. H. Wieland, *Chem. Ber. 40* (1907), 1667.
29. G. Ponzio, *Gazz. Chim. Ital. 53* (1923), 379.
30. T. Kubota; M. Yamakawa; M. Takasuka; K. Iwatani; H. Akazawa; I. Tanaka, *J. Phys. Chem. 71* (1967), 3597.
31. S. Califano; R. Moccia; R. Scarpati; G. Speroni, *J. Chem. Phys. 26* (1957), 1777.
32. M. Christl; J. P. Warren; B. L. Hawkins; J. D. Roberts, *J. Am. Chem. Soc. 95* (1973), 4392.

33. M. Shiro; M. Yamakawa; T. Kubota; H. Koyama, *Chem. Commun. 1968*, 1409.
34. M. Winnewisser; H. K. Bodenseh, *Z. Naturforsch. 22A* (1967), 1724.
35. H. K. Bodenseh; K. Morgenstern, *Z. Naturforsch. 25A* (1970), 150.
36. M. Winnewisser, *Chem. Phys. Lett. 11* (1971), 519.
37. B. P. Winnewisser; M. Winnewisser; F. Winther, *J. Mol. Spectrosc. 51* (1974), 65.
38. W. Beck; P. Swoboda; K. Feldl; R. S. Tobias, *Chem. Ber. 104* (1971), 533.
39. K. N. Houk; P. Caramella; L. L. Munchausen; Y.-M. Chang; A. Battaglia; J. Sims; D. C. Kaufman, *J. Electron. Spectrosc. Relat. Phenom. 10* (1977), 441.
40. P. Caramella; R. W. Gandour; J. A. Hall; C. G. Deville; K. N. Houk, *J. Am. Chem. Soc. 99* (1977), 385.
41. E. Beckmann, *Chem. Ber. 19* (1886), 993.
42. C. A. Lobry de Bruyn, *Chem. Ber. 19* (1886), 1370.
43. R. Philip, *Ges. Schiess-Sprengstoffw. 7* (1912), 109.
44. L. Wöhler; A. Weber, *Chem. Ber. 62* (1928), 2742.
45. P. Kurtz In Houben-Weyl: "Methoden der Organischen Chemie", *8* (1952), 355. G. Thieme Verlag, Stuttgart.
46. G. Adembri; P. Sarti-Fantoni; F. De Sio; P. F. Franchini, *Tetrahedron 23* (1967), 4697.
47. V. Bertini; A. De Munno; P. Pelosi; P. Pino, *J. Heterocycl. Chem. 5* (1968), 621.
48. R. Huisgen; M. Christl. *Chem. Ber. 106* (1973), 3291.
49. A. Quilico; G. Speroni, *Gazz. Chim. Ital. 69* (1939), 508; *70* (1940), 779.
50. A. Quilico; L. Panizzi, *Gazz. Chim. Ital. 72* (1942), 458.
51. R. Huisgen; M. Christl, *Angew. Chem. 79* (1967), 471.
52. C. F. Palazzo, *Gazz. Chim. Ital. 39* II (1909), 249.
53. A. Brandi; F. De Sarlo; A. Guarna; G. Speroni, *Synthesis 1982*, 719.
54. W. Beck; E. Schuierer, *Chem. Ber. 97* (1964), 3517; *J. Orgmet. Chem. 3* (1965), 55; *9* (1967), 5.
55. F. De Sarlo; A. Brandi; A. Guarna; A. Goti; S. Corezzi, *Tetrahedron Lett. 1983*, 1815; F. De Sarlo; A. Guarna; A. Brandi; A. Goti, *Tetrahedron 41* (1985), 5181.
56. H. Wieland, *Chem. Ber. 40* (1907), 418; *42* (1909), 803.
57. C. Wentrup; B. Gerecht; H. Briehl, *Angew. Chem. 91* (1979), 503.
58. T. Mukaiyama; T. Hoshino, *J. Am. Chem. Soc. 82* (1960), 5339.
59. R. Paul; S. Tchelitcheff, *Bull. Soc. Chim. 1963*, 140.
60. K. Torssell; O. Zeuthen, *Acta Chem. Scand. B 32* (1978), 118.
61. L. Birkenbach; K. Sennewald, *Liebigs Ann. 489* (1931), 7; *Chem. Ber. 65* (1932), 546.
62. G. Endres, *Chem. Ber. 65* (1932), 65.
63. I. De Paolini, *Gazz. Chim. Ital. 60* (1930), 700.
64. E. Gryszkiewiez-Trochimowski; D. Dymowski; E. Schmidt, *Bull. Soc. Chim. 1948*, 597. See also the procedure by W. Prandtl; K. Sennewald, *Chem. Ber. 62* (1929), 1758.
65. P. A. Wade; M. K. Pillay; S. M. Singh, *Tetrahedron Lett. 1982*, 4563; see also A. Werner; H. Buss, *Chem. Ber. 27* (1894), 2193.
66. J. E. Baldwin; C. Hoskins; L. Kruse, *J. Chem. Soc., Chem. Commun. 1976*, 795.
67. A. A. Hagedorn III; B. J. Miller; J. O. Nagy, *Tetrahedron Lett. 1980*, 229.
68. R. V. Stevens; R. P. Polniaszek, *Tetrahedron 39* (1983), 743.
69. D. G. Martin; D. J. Duchamp; C. G. Chidester, *Tetrahedron Lett.* (1973), 2549.
70. P. A. Wade; H. R. Hinney, *J. Am. Chem. Soc. 101* (1979), 1319.
71. P. A. Wade; M. K. Pillay, *J. Org. Chem. 46* (1981), 5425.
72. P. A. Wade; H. K. Yen; S. A. Hardinger; M. K. Pillay; N. V. Amin; P. D. Vail; S. D. Morrow, *J. Org. Chem. 48* (1983), 1796.
73. (a) C. Bellandi; M. De Amici; C. De Micheli; R. Gandolfi, *Heterocycles 22* (1984), 2187.
    (b) T. Shimizu; Y. Hayashi; M. Miki; K. Teramura, *Heterocycles 24* (1986) 889.
74. A. Werner; H. Buss. *Chem. Ber. 27* (1894), 2193.
75. G. Casnati; A. Ricca, *Tetrahedron Lett. 1967*, 327.
76. J. P. Declercq; G. Germain; M. Van Meerssche, *Acta Chrystallogr. B. 31* (1974), 2894.
77. A. F. Hegarty; M. Mullane, *J. Chem. Soc. Chem. Commun. 1984*, 229.
78. R. Huisgen; W. Mack, *Tetrahedron Lett. 1961*, 583.
79. R. Huisgen; W. Mack; E. Anneser, *Tetrahedron Lett. 1961*, 587; *Angew. Chem. 73* (1961), 656.
80. K. Bast; M. Christl; R. Huisgen; W. Mack, *Chem. Ber. 106* (1973), 3312.
81. H. Wieland; A. Höchtlen, *Liebigs Ann. 505* (1933), 237.
82. A. Quilico; G. Gaudiano; A. Ricca, *Gazz. Chim. Ital. 87* (1957), 638.
83. C. Grundmann; V. Mini; J. M. Dean; H. D. Frommeld, *Liebigs Ann. 687* (1965), 191.
84. R. H. Wiley; B. J. Wakefield, *J. Org. Chem. 25* (1960), 546.

85. J. Armand; J. P. Guetté; F. Valentini, *Comp. Rend. 263 C* (1966), 1388.
86. Y. Iwakura; K. Uno; S. Shiraishi; T. Hongu, *Bull. Chem. Soc. Japan 41* (1968), 2954.
87. C. Grundmann; J. M. Dean, *Angew. Chem. 76* (1964), 682.
88. H. Rheinboldt; M. Dewald; F. Jansen; O. Schmitz-Dumont, *Liebigs Ann. 451* (1927), 161.
89. G. Ponzio; G. Busti, *Gazz. Chim. Ital. 36* II (1906), 338.
90. C. Grundmann; R. Richter, *J. Org. Chem. 32* (1967), 2308.
91. C. Grundmann; S. K. Datta, *J. Org. Chem. 34* (1969), 2016.
92. G. A. Lee, *Synthesis 1982*, 508.
93. C. Grundmann; R. Richter, *J. Org. Chem. 32* (1967), 476.
94. M. Yamakawa; T. Kubota; H. Akazawa; I. Tanaka, *Bull. Chem. Soc. Japan 41* (1968), 1046.
95. R. V. Stevens, *Tetrahedron 32* (1976), 1599.
96. K. C. Liu; B. R. Shelton; R. K. Howe, *J. Org. Chem. 45* (1980), 3916.
97. K. E. Larsen; K. B. G. Torssell, *Tetrahedron 40* (1984), 2985.
98. K. B. G. Torssell; A. C. Hazell; R. G. Hazell, *Tetrahedron 41* (1985), 5569.
99. R. Fusco; L. Garanti; G. Zecchi, *Chim. Ind. (Milan) 57* (1975), 16.
100. L. Garanti; A. Sala; G. Zecchi, *J. Org. Chem. 40* (1975), 2403.
101. G. Just; K. Dahl, *Tetrahedron Lett. 1966*, 2441. *Tetrahedron 24* (1968), 5251.
102. G. Just; W. Zehetner, *Tetrahedron Lett. 1967*, 3389.
103. H. Kropf; R. Lambeck, *Liebigs Ann. 700* (1966), 18.
104. F. Eloy; R. Lenaers, *Bull. Soc. Chim. Belg. 72* (1963), 719.
105. R. Lenaers; F. Eloy, *Helv. Chim. Acta* (1963), 1067.
106. P. Vita Finzi; M. Arbasino, *Ric. Sci. 35 IIA* (1965), 1484.
107. T. Sasaki; T. Yoshioka, *Bull. Chem. Soc. Japan 40* (1967), 2604; *41* (1968), 2206.
108. P. Souchay; J. Armand, *Compt. Rend. 256* (1963), 4907.
109. J. Armand; J. P. Guetté; F. Valentini, *Bull. Soc. Chim. France 1968*, 4585.
110. W. R. Mitchell; R. M. Paton, *Tetrahedron Lett. 1979*, 2443.
111. R. A. Whitney; E. S. Nicholas, *Tetrahedron Lett. 1981*, 3371.
112. W. R. Mitchell; R. M. Paton, *J. Chem. Res. (S) 1984*, 58.
113. D. R. Brittelli; G. A. Boswell Jr., *J. Org. Chem. 46* (1981), 312, 316.
114. E. Coutouli-Argyropoulou, *Tetrahedron Lett. 1984*, 2029.
115. L. Horner; H. Oediger, *Chem. Ber. 91* (1958), 437.
116. Y. Uchida; S. Kozuka, *Bull. Soc. Chim. Japan 57* (1984), 2011.
117. S. Gabriel; M. Koppe, *Chem. Ber. 19* (1886), 1145.
118. K. Auwers; V. Meyer, *Chem. Ber. 22* (1889), 705.
119. A. Dondoni; G. Barbaro; A. Battaglia; P. Giorgianni, *J. Org. Chem. 37* (1972), 3196.
120. J. Ackrell; M. Altaf-ur-Rahman; A. J. Boulton; R. C. Brown, *J. Chem. Soc., Perkin I 1972*, 1587.
121. J. A. Chapman; J. Crosby; C. A. Cummings; R. A. C. Rennie; R. M. Paton, *J. Chem. Soc., Chem. Commun. 1976*, 240.
122. J. F. Barnes; M. J. Barrow; M. H. Harding; R. M. Paton; P. L. Ashcroft; J. Crosby; C. J. Joyce, *J. Chem. Res. (S) 1979*, 314.
123. T. Shimizu; Y. Hayashi; K. Teramura, *J. Org. Chem. 48* (1983), 3053.
124. (a) T. Shimizu; Y. Hayashi; T. Taniguchi; K. Teramura, *Tetrahedron 41*(1985), 727.
    (b) D. P. Curran; C. J. Fenk, *J. Am. Chem. Soc. 107* (1985), 6023.
125. M. Altaf-ur-Rahman; A. J. Boulton; D. M. Middleton, *Tetrahedron Lett. 1972*, 3469.
126. A. Casco; A. J. Boulton In "Advances in Heterocyclic Chemistry, Vol. 29, 1981, p. 251; W. Sliwa; A. Thomas, *Heterocycles 23* (1985), 399.
127. W. J. Middleton, *J. Org. Chem. 49* (1984), 919.
128. (a) R. F. Cunico; L. Bedell, *J. Org. Chem. 48* (1983), 2780.
    (b) R. D. Grant; J. T. Pinkey, *Aust. J. Chem. 37* (1984), 1231.
129. G. B. Bachmann; L. E. Strom, *J. Org. Chem. 28* (1963), 1150.
130. E. Kaji; K. Harada; S. Zen, *Chem. Pharm. Bull. Japan 26* (1978), 3254.
131. A. Rahman; M. Younas; N. A. Kahn, *J. Chem. Soc. Pak. 5* (1983), 243.
132. T. Shimizu; Y. Hayashi; K. Teramura, *Bull. Chem. Soc. Japan 57* (1984), 2531; T. Shimizu; Y. Hayashi; H. Shibafuchi; K. Teramura, *Bull. Chem. Soc. Japan 59* (1986) 2827.
133. S. D. Nelson, Jr.; D. J. Kasparian; W. F. Trager, *J. Org. Chem. 37* (1972), 2686.
134. A. Quilico; M. Simonetta, *Gazz. Chim. Ital. 76* (1946), 200; *77* (1947), 586.
135. E. Biekert; A. Kössel, *Liebigs Ann. 662* (1963), 93.
136. M. Jovitschitsch, *Chem. Ber. 28* (1895), 1213.
137. H. Wieland; L. Semper, *Chem. Ber. 39* (1906), 2522.
138. J. U. Nef; *Liebigs Ann. 280* (1894), 339.

139. R. Scholl, *Chem. Ber. 23* (1890), 3505.
140. G. Camels, *Compt. Rend. 99* (1884), 794.
141. A. F. Holleman, *Chem. Ber. 23* (1890), 2998.
142. H. Wieland; B. Rosenfeld, *Liebigs Ann. 484* (1930), 236.
143. P. Beltrame; G. Gelli; A. Loi; G. Saba, *Gazz. Chim. Ital. 113* (1983), 11.
144. M. H. Benn, *Can. J. Chem. 41* (1963), 2836; *42* (1964), 163, 2393; *43* (1965), 1, 1874.
145. M. H. Benn; L. J. Yelland, *Can. J. Chem. 45* (1967), 1595.
146. J. H. Davies; R. H. Davis; P. Kirby, *J. Chem. Soc.* (C) *1968*, 431.
147. A. Kjaer; S. R. Jensen, *Acta Chem. Scand. 22* (1968), 3324.
148. T. Bacchetti; A. Alemagna, *Rend. Ist. Lombardo Sci. Lett. A91* (1957), 30, 574.
149. C. Grundmann; H.-D. Frommeld, *J. Org. Chem. 31* (1966), 157.
150. M. G. Ettlinger; A. Kjaer In "Recent Advances in Phytochemistry" (T. J. Mabry; R. E. Alston; V. C. Runeckles, Eds.), Vol. 1, Appleton–Century–Crofts, New York, 1968, p. 53.
151. T. H. Keller; L. J. Yelland; M. H. Benn, *Can. J. Chem. 62* (1984), 437.
152. H. Wieland; L. Semper; E. Gmelin, *Liebigs Ann. 367* (1909), 52.
153. K. Thestrup-Pedersen; Marselisborg Hospital, personal communication.
154. S. A. Lang, Jr.; Y.-i Lin In "Comprehensive Heterocyclic Chemistry" (A. R. Katritzky and C. W. Rees, Eds.), Vol. 6, Pergamon Press, 1984, p. 1.
155. Y. Takeuchi; F. Furusaki, *Adv. Heterocyclic Chem. 21* (1977), 207.
156. B. J. Wakefield; D. J. Wright, *Adv. Heterocyclic Chem. 25* (1979), 147.
157. W. Kliegel, *Pharmazie 32* (1977), 643; *33* (1978), 331.
158. S. S. Ghabrial; I. Thomsen; K. B. G. Torssell, *Acta Chem. Scand. B.41* (1987), 426.

Chapter 3

The Nitrones

3.1 PHYSICOCHEMICAL PROPERTIES

Nitrones are generally rather stable compounds, easy to handle in air at ordinary temperature, but, under prolonged influence of light, rearrangements occur. They are cleaved into their components, a carbonyl compound and an N-substituted hydroxylamine, by treatment with acid or base.[1-3,22,23] In the presence of another carbonyl compound nitrones can exchange their carbonyl components in an acidic medium.

Structurally, a nitrone is a tautomer of an oxime, eq. (3.1). The two tautomers can be fixed by alkylation, the polar nitrones being N-alkylated oximes, eq. (3.2). Two geometric isomers (**1**) and (**2**) can be formed, which are interconvertible by heating.[4] The reaction in eq. (3.2) is actually a preparative method for nitrones, but they are always formed admixed with the oxime O-ethers **3** and **4**.

$$\text{Oxime} \rightleftharpoons \quad + \text{H}^+ \rightleftharpoons \text{Nitrone} \tag{3.1}$$

$$\tag{3.2}$$

Nitrones are slightly basic compounds. The ionization constants of the conjugate acids of most C^α-aromatic N-methyl nitrones have a pK_a range of ca. 7–9; e.g., $pK_a = 8.26$ for **5**, R = H,[5] eq. (3.3).

In analogy to keto-enol tautomerism nitrone-hydroxyenamine tautomerism is conceivable but not observed in unactivated compounds. For **6** and **7** the equilibrium is shifted entirely towards the nitrone structures **6a** and **7a**[6,7] but activation by a carbonyl group as in **8** results in the formation of the structures **8a,b**, eq. (3.4).[8] The hydroxyenamine tautomer can be fixed by silylation, eq. (3.5),[7] whereas benzoylation gives the rearranged α-benzoyloxyimine **9**, which after hydrolysis gives the α-benzoyloxyaldehyde.[9] This procedure constitutes a mild and efficient method for α-hydroxylation of carbonyl compounds, as has been observed previously.[10–13]

The slow dimerization of aliphatic nitrones proceeds via the N-hydroxyenamine form (**10**) with formation of isoxazolidines **11**[14–21a] that are in equilibrium with the nitrones, eq. (3.6). The aliphatic N-t-butyl nitrones appear to be fairly stable.[7–9] The formation of 1,4-dinitrones by lead tetraacetate oxidation of N,N-disubstituted hydroxylamines occurs via the N-hydroxyenamine form.[21b]

Infrared (IR) and ultraviolet (UV) data have been collected in previous reviews.[22,23] The N—O bond stretching frequency lies characteristically in the 1200–1280 cm^{-1} region for aromatic ketonitrones and close to 1100 cm^{-1} for aldonitrones.

$$\underset{\underset{10}{}}{RCH_{\!=\!N}\!\diagdown\!\underset{R^1}{\overset{O}{}}} \rightleftharpoons \underset{}{RCH_{\!=\!N}\!\diagdown\!\underset{R^1}{\overset{OH}{}}} \underset{10}{\rightleftharpoons} \underset{11}{\underset{HO\diagup N\diagdown R^1}{\overset{R \diagdown \diagup CH_2R}{\diagdown N \diagup O \diagdown R^1}}} \qquad (3.6)$$

The C=N stretching frequency appears in the 1560–1620 cm^{-1} region.[24-26] In the UV region a strong absorption is located at $\lambda_{max} = 244$ ($\varepsilon = 6000$, cyclohexane) for **1** ($R^1 = CH_3$, $R^2 = H$, $R^3 =$ cyclohexyl).[27] The λ_{max} value is shifted to a longer wavelength, and the intensities increase on conjugation, [**1**: $R^1, R^3 = Ph$, $R^2 = H$, $\lambda_{max} = 227, 236, 315$ ($\varepsilon = 9850, 9060, 14,000$, ethanol)].[28]

The ^{1}H and ^{13}C NMR shifts for some representative nitrones are shown in Table 3.1. The spectra of the geometric isomers **1** and **2** are often sufficiently different to allow the isomerization to be followed by NMR. The activation energy for the conversion of, e.g., **12** into **13** is 33.6 kcal mol^{-1}.[29] The Z and E isomers can be distinguished by their dipole moments.[30] In several instances the two isomers have been isolated and structurally assigned.[22,31-34]

$$\underset{12\ (Z)}{\text{CH}_3\text{-C}_6\text{H}_4\text{-C(Ph)=N(O)-CH}_2\text{Ph}} \rightleftharpoons \underset{13\ (E)}{\text{CH}_3\text{-C}_6\text{H}_4\text{-C(Ph)=N(CH}_2\text{Ph)-O}} \qquad (3.7)$$

Geminal ^{15}N coupling constants ($^2J_{^{15}N^1H}$) in cis- and trans-aldonitrones and vicinal ^{15}N coupling constants ($^3J_{^{15}N^1H}$) in cis- and trans-ketonitrones have been measured and calculated.[38] The C^α-carbon atom is located at δ 129.6 ppm in the ^{13}C NMR spectrum of **1**($R^1 = Ph$, $R^2 = H$, $R^3 = t$-Bu), which is slightly lower than the shift of imine or oxime carbon atoms (δ ca. 150–160 ppm). X-ray studies of p-chlorophenyl-N-methylnitrone **14**[39] show that the N—O bond has some double-bond character. It is considerably shorter (1.284 Å) than the N—O bond of the corresponding isomeric syn-oxime **15**, and the C=N distance of **14** is slightly longer than the C=N distance of **15**. In general, these measurements conform with the ^{13}C NMR data, which indicate a somewhat higher negative charge on the nitrone C^α-carbon atom relative to that of the oxime carbon atom.

14: Cl-C$_6$H$_4$-CH=N(O)CH$_3$ (N—O: 1.284, C=N: 1.309)

15: Cl-C$_6$H$_4$-CH=N-OCH$_3$ (N—O: 1.408, C=N: 1.260)

The mass spectra of nitrones have been studied by several groups.[40-45] Fragmentation-rearrangement reactions, such as loss of the N—O oxygen

Table 3.1. ^{1}H and ^{13}C Shifts for Nitrones

Compound	^{1}H NMR and ^{13}C NMR shifts, δ, TMS,[a] CDCl$_3$	Ref.
CH$_3$–CH=N(O)–C$_6$H$_{11}$ (H at C^1)	2.03 (CH$_3$, d), 3.28 (C^{1}H, m), 6.90 (H, q)	27
C$_6$H$_5$–CH=N(O)–CH$_3$	3.86 (CH$_3$, s), 7.37 (H, s), 7.38–7.45 (C^{3,4,5}H), 8.16–8.28 (C^{2,6}H)	27
C$_6$H$_5$–CH=N(O)–C$_6$H$_5$	7.88 (H, s), 7.36–7.46 (C^{3,4,5}H), 8.31–8.42 (C^{2,6}H), 7.56–7.78 (C$^{2',6'}$), 7.36–7.46 (C$^{3',4',5'}$H)	35
mesityl–CH=N(O)–CH$_3$ (H$_1$, H$_2$ on ring)	6.32 (H$_2$, d, J 8), 6.98 (H$_1$, d, J 8)	36
C$_6$H$_5$–CH=N(O)–C(CH$_3$)$_3$	130.8 (C^1), 129.6[b] (C$^\alpha$), 70.6 (N—$\underline{\text{C}}$), 28.2 (CH$_3$)	37
Me$_5$C$_6$–CH=N(O)–CH$_3$	3.89 (N—CH$_3$, s), 7.60 (H, s)	32
Me$_5$C$_6$–CH=N(CH$_3$)–O (isomer)	3.40 (N—CH$_3$, s), 7.92 (H, s)	32

[a]TMS = tetramethylsilane.
[b]Broad line, lw = 4 Hz.

(M$^+$-16) and formation of acylium ions by oxygen migration occur frequently, eq. (3.8), in the positive-ion spectra. The isomeric oxaziridines show mass spectral patterns similar to those of the nitrones, so interconversion is likely. The negative-ion mass spectra exhibit pronounced molecular ions and simpler fragmentation patterns. In contrast to the positive-ion spectra, no skeletal rearrangement fragments are produced upon electron impact. Since the mass spectra of the isomeric nitrones, *O*-ethers, and amides differ, this enables these types of isomers to be distinguished from each other.

Photoelectron spectra of nitrones have been measured and correlated with calculations of ionization potentials.[46–48]

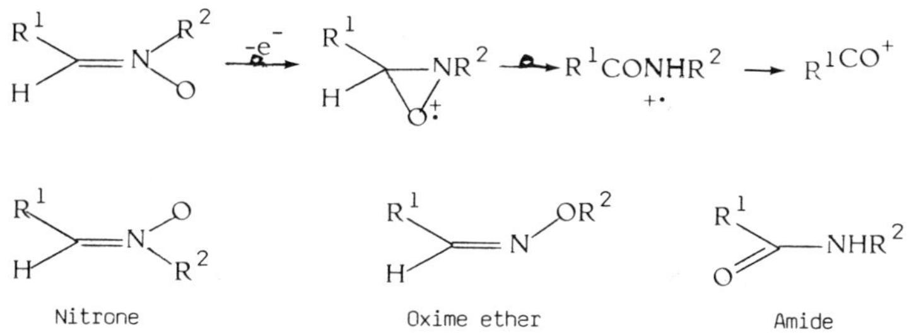

$$R^1\underset{H}{\diagdown}C=\underset{O}{\overset{R^2}{N}} \xrightarrow{-e^-} R^1\underset{H}{\diagdown}C-NR^2 \xrightarrow{} R^1CONHR^2 \xrightarrow{} R^1CO^+$$
$$\underset{\text{Nitrone}}{R^1\underset{H}{\diagdown}C=N\underset{R^2}{\diagdown}O} \qquad \underset{\text{Oxime ether}}{R^1\underset{H}{\diagdown}C=N\diagdown OR^2} \qquad \underset{\text{Amide}}{R^1\underset{O}{\diagdown}C-NHR^2}$$

(3.8)

3.2 REACTIONS OF NITRONES

A. The Nitrone-Amide Rearrangement

Nitrones are light sensitive[2,3,49-52] and rearrange stereospecifically[53] into oxaziridines following a disrotatory course, eq. (3.9). The oxaziridines react further thermally or photochemically to form amides but can also revert thermally to nitrones. This last reaction has occasionally been applied to the synthesis of nitrones, since oxaziridines are available by epoxidation of imines with peracids or hydrogen peroxide.[51,54-59] Nitrones are also found to undergo photochemical and acid-catalyzed *cis-trans*-isomerization.[53,60] These reactions are summarized in eq. (3.9). The nitrone-amide rearrangement, eq. (3.10), is catalyzed by acid in analogy to the Beckmann rearrangement[1-3,13,22,23,61-66] and has been used in synthesis. Reactions of the *N*-phenyl nitrone **16** with ^{18}O benzoyl chloride leads to the amide **17** with the label distributed equally between the two acyl groups. This indicates the intermediacy of a nitrilium ion,[65] eq. (3.11), and excludes a cyclic or sliding mechanism.[61,62] A nuclear chlorination has been observed in the reaction of *N*-aryl nitrones with phosgene or thionyl chloride, eq. (3.12).[67]

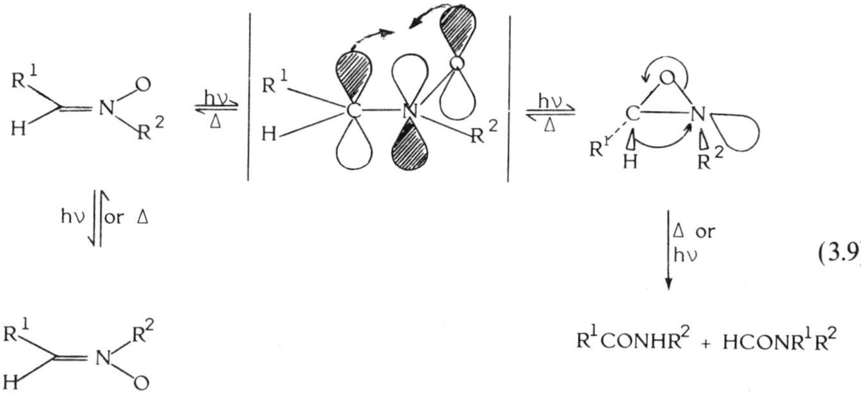

$$R^1CONHR^2 + HCONR^1R^2$$

(3.9)

$$R^1CH=N^+(R^2)OH \xrightarrow{H^+, H_2O} R^1C\equiv\overset{+}{N}-R^2 \longrightarrow R^1CONHR^2 \quad (3.10)$$

$$\underset{\mathbf{16}}{R^1CH=N^+(Ph)(OCOPh)} \longrightarrow R^1C\equiv\overset{+}{N}-Ph \longrightarrow R^1C(OCOPh)=N-Ph \downarrow \underset{\mathbf{17}}{R^1CO-N(COPh)(Ph)} \quad (3.11)$$

$$RCH=N^+(Ph)(OCOCl) \xrightarrow{Cl^-} RCH=N-C_6H_4Cl \quad (3.12)$$

B. The Behrend Rearrangement

Another type of tautomerism is attributed to Behrend, who observed the base-catalyzed rearrangement of **18** as shown in eq. (3.13).[68] The isomer of lowest energy is favored, as reflected by the effect of para-substitution. The favored isomer is that in which the substituent can interact conjugatively.[26,69]

$$\underset{\mathbf{18}}{X\text{-}C_6H_4\text{-}CH_2\text{-}N(O)=CH\text{-}C_6H_5} \;\rightleftharpoons\; \underset{\mathbf{19}}{X\text{-}C_6H_4\text{-}CH=N(O)\text{-}CH_2\text{-}C_6H_5} \quad (3.13)$$

C. The Nitrone-Oxime O-Ether Rearrangement

Thermolysis at ca. 150–200°C of certain N-diphenylmethyl nitrones gives the corresponding oxime O-diphenylmethyl ethers via isomerizing iminoxy radicals, eq. (3.14).[2,3,34,70,71] The radical character and the intermolecularity of the reaction is proved by the observation of CIDNP effects[72] and the production of crossover products.[73]

D. Thermolytic Alkene Elimination

At 80°C N-alkylfluoren-9-yl nitrones undergo a Cope-type elimination with formation of alkene, eq. (3.15).[74-76] There is evidence that the elimination is not synchronous, but that the C—N cleavage precedes the O—H bond formation.

E. Reactions of Nitrones with Nucleophiles

Several nucleophiles are found to attack the carbon end of the dipole and form α-substituted hydroxylamines, eq. (3.16). Compound **20** usually undergoes secondary reactions, depending on the structure of Nu$^-$. The acid- or base-catalyzed hydrolytic cleavage represents the simplest reaction of this class: Nu$^-$ = OH$^-$. In case Nu$^-$ is a thiocarboxylic acid, the nitrone is converted into a thione (**21**), eq. (3.17).[77]

The cleavage of the nitrone can also be performed with hydroxylamine and hydrazine, thus giving the corresponding oximes and hydrazones directly.[1,22,23] Hydrogen cyanide adds to the nitrones, giving 1-cyanoalkylhydroxylamines,[24,25] which often undergo further transformations, such as elimination of water[78-80] and cyclization with a second molecule to produce imidazoles **22**, eqn. (3.18).[81] Adding a thiol to N-α-cyanoalkyl nitrones promotes the cyclization to imidazole **23**, eq. (3.19).[82] N-2-Hydroxyalkyl or N-3-hydroxylalkyl nitrones are in equilibrium with their hydroxylamine form **24**, eq. (3.20).[83-91] A variety of nucleophiles have been added to four-membered cyclic nitrones.[92-96]

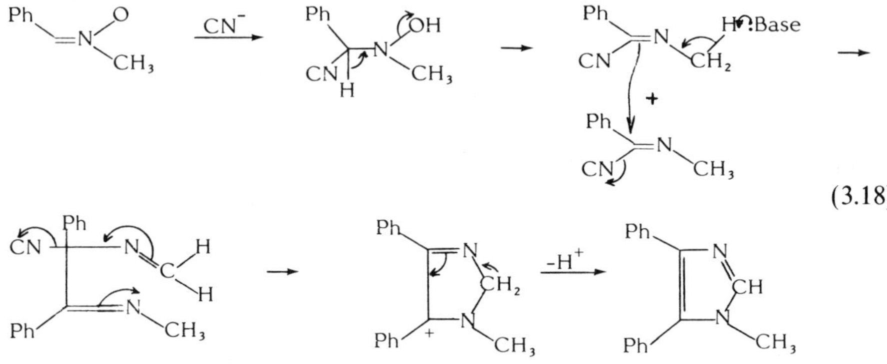

(3.18)

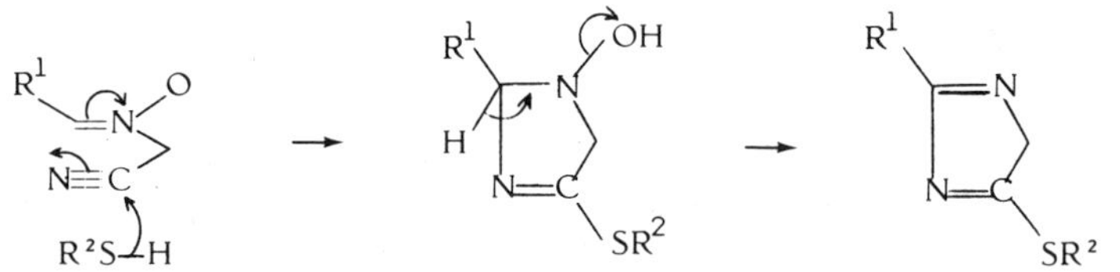

(3.19)

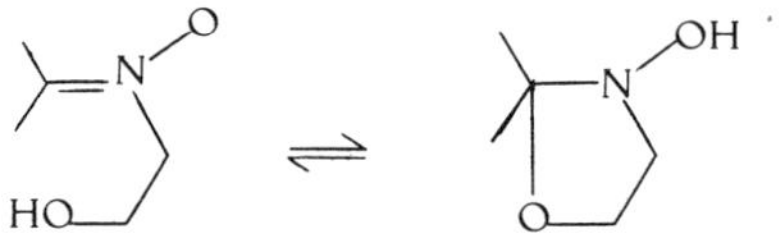

(3.20)

Carbanions,[1-3] active methylene compounds,[36,97-103] ylides,[104-108] and organometallic compounds[109-115] add to nitrones, eqs. (3.21)–(3.23). The reaction of nitrones with the carbanion from malonate proceeds to the isoxazolidone **25**.[116] The reaction of nitrones with phosphorous ylides opens a route to aziridines, eq. (3.22).[117] Compound **27** is formed by adding ylides to N-benzyl nitrone obtained via the Behrend rearrangement. Aziridines are also one of the products from the reaction of nitrones with α-trimethylsilyl carbanions.[118,119] An example of adding an organometallic compound is demonstrated in eq. (3.23), where the Reformatzky reagent gives isoxazolidones (**28**).[120]

[Equation (3.21): reaction of pyrrolidine nitrone with HC(COOEt)=C(OEt) giving bicyclic product 25]

[Equation (3.22): PhCH=N(O)CH₃ ⇌ Ph-CH₂-N(O)= ; with (EtO)₂PCH₂COOEt, NaH → aziridines 26 and 27]

[Equation (3.23): ArCH=N(O)R + R¹R²C(Br)COOEt, Zn → isoxazolidinone 28]

R = Alkyl   R¹, R² = H, Alkyl, COOR

F. Oxidations of Nitrones

Lead tetraacetate (LTA) oxidizes aldonitrones to $(N)O$-acetylhydroxamic acids (**29**), eq. (3.24), and ketonitrones are cleaved to the acylals (**30**), eq. (3.25).[121] N-Alkyl groups are also attacked by LTA as in eq. (3.26), and if one of the phenyl groups is labeled, a scrambled product **31** is obtained. Cleavage of nitrones is effected by ozone, eq. (3.27)[122] and by periodate.[123] The actions of various oxidants, such as Fe^{3+},[124,125] NBS (N-bromosuccinimide),[126,127] NCS (N-chlorosuccinimide),[128] SeO_2,[129] HOBr,[130] and Br_2[131] have been investigated. Three types of products can occur:

1. Hydroxamic acids are formed, as shown in eq. (3.24).
2. Cleavage occurs as demonstrated in eqs. (3.25) or (3.27).
3. The β-position in alkyl nitrones is attacked, e.g., by SeO_2, alkyl nitrites,[132] or halogens, as depicted in compound **32**.

[Equation (3.24): PhCH=N(O)-t-Bu → LTA → Ph-C(=O)-N(OAc)-t-Bu, 94% **29**]

[Equation (3.25): Ph₂C=N(O)Ph → LTA → Ph₂C(OAc)₂ + PhNO, 87% **30**]

$$\text{Ph}^*\text{-CH=N(O)-Ph} \xrightarrow{\text{LTA}} \text{Ph}^*\text{-C(=O)-N(OAc)-CH(OAc)-Ph}^* \quad (3.26)$$

31

$$R^1\text{-CH=N(O)R}^2 \xrightarrow{O_3} R^1\text{-CHO} + R^2\text{NO} \quad (3.27)$$

$$R^1\text{-CH}_2 \xrightarrow{[\text{ox}]} R^1\text{-CH=N(O)R}^2$$

32

G. Reductions of Nitrones

The reduction of nitrones can be directed toward selective (a) deoxygenation with formation of imines, (b) reduction of the double bond to give *N*-substituted hydroxylamines, and (c) complete reduction to amines.

Deoxygenation of nitrones is most effectively carried out by phosphines or phosphites, eq. (3.28).[133–135] Adding phosphite to the double bond has been observed.[136] Lithium aluminum hydride[25,137] and sodium borohydride[25,79] are normally used for reducing the nitrones to hydroxylamines (**33**), and the reduction usually stops at this level, eq. (3.29). For total reduction to the amine **34**, catalytic reduction is the best method, and Raney-Ni appears to be the most frequently used catalyst.[1] With Pt as catalyst it is possible to stop the reduction at the hydroxylamine stage.[138a] Selective reduction to imine is accomplished with sodium hydrogentelluride at pH 10–11. The amine is obtained in refluxing ethanol at pH 6.[138b]

$$R^1\text{-CH=N(O)R}^2 \xrightarrow{R_3P} R^1\text{-CH=NR}^2 + R_3PO \quad (3.28)$$

$$R^1\text{-CH=N(O)R}^2 \xrightarrow[\text{NaBH}_4]{\text{LiAlH}_4} R^1\text{-CH}_2\text{-N(OH)R}^2 \quad \textbf{33} \quad (3.29)$$

$$\xrightarrow{\text{H}_2/\text{Ni}} R^1\text{-CH}_2\text{-NHR}^2$$

34

H. Nitrones as Radical Scavengers

Nitrones serve as excellent scavengers for several types of radicals,[139] carbon centered as well as heteroatom centered (O, N, S, metal), eq. (3.30). In that respect they are superior to nitroso compounds, which normally only trap carbon- and oxygen-centered radicals. A drawback is that the radical center of the nitroxide radical formed **35** is located too far from the original radical R˙, so little can be concluded about its structure. The electron spin resonance (ESR) spectrum of the rather stable nitroxide radical can conveniently be recorded. The g value and the triplet N and doublet H splittings give some information on the nature of the trapped radical R˙. The second-order rate constant, k, of the trapping reaction (3.30) is about 10^5 L mol^{-1} sec^{-1}.

$$\text{PhCH=N(O)X} + \text{R}^\cdot \xrightarrow{k} \text{Ph-CR(H)-N(O}^\cdot\text{)X} \quad \underline{35} \tag{3.30}$$

I. 1,3-Dipolar Cycloaddition

The mechanism and regiochemistry of the 1,3-dipolar cycloaddition is discussed in Sections 1.5–1.8, and the synthetic applications in Chapter 5. References to reviews are given in Section 1.1. It is a high-yielding reaction proceeding smoothly at room temperature with reactive olefins. Highly substituted olefins require heating to 100–150°C for 1–24 h. Elevated temperatures have the drawback that cycloreversions, eq. (3.31)[140] (cf. Section 1.6) and (Z)-(E)-isomerization, eq. (3.32), take place extensively, leading to isomerization of the cycloaddition product.

$$R^1R^2C=N(O)R^3 + CH_2=CHX \underset{\Delta}{\rightleftharpoons} \text{isoxazolidine} \tag{3.31}$$

J. Z-E-Isomerization

The configurational stability of nitrones is of interest in connection with the cycloaddition, since it will affect the stereoisomeric distribution of the products. Ordinarily, the rotation barrier for most nitrones, eq. (3.32), is so high, $E_a \geq 30$ kcal mol^{-1}, that the nitrones are completely stable at ordinary reaction temperatures less than 80°C and reaction times less than 24 h. Higher temperatures, e.g., in refluxing xylene, cause rapid equilibration. Electron withdrawing groups at C_α (e.g., acyl, COOR, CN) lower the barrier for rotation,[141,142] which for N,α-diphenyl-α-cyanonitrone is 24.6 kcal mol^{-1} [141] as compared with the barrier for the C_α-diaryl, N-alkyl, or C_α-aryl, -H, N-alkyl nitrones, which is 29–34 kcal mol^{-1}.[29,32,145] Isomerization is catalyzed by light and acids.[53,60,143,144]

$$R^1R^2C_\alpha=N(R^3)(O) \rightleftharpoons R^1R^2C_\alpha=N(O)(R^3) \tag{3.32}$$

3.3 SYNTHESIS OF NITRONES

Preparatory aspects for nitrone synthesis have been comprehensively reviewed.[1-3,22,23,51] Considering the importance of nitrones as an efficient C—C coupling device for the isoxazolidine moiety, we give a summary of the earlier work together with comments on recent development. There are several significant classical preparative methods available.

1. Condensation of *N*-monosubstituted hydroxylamines with carbonyl compounds
2. Dehydrogenation of *N,N*-disubstituted hydroxylamines
3. *N*-Alkylation of oximes
4. Reaction of aromatic nitroso compounds with active methylene compounds; the Kröhnke reaction
5. *N*-Oxidation of imines (Schiff's bases)

In addition there are a number of published procedures of more limited generality.[146-156]

A. Condensation of N-Monosubstituted Hydroxylamines with Carbonyl Compounds

Using carbonyl compounds is the most important method for our purpose because it embodies generality and experimental simplicity with the basic synthetic strategy of using readily available carbonyl compounds as building blocks. For the subsequent 1,3-dipolar addition the *N*-monosubstituted hydroxylamine activates the carbonyl carbon for reacting with the olefin or acetylene. The nature of the *N*-substitutent R^1 influences only marginally the efficiency of the addition step, eq. (3.33), but it affects the cis-trans ratio of **36** to a certain extent. However, certain restrictions are placed on the substituent R^1, because usually the unsubstituted amino function is desired in the amino alcohol **37** ($R^1 = H$). Benzylhydroxylamine may serve the purpose because it can be removed by catalytic hydrogenation simultaneously with the cleavage of the N—O bond of the ring. Another possibility of solving the problem is to use *N*-α-*O*- or *S*-alkyl substituted nitrones,[157-161,188] as practiced, e.g., in some carbohydrate-based oximes, eq. (3.34). Optical activity is simultaneously induced into the isoxazolidine **38**. However, suitable R^1-substituents that can easily be removed by various methods are still needed.

The preparation of oxazoline[198] (**39**) and imidazolinone N-oxides,[199] (**40**), and their cycloadditions to olefins and acetylenes has been described, eqs. (3.35), (3.36). Use of a chiral hydroxylamino alcohol could conceivably introduce asymmetry in the β-hydroxyketone obtained on reduction and hydrolysis of the isoxazolidine, eq. (3.37).

Aldehydes react rapidly and quantitatively with the monosubstituted hydroxylamines at room temperature, whereas ketones or acetals require several hours of heating. Cyclic nitrones are preferentially prepared by zinc reduction of nitro-substituted carbonyl compounds, eq. (3.38).[79] Imines have also been used as starting material for nitrone synthesis.[80]

B. Dehydrogenation of N,N-Disubstituted Hydroxylamines

A number of oxidants have been applied to the dehydrogenation of hydroxylamines, eq. (3.39): O_2/Cu^{2+},[79] peroxides,[153,162-165] Fe^{3+},[111,132,166] Cu^{2+},[167,172] Ag^+,[111] $KMnO_4$,[111,162] quinone,[168] PbO_2,[169,170] $NaIO_4$,[123] $K_2Cr_2O_7$,[171] HgO,[70,138,173,174] Pd,[175] and nitrosobenzene.[177-179] The dehydrogenation is a two-step reaction: the first step is an oxidation of the hydroxylamine anion to the radical; the second step is a hydrogen abstraction, eq. (3.37). Pd[175] acts as a catalyst and hydrogen is evolved in the reaction. Secondary amines have been transformed to nitrones by tungstate-catalyzed hydrogen peroxide oxidation.[176]

$$\underset{R^1}{\overset{R\,H_2C}{>}}N-OH \xrightarrow{[O]} \underset{R^1}{\overset{R\,H_2C}{>}}N-O^{\cdot} \xrightarrow{[O]} \underset{R^1}{\overset{R}{>}}N-O \qquad (3.39)$$

C. N-Alkylation of Oximes

Alkylation of oximes with alkyl halides gives a mixture of nitrone and O-alkylated oxime, eq. (3.40). The counter ion, e.g. Li^+, Na^+, K^+, $N(CH_3)_4^+$, and the alkylating agent (RBr, RI, dialkyl sulfate) have comparatively small effects on the product ratio, which for aldoximes favors the nitrone formation[180] but for ketoximes, as a result of steric hindrance, favors O-alkylation.[182] The problem with oximes being ambident nucleophiles can be circumvented by converting them first into O-trimethylsilyl oximes. They are conveniently N-alkylated to nitrones with trialkyloxonium tetrafluoroborate.[197] A route to cyclic nitrones involving silver ion–catalyzed cyclization of allenic oximes has been reported.[200] N-Methylation of the oxime of diethyl 2-ketomalonate has been accomplished with diazomethane.[181] Silver salts of oximes give essentially O-alkylation.[183-185] The geometrical isomerism of aldoximes exerts control over the course of the alkylation in that (E)-benzaldoxime gives predominantly O-alkylation and (Z)-benzaldoxime gives the nitrone.[185-187] Michael addition of oximes to acrylates or acrylonitrile gives a high yield of the intermediary nitrone, which subsequently reacts with a second molecule of the olefin to give the isoxazolidine, eq. (3.41).[188-191] Apart from a few cyclic derivatives,[7,192,193] N-acylnitrones are practically unknown. They appear as short-lived inter-

$$>=N-OH \xrightarrow[\text{Base}]{RX} >=N\overset{O}{\underset{R}{\diagdown}} + >=N-OR \qquad (3.40)$$

(3.41)

mediates in the oxidation of N-alkylhydroxamic acids.[194] Fairly stable carbamoylnitrones are formed by reacting Z-benzaldoxime with methyl isocyanate or p-chlorophenylisocyanate in ether at −30°C, eq. (3.42).[195]

$$\text{PhCH=N}^{OH} \xrightarrow{p\text{-ClC}_6\text{H}_4\text{NCO}} \text{PhCH=N}\begin{array}{c}O\\ \diagdown\\ =O\\ \text{HNC}_6\text{H}_4\text{Cl}-P\end{array} \quad (3.42)$$

D. Reaction of Aromatic Nitroso Compounds with Active Methylene Compounds: The Kröhnke Reaction[1]

The Kröhnke reaction is restricted to the preparation of N-arylnitrones and requires in principle an activated methylene group (occasionally a methine or methyl group) possessing a good leaving group, as shown in eq. (3.43).

$$\text{ArNO} + \text{CH}_2 \begin{array}{c}x\\ \diagdown\\ y\end{array} \xrightarrow{\text{Base}} \text{Ar} \overset{O^-}{\underset{}{N}}-CH\begin{array}{c}x\\ \diagdown\\ y\end{array} \longrightarrow \text{Ar} \overset{O}{\underset{}{N}}=\text{CHX} \quad (3.43)$$

x = activating group: aryl, CN, RCO

y = leaving group : halogen, pyridinium, sulfonium

E. N-Oxidation of Imines (Schiff's Bases)

Oxidation of the imine function, whether it is part of a heteroaromatic ring or an isolated C=N group, is most effectively carried out with peracids.[51,135] In rare cases hydrogen peroxide or other reagents have been used. N-Heteroaromatics normally give the corresponding N-oxides (= nitrones) directly. Imines or Schiff's bases give initially oxaziridines, which under controlled conditions thermally rearrange to nitrones,[58,135,196] eq. (3.44); cf. the nitrone-amide rearrangement in Section 3.2. Occasionally the nitrones are formed directly.[32,58]

$$\begin{array}{c}R^1\\ \diagup\\ R^2\end{array}C=NR^3 \xrightarrow{\text{Peracid}} \begin{array}{c}R^1\\ \diagup\\ R^2\end{array}\overset{O}{\triangle}NR^3 \longrightarrow \begin{array}{c}R^1\\ \diagup\\ R^2\end{array}C=\underset{O}{NR^3} \quad (3.44)$$

(2 isomers)   (2 isomers)

R^1, R^2, R^3 = Alk, Ar, H.

REFERENCES

1. W. Rundel In Houben-Weyl: "Methoden der Organischen Chemie" *10:4* (1968), 309. (E. Müller, Ed.) G. Thieme Verlag, Stuttgart.
2. E. Breuer In "The Chemistry of Amino, Nitroso, and Nitro Compounds and their Derivatives", Supplement F (S. Patai, Ed.), Wiley, Chichester, 1982, p. 459.

3. J. J. Tuffariello In "1,3-Dipolar Cycloaddition Chemistry" (A. Padwa, Ed.), Vol. 2, Wiley, New York, 1984, p. 83.
4. F. Barrow; F. J. Thorneycroft, *J. Chem. Soc. 1939*, 773.
5. V. A. Bren; E. A. Medyantseva; I. M. Andreeva; V. I. Minkin, *Zh. Org. Khim. 9* (1973), 767.
6. R. Bonnett; D. E. McGreer, *Can. J. Chem. 40* (1962), 177.
7. K. Torssell; O. Zeuthen, *Acta. Chem. Scand. B 32* (1978), 118.
8. D. St. C. Black; V. M. Clark; B. G. Odell; A. Todd, *J. Chem. Soc., Perkin I 1976*, 1944.
9. C. H. Cummins; R. M. Coates, *J. Org. Chem. 48* (1983), 2070.
10. D. H. R. Barton; N. J. G. Gutteridge; R. H. Hesse; M. M. Pechet, *J. Org. Chem. 34* (1969), 1473.
11. N. J. G. Gutteridge; J. R. M. Dales, *J. Chem. Soc.* (C) *1971*, 122.
12. J. P. Alazard; B. Khemis; X. Lusinchi, *Tetrahedron 31* (1975), 1427; *Tetrahedron Lett. 1972*, 4795.
13. M. Cherest; X. Lusinchi, *Tetrahedron 38* (1982), 3471; *Bull. Soc. Chim. France 1984*, 227.
14. R. Foster; J. Iball; R. Nash, *Chem. Commun. 1968*, 1414; *J. Chem. Soc., Perkin II 1974*, 1210.
15. A. D. Baker; J. E. Baldwin; D. P. Kelly; J. De Bernardis, *Chem. Commun. 1969*, 344.
16. W. Kliegel, *Tetrahedron Lett. 1969*, 2627.
17. B. Princ; O. Exner, *Coll. Czech. Chem. Commun. 44* (1979), 2221.
18. R. A. Reamer; M. Sletzinger; I. Shinkai, *Tetrahedron Lett. 1980*, 3447.
19. F. De Sarlo; A. Brandi; A. Guarna, *J. Chem. Soc., Perkin I 1982*, 1395.
20. H. Stamm; H. Steudle, *Arch. Pharm. 309* (1976), 1014.
21. (a) F. De Sarlo; A. Brandi; P. Mascagni, *Synthesis 1981*, 561.
    (b) H. G. Aurich; M. Schmidt; T. Schwerzel, *Chem. Ber. 118* (1985), 1105.
22. J. Hamer; A. Macaluso, *Chem. Rev. 64* (1964), 473.
23. G. R. Delpierre; M. Lamchen, *Quart. Rev. 19* (1965), 329.
24. M. Masui; C. Yijima, *J. Chem. Soc.* (C) *1967*, 2022.
25. M. L. M. Pennings; D. N. Reinhoudt; S. Harkema; G. J. van Hummel, *J. Org. Chem. 47* (1982), 4419.
26. P. A. S. Smith; S. E. Gloyer, *J. Org. Chem. 40* (1975), 2504.
27. K. Koyano; H. Suzuki, *Tetrahedron Lett. 1968*, 1859.
28. O. H. Wheeler; P. H. Gore, *J. Am. Chem. Soc. 78* (1956), 3363.
29. T. S. Dobashi; M. H. Goodrow; E. J. Grubbs, *J. Org. Chem. 38* (1973), 4440.
30. O. Exner In "The Chemistry of Double-Bonded Functional Groups", Part I. Wiley, New York, 1977, p. 1.
31. P. M. Weintraub; P. L. Tiernan, *J. Org. Chem. 39* (1974), 1061.
32. J. Bjørgo; D. R. Boyd; D. C. Neill; W. B. Jennings, *J. Chem. Soc., Perkin I 1977*, 254.
33. Y. Yoshimura; Y. Mori; K. Tori, *Chem. Lett. 1972*, 181.
34. T. S. Dobashi; D. R. Parker; E. J. Grubbs, *J. Am. Chem. Soc. 99* (1977), 5382.
35. K. Koyano; H. Suzuki, *Bull. Chem. Soc. Japan 42* (1969), 3306.
36. J. E. Baldwin; A. K. Qureshi; B. Sklarz, *J. Chem. Soc.* (C) *1969*, 1073.
37. K. Torssell; H. J. Jakobsen, unpublished observation.
38. D. R. Boyd; M. E. Stubbs; N. J. Thompson; H. J. C. Yeh; D. M. Jerina; R. E. Wasylichen, *Org. Magn. Res. 14* (1980), 528.
39. K. Folting; W. N. Lipscomb; B. Jerslev, *Acta Chryst. 17* (1964), 1263.
40. R. Grigg; B. G. Odell, *J. Chem. Soc. B 1966*, 218.
41. L. A. Neiman; V. I. Maimind; M. M. Shemyakin; V. A. Pushkov; V. N. Bucharev; Yu. S. Nekrasov; N. S. Vulfson, *Zh. Obshch. Khim. 37* (1967), 1600.
42. J. H. Bowie; S.-O. Lawesson; B. S. Larsen; G. E. Lewis; G. Schroll, *Austr. J. Chem. 21* (1968), 2031.
43. R. T. Coutts; G. R. Jones; A. Benderley; A. L. C. Mak, *Can. J. Pharm. Sci. 13* (1978), 61.
44. R. G. Kostyanovskii; A. P. Pleshkova; V. N. Voznesenskii; V. I. Khovstenko; A. Sh. Sultanov; V. K. Mavrodiev; A. I. Mishchenko; A. V. Prosyanik, *Izvest. Akad. Nauk SSSR, Ser. Khim. Nauk 1979*, 1388.
45. R. G. Kostyanovskii; A. P. Pleshkova; V. N. Voznesenskii; A. I. Mishchenko; A. V. Prosyanik; V. I. Markov, *Izvest. Akad. Nauk, SSSR, Ser. Khim Nauk 1980*, 322.
46. K. N. Houk; P. Caramella; L. L. Munchausen; Y.-M. Chong; A. Battaglia; J. Sims; D. C. Kaufman, *J. Electron Spectrosc. 10* (1977), 441.
47. D. Mukherjee; L. W. Domelsmith; K. N. Houk, *J. Am. Chem. Soc. 100* (1978), 1954.
48. J. Bastide; J. P. Meier; T. Kubota, *J. Electron Spectrosc. 9* (1976), 307.
49. L. Alessandri, Atti Accad. Naz. Lincei, *Rend Classe Sci. Fis. Mat. Nat.* [II] *19* (1910), 122; *Chem. Abstr. 5* (1911), 276.

References

50. F. Kröhnke, *Liebigs Ann. 604* (1957), 203.
51. W. Rundel In Houben-Weyl, "Methoden der Organischen Chemie", *10:4* (1968), 309, 449. (E. Müller, Ed.), G. Thieme Verlag, Stuttgart.
52. G. G. Spence; E. C. Taylor; O. Buchardt, *Chem. Rev. 70* (1970), 231.
53. J. S. Splitter; T.-M. Su; H. Ono; M. Calvin, *J. Am. Chem. Soc. 93* (1971), 4075.
54. W. D. Emmons, *J. Am. Chem. Soc. 79* (1957), 5739.
55. L. Horner; E. Jürgens, *Chem. Ber. 90* (1957), 284.
56. H. Krimm, *Chem. Ber. 91* (1958), 1057.
57. R. Bonnett; V. M. Clark; A. Todd, *J. Chem. Soc. 1959*, 2101.
58. D. R. Boyd; P. B. Coulter; N. D. Sharma; W. B. Jennings; V. E. Wilson, *Tetrahedron Lett. 1985*, 1673.
59. J. B. Hendrickson; D. A. Pearson, *Tetrahedron Lett. 1983*, 4657.
60. L. L. Rodina; I. Kurutz; A. I. Shcherban; J. K. Korobitsyna, *Zh. Org. Khim. 14* (1978), 889.
61. S. Tamagaki; S. Kozuka; S. Oae, *Tetrahedron 26* (1970), 1795.
62. S. Tamagaki; S. Oae; *Bull. Chem. Soc. Japan 44* (1971), 2851.
63. D. H. R. Barton; M. J. Day; R. H. Hesse; M. M. Pechet, *J. Chem. Soc., Perkin I 1975*, 1764.
64. P. W. Jeffs; G. Molina, *J. Chem. Soc., Chem. Commun. 1973*, 3.
65. H. W. Heine; R. Zibuck; W. J. A. van den Heufel, *J. Am. Chem. Soc. 104* (1982), 3691.
66. J. E. McMurry; V. Farina, *Tetrahedron Lett. 1983*, 4653.
67. D. Liotta; A. D. Baker; F. Weinstein; D. Felsen; R. Engel; N. L. Goldman, *J. Org. Chem. 38* (1973), 3445; cf. also J. Meisenheimer, *Chem. Ber. 59* (1926), 1848.
68. R. Behrend; E. König, *Liebigs Ann. 263* (1891), 339.
69. M. Michalska; I. Orlich, *Bull. Akad. Pol. Sci. Ser. Chim. 23* (1975), 655.
70. A. C. Cope; A. C. Haven, Jr., *J. Am. Chem. Soc. 72* (1950), 4896.
71. J. A. Villarreal; T. S. Dobashi; E. J. Grubbs, *J. Org. Chem. 43* (1978), 1890.
72. D. G. Morris, *J. Chem. Soc., Chem. Commun. 1971*, 221.
73. J. A. Villarreal; E. J. Grubbs, *J. Org. Chem. 43* (1978), 1896.
74. D. R. Boyd; D. C. Neill, *J. Chem. Soc., Perkin I 1977*, 1308.
75. D. R. Boyd; D. C. Neill; M. E. Stubbs, *J. Chem. Soc., Perkin II 1978*, 30.
76. W. M. Leyshon; D. A. Wilson, *J. Chem. Soc., Perkin I 1975*, 1920.
77. K. Kimura; H. Niwa; S. Motoki, *Bull. Chem. Soc. Japan 50* (1977), 2751.
78. V. Bellavita, *Gazz. Chim. Ital. 70* (1940), 584.
79. R. Bonnetti; R. F. C. Brown; V. M. Clark; I. O. Sutherland; A. Todd, *J. Chem. Soc. 1959*, 2094.
80. M. Abou-Gharbia; M. M. Joullie, *Synthesis 1977*, 318.
81. N. G. Clark; E. Cawkill, *Tetrahedron Lett. 1975*, 2717.
82. M. Masui; K. Suda; M. Yamauchi; C. Yijima, *J. Chem. Soc., Perkin I 1972*, 1955.
83. L. B. Volodarskii; A. Ya. Tikhonov, *Izvest. Akad. Nauk SSSR, Ser. Khim. Nauk 1972*, 1218.
84. W. Kliegel, *Liebigs Ann. 733* (1970), 192.
85. T. Poloński; A. Chimiak, *J. Org. Chem. 41* (1976), 2092.
86. H. Moerli; M. Lappenberg, *Arch. Pharm. 311* (1978), 806.
87. E. G. Janzen; R. C. Zawalski, *J. Org. Chem. 43* (1978), 1900.
88. L. B. Volodarskii; T. K. Sevastyanova, *Zh. Org. Khim. 7* (1971), 1687.
89. W. Kliegel; H. Becker, *Chem. Ber. 110* (1977), 2067.
90. W. Kliegel; L. Preu, *Chem. Ztg. 108* (1984), 283.
91. W. Kliegel; J. Graumann, *Liebigs Ann. 1984*, 1545.
92. M. L. M. Pennings; D. N. Reinhoudt; S. Harkema; G. J. van Hummel, *J. Org. Chem. 47* (1982), 4419.
93. M. L. M. Pennings; D. N. Reinhoudt; S. Harkema; G. J. van Hummel, *J. Org. Chem. 48* (1983), 486.
94. M. L. M. Pennings; D. N. Reinhoudt, *J. Org. Chem. 48* (1983), 4043.
95. M. L. M. Pennings; D. N. Reinhoudt, *J. Org. Chem. 47* (1982), 1816.
96. M. L. M. Pennings; D. N. Reinhoudt, *Tetrahedron Lett. 1982*, 1003.
97. N. A. Akmanova; D. Ya. Mukhametova; Kh. F. Sagitdinova; F. A. Akbutina, *Zh. Org. Khim. 15* (1979), 2060.
98. K. Krishnan; N. Singh, *J. Indian Chem. Soc. 51* (1974), 802.
99. S. Tomoda; Y. Takeuchi; Y. Nomura, *Chem. Lett. 1982*, 1787.
100. H. Stamm; J. Hoenicke, *Liebigs Ann. 748* (1971), 143; *Synthesis 1971*, 145.
101. H. Stamm; H. Steudle, *Arch. Pharm. 309* (1976), 935; *310* (1977), 873.
102. S. Zbaida; E. Breuer, *Tetrahedron 34* (1978), 1241.
103. V. S. Velezhava; I. S. Yaroslavskii; L. N. Kurkovskaya; N. N. Suvorov, *Zh. Org. Khim USSR 19* (1983), 1518.

104. R. Huisgen; J. Wulff, *Chem. Ber. 102* (1969), 746.
105. D. St. C. Black; V. C. Davis, *J. Chem. Soc. Chem. Commun. 1975*, 416.
106. E. Breuer; S. Zbaida, *J. Org. Chem. 42* (1977), 1904.
107. E. Breuer; S. Zbaida; J. Pesso; I. Ronen-Braunstein, *Tetrahedron 33* (1977), 1145.
108. E. Breuer; S. Zbaida; J. Pesso; S. Levi, *Tetrahedron Lett. 1975*, 3103.
109. A. Angeli; L. Alessandri; M. Aiazzi-Mancini, *Atti Acad. Lincei 20* 1 (1910), 546.
110. A. Dornov; H. Gehrt; F. Ische, *Liebigs Ann. 585* (1954), 220.
111. G. E. Utzinger; R. A. Regenass, *Helv. Chim. Acta 37* (1954), 1885, 1892.
112. D. St. C. Black; V. M. Clark; R. S. Thakur; Lord Todd, *J. Chem. Soc., Perkin I 1976*, 1951.
113. D. St. C. Black; N. A. Blackman; L. M. Johnstone, *Austr. J. Chem. 32* (1979), 2025.
114. C. Berti; M. Colonna; L. Greci; L. Marchetti, *Tetrahedron 31* (1975), 1745; *32* (1976), 2147; *J. Heterocycl. Chem. 16* (1979), 17.
115. H. Stamm; J. Hoenicke, *Liebigs Ann. 749* (1971), 146.
116. L. S. Kaminski; M. Lamchen, *J. Chem. Soc. (C) 1967*, 1683.
117. E. Breuer; I. Ronen-Braunstein, *J. Chem. Soc., Chem. Commun. 1974*, 949.
118. O. Tsuge; K. Sone; S. Urano; K. Matsuda, *J. Org. Chem. 47* (1982), 5171.
119. O. Tsuge; K. Sone; S. Urano, *Chem. Lett. 1980*, 977.
120. H. Stamm; H. Steudle, *Tetrahedron 35* (1979), 647.
121. L. A. Neiman; S. V. Zhukova, *Zh. Org. Khim. 10* (1974), 2175; *Tetrahedron Lett. 1973*, 499.
122. R. E. Erickson; T. M. Myszkiewicz, *J. Org. Chem. 30* (1965), 4326.
123. A. K. Qureshi; B. Sklarz, *J. Chem. Soc. (C) 1966*, 412.
124. J. F. Elsworth; M. Lamchen, *J. Chem. Soc. (C) 1966*, 1477; *J. S. Afr. Chem. Soc. 25* (1972), 1.
125. A. R. Forrester; R. H. Thomson, *J. Chem. Soc. 1963*, 5632.
126. D. St. C. Black; N. A. Blackman, *Austr. J. Chem. 32* (1979), 1795.
127. D. St. C. Black; N. A. Blackman; R. F. C. Brown, *Austr. J. Chem. 32* (1979), 1785.
128. U. M. Kempe; T. K. Das Gupta; K. Blatt; P. Gygax; D. Felix; A. Eschenmoser, *Helv. Chim. Acta 55* (1972), 2187.
129. D. St. C. Black; A. B. Boscacci, *Austr. J. Chem. 29* (1976), 2511.
130. G. W. Alderson; D. St. C. Black; V. M. Clark; Lord Todd, *J. Chem. Soc., Perkin I 1976*, 1955.
131. J. K. Sutherland; D. A. Widdowson, *J. Chem. Soc. 1964*, 3495.
132. D. St. C. Black; V. M. Clark; R. S. Thakur; Lord Todd, *J. Chem. Soc., Perkin I 1976*, 1951.
133. M. Hamana, *J. Pharm. Soc. Japan 75* (1955), 139.
134. E. Howard, Jr.; W. F. Olszewski, *J. Am. Chem. Soc. 81* (1959), 1483.
135. A. R. Katritzky; M. Lagowski In "Chemistry of the Heterocyclic N-Oxides", Academic Press, London, 1971.
136. P. Milliet; X. Lusinchi, *Tetrahedron 35* (1979), 43.
137. O. Exner, *Coll. Czech. Chem. Commun. 20* (1955), 202.
138. (a) J. Thesing; H. Mayer, *Liebigs Ann. 609* (1957), 46.
    (b) D. H. R. Barton; A. Fekih; X. Lusinchi, *Tetrahedron Lett. 1985*, 4603.
139. (a) E. G. Janzen, *Accts. Chem. Res. 4* (1971), 31.
    (b) C. Lagercrantz, *J. Phys. Chem. 75* (1971), 3466.
140. G. Bianchi and R. Gandolfi In "1,3-Dipolar Cycloaddition Chemistry" (A. Padwa, Ed.), Vol. 2, Wiley, New York, 1984, p. 451.
141. K. Koyano; I. Tanaka, *J. Phys. Chem. 69* (1965), 2545.
142. Y. Inouye; J. Hara; H. Kakisawa, *Chem. Lett. 1980*, 1407.
143. J. Bjørgo; D. R. Boyd; D. C. Neill, *J. Chem. Soc., Chem. Commun. 1974*, 478.
144. W. B. Jennings; D. R. Boyd; L. C. Waring, *J. Chem. Soc., Perkin II 1976*, 610.
145. D. R. Boyd; D. C. Neill, *J. Chem. Soc., Perkin I 1977*, 1308.
146. Y. Inouye; Y. Watanabe; S. Takahashi; H. Kakisawa, *J. Chem. Soc. Japan 52* (1979), 3763.
147. N. S. Ooi; D. A. Wilson, *J. Chem. Res (S) 1980*, 394.
148. T. Markowicz; J. Skolimowski; R. Skowronski, *Pol. J. Chem. 55* (1981), 2505.
149. A. H. M. Kayen; T. J. de Boer, *Rec. Trav. Chim. 96* (1977), 8.
150. C. Schenk; M. L. Beekes; J. A. M. van der Drift; T. J. de Boer, *Rec. Trav. Chim. 99* (1980), 278.
151. M. Abou-Gharbia; M. M. Joullie, *Synthesis 1977*, 318.
152. H. Suginome; N. Sato; T. Masamune, *Tetrahedron 27* (1971), 4861.
153. R. Kreher; H. Morgenstern, *Chem. Ber. 115* (1982), 2679.
154. G. Tacconi; P. P. Righetti; G. Desimoni, *J. Prakt. Chem. 322* (1980), 679.
155. N. S. Ooi; D. A. Wilson, *J. Chem. Res (S) 1981*, 18.
156. J. A. Damavandy; R. A. Y. Jones, *J. Chem. Soc., Perkin I 1981*, 712.
157. A. Vasella, *Helv. Chim. Acta 60* (1977), 426, 1273.

158. A. Vasella; R. Voeffray; J. Pless; R. Huguenin, *Helv. Chim. Acta 66* (1983), 1241.
159. S. Mzengeza; R. A. Whitney, *J. Chem. Soc., Chem. Commun. 1984*, 606.
160. D. A. Kerr; D. A. Wilson, *J. Chem. Soc. (C) 1970*, 1718.
161. W. M. Leyshon; D. A. Wilson, *J. Chem. Soc., Perkin I 1975*, 1920, 1925.
162. G. E. Utzinger, *Liebigs Ann. 556* (1943), 50.
163. H. E. De La Mare; G. M. Coppinger, *J. Org. Chem. 28* (1963), 1068.
164. J. Thesing; A. Müller; G. Michel, *Chem. Ber. 88* (1955), 1027.
165. W. H. Rastetter; T. R. Gadek; J. P. Tane; J. W. Frost, *J. Am. Chem. Soc. 101* (1979), 2228.
166. R. Behrend; K. Leuchs, *Liebigs Ann. 257* (1890), 223.
167. H. Rupe; R. Wittwer, *Helv. Chim. Acta 5* (1922), 217.
168. E. J. Alford; J. A. Hall; M. A. T. Rogers, *J. Chem. Soc. (C) 1966*, 1103.
169. J. Thesing, *Chem. Ber. 87* (1954), 507.
170. A. Treibs; G. Fritz, *Liebigs Ann. 611* (1958), 162.
171. R. Behrend; E. König, *Liebigs Ann. 263* (1891), 191.
172. J. Thesing; H. Mayer, *Chem. Ber. 89* (1956) 2159.
173. J. Thesing; W. Stirrenberg, *Chem. Ber. 92* (1959), 1748.
174. A. H. Wragg; T. S. Stevens, *J. Chem. Soc. (C) 1959*, 462.
175. S.-I. Murahashi; H. Mitsui; T. Watanabe; S.-i. Zenki, *Tetrahedron Lett. 1983*, 1049.
176. H. Mitsui; S-i. Zenki; T. Shiota; S-I. Murahashi, *J. Chem. Soc., Chem. Commun. 1984*, 874.
177. C. E. Griffin; N. F. Hepfinger; B. L. Shapiro, *Tetrahedron 21* (1965), 2735.
178. P. Ehrlich; F. Sachs, *Chem. Ber. 32* (1899), 2341.
179. F. Kröhnke, *Chem. Ber. 71* (1938), 2584.
180. F. Nerdel; I. Huldschinsky, *Chem. Ber. 86* (1953), 1005.
181. R. G. Kostyanovskii; V. I. Markov; A. I. Mischenko; A. V. Prosyanik, *Izvest. Akad. Nauk SSSR, Ser. Khim. Nauk 1977*, 250.
182. P. A. S. Smith; J. E. Robertson, *J. Am. Chem. Soc. 84* (1962), 1197.
183. H. Lindemann; K. T. Tschang, *Chem. Ber. 60* (1927), 1725.
184. O. L. Brady; F. B. Dunn; R. F. Goldstein, *J. Chem. Soc. 1926*, 2386.
185. E. Buehler, *J. Org. Chem. 32* (1967), 261.
186. E. Buehler; G. B. Brown, *J. Org. Chem. 32* (1967), 265.
187. E. Beckmann, *Chem. Ber. 22* (1889), 429.
188. N. K. A. Dalgaard; K. E. Larsen; K. B. G. Torssell, *Acta Chem. Scand. B 38* (1984), 423.
189. M. Ochiai; M. Obayashi; K. Morita, *Tetrahedron 23* (1967), 2641.
190. A. Lablache-Combier; M.-L.Villaume; R. Jacquesy, *Tetrahedron Lett. 1967*, 4959.
191. R. Grigg; M. Jordan; A. Tangthongkum; F. W. B. Einstein; T. Thomas, *J. Chem. Soc., Perkin I 1984*, 47.
192. H. Gnichtel; R. Walentowski; K. E. Schuster, *Chem. Ber. 105* (1972), 1701.
193. N. V. Abbakumova; A. F. Vasilev; E. B. Nazarova, *Khim. Geterosikl. Soed. 1981*, 968.
194. S. A. Hussain; A. H. Sharma; M. J. Perkins; D. Griller, *J. Chem. Soc., Chem. Commun. 1979*, 289.
195. V. P. Tashchi; T. Orlova; Yu. G. Putsykin; A. F. Rukasov, *Dokl. Akad. Nauk SSSR 266* (1982), 1167.
196. T. D. Lee; J. F. W. Keana, *J. Org. Chem. 41* (1976), 3237.
197. N. A. LeBel; N. Balasubramanian, *Tetrahedron Lett. 1985*, 4331.
198. S. P. Ashburn; R. M. Coates, *J. Org. Chem. 50* (1985), 3076.
199. M. Chmielewski; C. Belzecki, *Bull. Pol. Acad. Sci. Chem. 32* (1984), 195.
200. D. Lathbury; T. Gallagher, *Tetrahedron Lett. 1985*, 6249.

Chapter 4

Alkyl and Silyl Nitronates

4.1 INTRODUCTION. THE NITRONIC ACIDS[1,2]

Aliphatic nitro compounds with an α-hydrogen can exist in two tautomeric forms: the nitro form **1** and the aci-nitro form **2**, the nitronic acid, eq. (4.1). They were observed conductometrically by Holleman[3] in 1895, and shortly after that the two tautomeric forms of phenylnitromethane were synthesized.[4] At about the same time the first nitronic ester,

$$H_2NCOC(CN)=N\begin{smallmatrix}O\\OC_2H_5\end{smallmatrix}$$

was prepared.[5] The term "nitronic acid" (suffix) was coined by Bamberger[6] for the

$$=N\begin{smallmatrix}OH\\O\end{smallmatrix}$$

group and is widely used as well as the term "nitronate" for its esters. As a prefix the name *aci-nitro* has been suggested. The equilibrium is strongly shifted towards the nitro structure **1**, which is a considerably weaker acid than **2**. In Table 4.1 the ionization constants for a few nitro compounds are collected. In a few cases, when R^1 and R^2 are electron withdrawing groups, e.g., nitro groups in trinitromethane, $HC(NO_2)_3$, the C proton is sufficiently activated to be dissociated in aqueous solution and a noticeable amount of **2** ($R^1, R^2 = NO_2$) is formed.

$$\underset{\underline{1}}{\overset{R^1}{\underset{R^2}{>}}CHNO_2} \underset{k_{-1}}{\overset{k_1}{\rightleftarrows}} \overset{R^1}{\underset{R^2}{>}}\overset{-}{C}-N\begin{smallmatrix}O\\O\end{smallmatrix} \leftrightarrow \overset{R^1}{\underset{R^2}{>}}C=N\begin{smallmatrix}O^-\\O\end{smallmatrix} + H^+ \underset{k_{-2}}{\overset{k_2}{\rightleftarrows}}$$

$$\underset{\underline{2}}{\overset{R^1}{\underset{R^2}{>}}C=N\begin{smallmatrix}OH\\O\end{smallmatrix}} \qquad K_a^{NO_2} = \frac{k_1}{k_{-1}} \qquad K_a^{aci} = \frac{k_{-2}}{k_2}$$

(4.1)

Table 4.1. Ionization Constants of Nitroalkanes and Corresponding Nitronic Acids in Water at 25°C

Nitro compound	$pK_a^{NO_2}$	pK_a^{aci}	Ref.
CH_3NO_2	10.21	3.25^a	7
$CH_3CH_2NO_2$	8.46	4.40	7
$C_6H_5CH_2NO_2$	6.8	3.9	8
$(CH_3)_2CHNO_2$	7.68	5.11	7
cyclohexyl-NO_2	8.3	6.35	1, 9
$CH_2(NO_2)_2$	3.60^b	1.86	1, 10
$CH_3CH_2CH(NO_2)_2$	5.53^b	4.1	10, 11
$CH(NO_2)_3$	0.18^b	~0	1, 12

^{a} 0°C.
^{b} 20°C.

Likewise, chlorine and bromine show an acid-strengthening effect,[13] but fluorine is decidedly acid weakening.[14] The effect of alkyl groups is somewhat unpredictable. Contrary to expectations, 2-nitropropane ($pK_a^{NO_2} = 7.68$) is a stronger acid than 1-nitropropane ($pK_a^{NO_2} = 8.98$) or nitroethane ($pK_a^{NO_2} = 8.46$). However, the pK_a behavior became understandable when dissociation rates k_1 and k_{-2} and protonation rates k_{-1} and k_2 were measured. Indeed, 2-nitropropane dissociates slower than 1-nitropropane, but the rate of reprotonation of the more stable 2-nitronate is much slower than that of 1-nitronate, which results in the higher equilibrium constant (lower pK_a) for 2-nitropropane. The mechanism of tautomerization of nitro compounds and the determination of reaction rates have been investigated by several research groups and are covered by earlier reviews.[1,2]

Primary and secondary nitro compounds are rapidly deprotonated by base. The alkali metal salts are soluble in water but poorly soluble in alcohol-ether. Basification of an alcoholic solution of the nitro compounds with sodium hydroxide and subsequent addition of ether will precipitate the salts. The salts are unstable, and it is advisable not to isolate them. **Caution! The sodium salt of nitromethane undergoes violent decomposition or detonation on heating and drying!**

Careful acidification at 0°C with an acid of low nucleophilicity leads to a kinetically favored formation of the nitronic acid, provided $k_2 > k_{-1}$, eq. (4.1). Since the nitro form is more stable than the tautomeric *aci*-nitro form, it reverts slowly into the nitro compound. This is a rapid process for most nitro compounds (minutes to hours), but in certain cases the nitronic acid can be isolated and stored in a refrigerator. The half-life exceeds several weeks or months, when the *aci*-nitro form can gain extra stability by conjugation as in **3** or by steric hindrance as in **4** (Table 4.2). A competing reaction to the reformation of the nitro compound on acidification is Nef's reaction,[20] in which

Introduction. The Nitronic Acids

Table 4.2. Stability of Nitronic Acids

Compound	Mp (°C)	Half-life[a] at ca. 25°C	Ref.
$CH_3CH=NO_2H$	Liquid	Min	15
$C_6H_5CH=NO_2H$	84	Days	4
$(o\text{-}CH_3\text{-}C_6H_4)(H)C=NO_2H$	45	Min	16
$(4\text{-}Br\text{-}C_6H_3)(NC)C=NO_2H$	64	Days	17
3	ca. 145–150	Weeks	18
4	ca. 101–118	Months	19

[a] Approximate time for exhibiting evident decomposition.

3 (fluorenone nitronic acid structure)

4 (aryl methyl nitronic acid structure with Cl and Br substituents)

the nitronate ion is converted into a carbonyl compound (route 4.2a). The Nef reaction is favored under conditions of strong acids and where addition of the nitronate salt takes place in excess of acid. Use of anhydrous hydrochloric acid converts primary nitronate salts to hydroxamic chlorides, eq. (4.3).[1,21] A transient blue color often appears on acidification of the nitronate ion, indicating the formation of an intermediate nitroso compound as shown in eqs. (4.2),(4.3).

$$R^1R^2C=N(OH)O \xrightarrow{H^+, H_2O} R^1R^2C(OH)N(OH)_2 \xrightarrow{a} R^1R^2C(OH)NO \xrightarrow{} R^1R^2C=O + HNO \quad (4.2)$$

$$\xrightarrow{b} R^1CONHOH \text{ (Hydroxamic acid, } R^2 = H\text{)}$$

$R^1, R^2 = \text{Alk, Ar, H}$

$$R^1\text{-}CHCl\text{-}N(OH)(O\text{-}H^+) \rightarrow R^1\text{-}CHCl\text{-}N(OH)_2 \rightarrow R^1\text{-}CCl=NOH \quad (4.3)$$

At low pH, Nef conditions help to fix the undissociated nitronic acid structure, rendering it susceptible to an attack of a nucleophilic water molecule or chloride ion at the carbon atom with concomitant rearrangement into ketones and aldehydes (route 4.2a), hydroxamic acids (route 4.2b), or hydroximic acid chlorides, eq. (4.3). At very low pH, formation of hydroxamic acids is favored.[22,23] The intermediacy of nitrile oxides has been invoked in route 4.2b.[24] At moderate pH (ca. 5) protonation of the carbon atom competes favorably with O-protonation, and the nitronic acid reverts to the nitro compound. Nitronic acids are unstable compounds, and none of their transformations goes entirely to one single product even under the most favorable conditions. In addition to the main products (nitro compounds or Nef products) formation of oximes, nitrolic acids, or geminal nitrosonitro derivatives and nitrite ions have been observed. These products can occasionally appear in main products, as in the case of the decomposition of cyclohexanenitronic acid at pH 3.05, eq. (4.4).[25]

$$\text{C}_6\text{H}_{10}\text{=NO}_2\text{H} \xrightarrow{\text{pH 3.05}} \text{cyclohexanone} + \text{cyclohexyl-NO}_2 + \text{cyclohexyl-NOH} + \text{cyclohexyl(NO)(NO}_2\text{)} \quad (4.4)$$

32%   6%   31%   31%

Reaction of nitronate salts with halogens gives α-halonitro compounds.[26] With nitrous acid geminal nitrosonitro derivatives are formed, which for $R^2 = H$ rearrange to nitrolic acids, eq. (4.5).[27,28] Oxidation of nitronic acids or nitronates by alkaline potassium permanganate[29-35] as well as reduction by HI,[36] Ti^{3+},[37,38] Va^{2+},[39,40] and Cr^{2+} [41] gives carbonyl compounds, often in better yields than the classical Nef reaction. It is observed that nitronates occasionally undergo dimerization, probably via a radical process,[42] eqs. (4.6)–(4.10).

$$R^1R^2C=N(O^-)(O) \xrightarrow{E^+} R^1R^2C(E)\text{-NO}_2 \rightarrow R^1\text{-C}(=NOH)(NO_2) \quad (4.5)$$

$E^+ = NO^+, CL_2, Br_2, I_2,$ etc.   (for $E^+ = NO^+$ $R^2 = H$)

Nitrolic acid

$$R_2C=N(OH)(O) + R_2C=N(O^-)(O) \rightarrow R_2C=N(OH^\bullet)(O) + R_2C=N(O^\bullet)(O) \quad (4.6)$$

$$R_2C=N(O^\bullet)(O) + R_2C=N(O^-)(O) \rightarrow R_2C(NO_2)\text{-CR}_2\text{-N}(O^{\bullet-})(O) \quad (4.7)$$

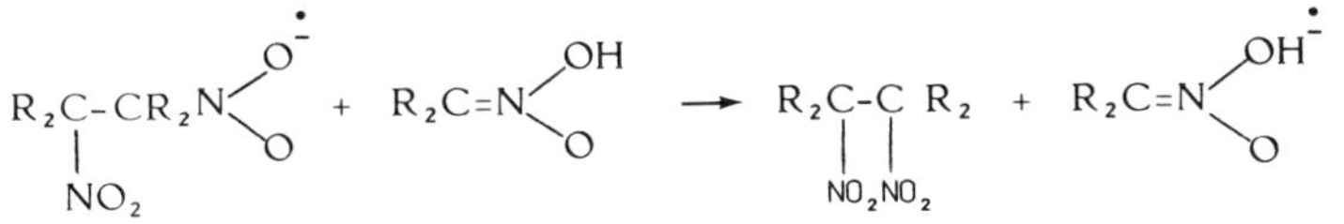

$$R_2C-CR_2-N\begin{matrix}O^-\\O\end{matrix} + R_2C=N\begin{matrix}OH\\O\end{matrix} \longrightarrow R_2C-CR_2 + R_2C=N\begin{matrix}OH^-\\O\end{matrix} \quad (4.8)$$
$$|||$$
$$NO_2NO_2\,NO_2$$

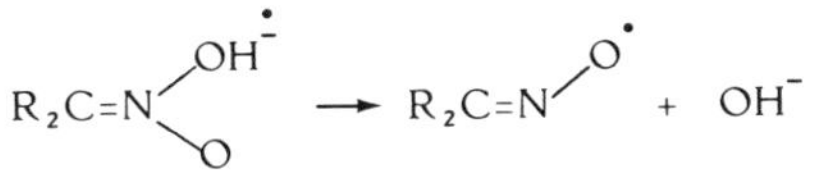

$$R_2C=N\begin{matrix}OH^-\\O\end{matrix} \longrightarrow R_2C=N\begin{matrix}O^-\\O\end{matrix} + OH^- \quad (4.9)$$

$$R_2C=N\begin{matrix}O^-\\O\end{matrix} + R_2C=N\begin{matrix}O^-\\O\end{matrix} \longrightarrow R_2C=N\begin{matrix}O^-\\O\end{matrix} + R_2C=N\begin{matrix}O^-\\O\end{matrix} \quad (4.10)$$

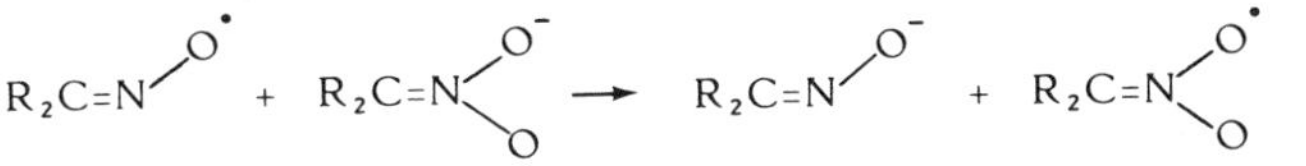

The α-ketonitronic acids present a special problem because three tautomeric forms are possible: they have been identified by IR, UV, and NMR spectroscopy, eq. (4.11).[43,44] Their relative amounts are strongly solvent dependent.

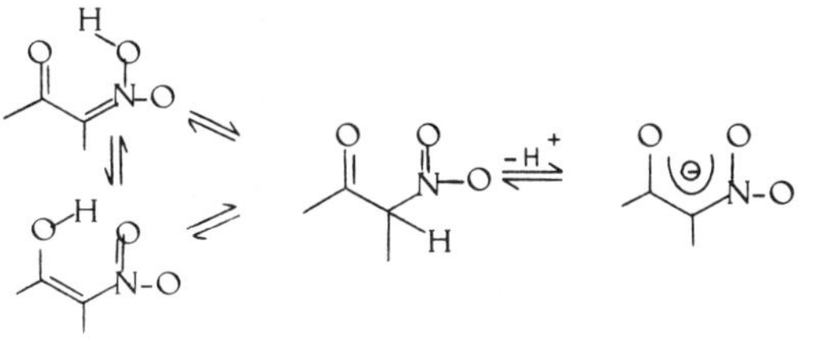

(4.11)

Certain highly substituted γ-ketonitronic acids occur in an unstable ring tautomeric dihydro-1,2-oxazin-N-oxide form **5**, eq. (4.12).[45–48] The IR spectrum of compound **5** exhibits OH-absorption at ca. 3100 cm^{-1}, and the carbonyl stretching in the 1650–1800 cm^{-1} region is missing. The oxazine **5** decomposes slowly into the dicarbonyl compound. The sterically unhindered 1-nitro-4-pentanone does not show any sign of a similar equilibrium.

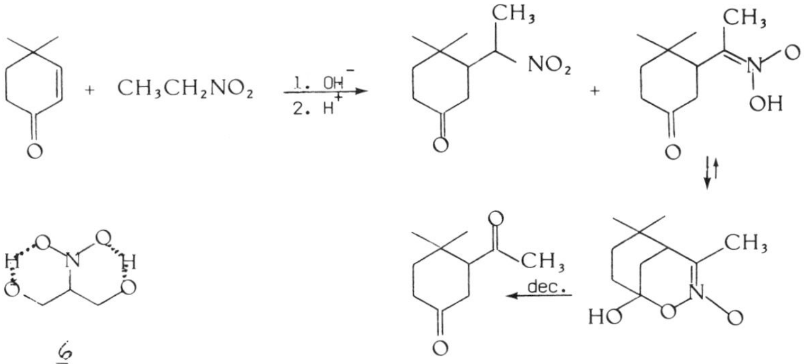

(4.12)

Nitronic acids and nitronates are also formed by Lewis acid–catalyzed (TiCl$_4$, Ti(Oi–Pr)$_2$Cl$_2$, SnCl$_4$, AlCl$_3$) addition of trimethylsilylenol ethers to nitroolefins, eq. (4.13).[48–51a] It is likely that silyl nitronates appear as intermediates in the aluminum chloride–catalyzed addition of allylsilane to α-nitroolefins, which leads to 1,4-diones by subsequent reduction with titanous ions.[51b] The Michael reactions, eqs. (4.12) and (4.13) represent useful preparative methods for 1,4-dicarbonyl compounds. 2-*aci*-Nitro-1,3-propanediol (**6**) is stabilized by intramolecular hydrogen bonds.[52]

There is no report in the literature that nitronic acids or salts undergo 1,3-dipolar addition with olefins in analogy with the silyl or alkyl nitronates. It is, however, reported that methyl acrylate and potassium trinitromethide at pH ∼ 5 give substantial amounts of methyl 4,4-dinitro-2-hydroxybutyrate, eq. (4.14).[53] The formation of this compound can either be explained by the 1,3-dipolar addition (route 4.14a) or by the two-step mechanism (route 4.14b).

The nitronate group is planar.[54,55] The C—N bond is shorter than a single C—N bond (ca. 1.45 Å) but longer than an imine bond (ca. 1.23 Å), and the N—O bond is slightly longer than that of the nitro group, indicating that the nitronate function has a considerable C—N double-bond character and that the charge density is located on the oxygen atoms.

Aliphatic nitro compounds are characterized by an intense UV absorption near 210 nm (π, π^* transition, $\varepsilon \sim 10{,}000$) and a low-intensity band at 270–280 nm (n, π^* transition, $\varepsilon \sim 20$).[56,57] The long-wave λ_{max} value is not affected significantly by the presence of geminal nitro or halogen groups. The anion of aliphatic mononitro compounds shows an intense absorption band in the 230–240-nm region ($\varepsilon \sim 10{,}000$). It moves into the visible region for the geminal dinitroalkane ion, $\lambda_{max} \approx 380$ nm.[58] Ethanenitronate, $CH_3CH=NO_2^-$ absorb at λ_{max} 229 ($\varepsilon = 9000$) and 1-nitroethanenitronate $CH_3C(NO_2)NO_2^-$ at λ_{max} 381, $\log \varepsilon = 4.21$.[59]

Conjugation causes the expected shift in absorption toward longer wavelengths; $C_6H_5CH=NO_2H$ absorbs at $\lambda_{max}^{H_2O}$ 284 nm ($\varepsilon = 20{,}000$), and its anion $C_6H_5CH=NO_2^-$ absorbs at $\lambda_{max}^{H_2O}$ 293 nm ($\varepsilon = 20{,}000$).[60] The UV spectra of the nitronic acids are very similar to those of their anions and esters.

In the IR region the nitronic acids show a broadband absorption around 2500–3000 cm^{-1} emanating from OH stretching similar to that of carboxylic acids. A characteristic C=N absorption is located in the 1620–1680 cm^{-1} region (s); this absorption is shifted to lower frequencies for the salts, 1585–1605 cm^{-1} (s).[61–63] The corresponding aliphatic nitro group stretching frequency is located at 1550–1560 cm^{-1}.

4.2 ALKYLATION OF NITRO COMPOUNDS. SYNTHESIS OF ALKYL NITRONATES (NITRONIC ESTERS)

The synthesis of alkyl nitronates has been reviewed comprehensively up to the end of 1979.[1,2,64] The ambident nitronate ion can undergo C- and O-alkylation, eq. (4.15), and the ratio is determined by the following factors:

1. The structure of the nitronate salt and the counter ion
2. The structure of the alkylating agent
3. The leaving group of the alkylating agent (X^-)
4. The heterogeneity of the medium, the solvent and the temperature

$$M^+ \quad \underset{R^2}{\overset{R^1}{>}}N\underset{O}{\overset{O^-}{<}} \quad \xrightarrow{R^3X} \quad \underset{R^2}{\overset{R^1}{>}}\underset{R^3}{\underset{|}{C}}-NO_2 \quad + \quad \underset{R^2}{\overset{R^1}{>}}N\underset{O}{\overset{OR^3}{<}} \quad + \quad X^- \quad (4.15)$$

$X = Cl, Br, I, RSO_3, O^+R_2, N_2^+, N^+(CH_3)_3$

C-Alkylation gives a nitro compound, and O-alkylation gives an alkyl nitronate (nitronic ester).

As a general rule, alkylations take place predominantly—in several cases exclusively—on the oxygen, giving initially unstable nitronic esters in fair to good yields. There are some noteworthy exceptions to this rule:

1. C-alkylations of nitronate salts by o- and p-nitrosubstituted benzyl halides, eq. (4.16).[65,66]
2. C-Alkylation of polynitro compounds by methyl iodide, eq. (4.17)[67-72]
3. C-Arylation with iodonium salts, eq. (4.18)[73]
4. C-Alkylation with N-substituted pyridinium salts, eq. (4.19)[74-76]
5. C-Alkylation by organomercurials, eq. (4.20)[77,78]
6. C-Alkylation of methyl α-nitroacetate in aprotic solvents, eq. (4.21)[79,80]
7. Pd-catalyzed C-alkylation[81,82]
8. The Michael addition, eq. (4.12)[26]
9. The nitro-aldol reaction, eq. (4.30)[26]
10. C-Alkylation and acylation by double deprotonation, eq. (4.45)[83-87]

$$p\text{-}O_2N\text{-}C_6H_4\text{-}CH_2Cl + (CH_3)_2C=N(O^-)(O) \rightarrow p\text{-}O_2N\text{-}C_6H_4\text{-}CH_2C(CH_3)_2NO_2 \quad (4.16)$$

$$CH_3I + (NO_2)_2C=N(O^-)(O) \rightarrow CH_3C(NO_2)_3 \quad (4.17)$$

$$Ph_2I^+ + (CH_3)_2C=N(O^-)(O) \rightarrow PhC(CH_3)_2NO_2 + PhI \quad (4.18)$$

$$\text{(2,4,6-triphenyl-N-R-pyridinium)} + (CH_3)_2C=N(O^-)(O) \rightarrow (CH_3)_2CRNO_2 + \text{(2,4,6-triphenylpyridine)} \quad (4.19)$$

$$RHgCl + (CH_3)_2C=N(O^-)(O) \xrightarrow{h\nu} RC(CH_3)_2NO_2 + Hg^\circ + Cl^- \quad (4.20)$$

$$RI(Br) + R^1OOCCH=N(O^-)(O) \rightarrow R^1OOCCHRNO_2 \quad (4.21)$$

A characteristic feature of reactions (4.16) and (4.20) is that they predominantly proceed via radical intermediates in a chain process, termed $S_{RN}1$. The mechanism is well investigated for the reaction of p-nitrobenzyl chloride with 2-propanenitronate, (4.16), and it is formulated in eqs. (4.22)–(4.25).[88-91] Adding 1,4-dinitrobenzene as an electron trap stops the reaction and increases the O-alkylation from 6% to 88% and strongly supports the suggested electron transfer mechanism.[92] In eqs. (4.17)–(4.20) charge transfer or electron transfer from the nitronate ion to the alkylating agent, as in eq. (4.22), could give the intermediate radicals **7–10**, which fragment according to eqs. (4.26)–(4.29). The alkylation, eq. (4.16), shows a characteristic effect on the leaving group (Table 4.3).[93]

$$CH_3I^{-\bullet} \longrightarrow CH_3^{\bullet} + I^- \qquad (4.26)$$
$$\underline{7}$$

$$Ph_2I^{\bullet} \longrightarrow PhI + Ph^{\bullet} \qquad (4.27)$$
$$\underline{8}$$

(4.28)

$$\underline{9}$$

$$RHgCl^{-\bullet} \longrightarrow R^{\bullet} + Hg^{\circ} + Cl^- \qquad (4.29)$$
$$\underline{10}$$

Table 4.3. Effect of the Leaving Group in the Reaction of p-NO$_2$-benzyl-X with the Lithium Salt of 2-Nitropropane in DMF^a at $-16°C$, Eq. (4.16)[93]

O$_2$N—⟨⟩—CH$_2$—X (X = leaving group)	C-Alkylation (%)	O-Alkylation (%)
N(CH$_3$)$_3$ (25°C)	93	0
Cl	93	1
O—Ts	40	32
Br	17	65
I	9	81

^{a}DMF = dimethylformamide.

The experimental results are explained by the supposition that C-alkylation is preceded by an electron transfer as in (4.22) and that the direct S_N2 displacement of N(CH$_3$)$_3$, Cl, and OTs is slow. For Br and I the S_N2 displacement by the negatively charged nitronate oxygen competes effectively, leading to O-alkylation. In the unsubstituted benzyl halide series there is no leaving group effect. The yields of O-alkylation products are about 82–84%.

Equation (4.21) shows a distinct solvent dependence. It has been argued that C-alkylation predominates in aprotic polar solvents because the negative charge density shifts to the carbon atom.[94] Two C-alkylation reactions are exceptional: (1) the Michael addition (4.12)[26] and (2) the nitro-aldol reaction (Henry's reaction) (4.30).[26] These two reactions give only C-alkylation and have found vast application in organic synthesis as a carbon-carbon coupling procedure. The reversible character of these reactions has to be considered when the most favorable experimental conditions are investigated. Normally only catalytic amounts of base are required.

$$\begin{matrix} R^1 \\ R^2 \end{matrix} C=N \begin{matrix} O^- \\ O \end{matrix} + \begin{matrix} R^3 \\ R^4 \end{matrix} C=O \rightleftharpoons R^1 R^2 \underset{NO_2}{C} - \underset{OH}{C} R^3 R^4 \qquad (4.30)$$

$$R^1, R^2, R^3, R = Alk, Ar, H$$

Excellent yields of nitronic esters are obtained by alkylation of the sodium nitronate with trimethyl and triethyloxonium tetrafluoroborate[95–97,106] at 0°C in methylene chloride (Table 4.4). Two isomers are formed, (E) and (Z), which have been observed spectroscopically and also have been separated.

O-Alkylation of nitronic acids has been successfully carried out with diazomethane and diazoethane, eq. (4.31).[95,98–118] To be effective the acidity of the α-proton should be high (pK_a < 8), as in α-nitroacetates and similar compounds. Thus, 1-nitropropane does not react with diazomethane, but this drawback can be circumvented by first transforming the nitro compound into its aci-form.[95] No C-alkylation was observed with diazomethane or trialkyloxonium salts, but α-nitroketones gave minor amounts of the isomeric methyl enol ether.[113]

Table 4.4. Nitronic Esters Prepared by Alkylation of Nitro Compounds with Oxonium Tetrafluoroborate or Diazomethane (^{1}H NMR Data)

Nitronic ester	Yield (%) E+Z	^{1}H NMR vinyl H Z	^{1}H NMR vinyl H E	Ratio Z/E	Ref.
H$_3$C, O / N / H O CH$_3$ (E) and H$_3$C, OCH$_3$ / N / H O (Z)	90–95	5.91	6.25	1:5	95
CH$_3$CH=NO$_2$C$_2$H$_5$	94	5.57	6.27	1:2	95
C$_2$H$_5$CH=NO$_2$C$_2$H$_5$	79	6.05	6.11	1.2:1	95
C$_3$H$_8$CH=NO$_2$C$_2$H$_5$	90–95	5.75 2.26 (CCH$_2$) 4.14 (OCH$_2$)	6.04 2.26 (CCH$_2$) 4.10 (OCH$_2$)	1:7	95
p-BrPhCH=NO$_2$C$_2$H$_5$	95	6.80	7.03	1:3	95
p-NO$_2$PhCH=NO$_2$C$_2$H$_5$	92	6.95	7.22	1:4	95
(CH$_3$)$_2$C=NO$_2$C$_2$H$_5$	75–80				
CH$_3$CH$_2$C=NO$_2$C$_2$H$_5$ | CH$_3$	90–95	1.94 (CH$_3$)	2.00 (CH$_3$)	1:1	95
CNCH=NO$_2$CH$_3$		6.00	6.43	57:43	112
CH$_3$OOCCH=NO$_2$CH$_3$		6.46	6.79	60:40	112
CH$_3$COCH=NO$_2$CH$_3$		6.58	6.85	45:55	112

$$R^1R^2CH-NO_2 \rightleftharpoons R^1R^2C=N(OH)O \xrightarrow{CH_2N_2} R^1R^2C=N(OCH_3)O \text{ (11a)} + R^1R^2C=N(O)(OCH_3) \text{ (11b)} \quad (4.31)$$

R^1	R^2	11a + b Ref.
NO$_2$	NO$_2$	103, 106, 108, 109
H	NO$_2$	105, 108
COOCH$_3$	COOCH$_3$	104, 108
H	COOCH$_3$	102, 103, 112
H	Ph	101, 107
H	CN	112
H	COCH$_3$	112
H	p-NO$_2$Ph	112
H	PhSO$_2$	98, 113, 118

O-Methylation of nitronate salts has also been performed with dimethyl sulfate,[98,119,120] producing oximes and carbonyl compounds as principal products.

Alkyl nitronates are like the nitronic acid unstable compounds, with half-lives from a few minutes to several hours or days at 25°C. In a few cases they remain unchanged for several weeks. They fragment into an oxime and a carbonyl compound (13). Equation (4.32) has been optimized and can actually be employed as a preparative method for aldehydes and ketones from alkyl

$$\ce{>=N(O)OC(H)<} \rightarrow \ce{>=NOH} + \ce{>=O} \qquad (4.32)$$
$$\qquad\qquad\qquad \underline{12} \quad\;\; \underline{13}$$

halides[119-123] and also from alcohols, according to the Mitsunobu procedure.[124] The fragmentation products **12** and **13** are directly produced in the reaction of alkali metal nitronates with alkyl halides. With silver nitronate it is occasionally possible to isolate the initially formed nitronic ester.[1,64]

The cyclic nitronic esters of the general structure **14** are more stable than the acyclic ones. These derivatives are produced in good yields from 1,3-halonitro or 1,3-dinitroalkanes, eq. (4.33).[125-133]

$$\text{1,3-halonitro} \xrightarrow{\text{KOAc}} \underline{14} \qquad (4.33)$$

X = Cl, Br, I, NO$_2$

The 1,3-halonitro derivatives can be obtained via Michael addition of a nitroalkane to an electron-deficient olefin, eq. (4.34). Subsequent bromination and treatment with base yield the nitronic ester (2-isoxazoline-*N*-oxide). The primary nitro compound **15** gives, apart from the nitronic ester **16**, substantial amounts of the nitrocyclopropane **17**. Secondary 1,3-nitrohalo compounds in many cases give no cyclopropanes. The same cyclization has been applied to 1,1-dinitro-3-bromo compounds, the preparation of which is exemplified by eqs. (4.35)–(4.37).[134-140] It is noteworthy that the alkylation of mercury tri-

$$\text{PhCH=C(Ph)O} \xrightarrow[\text{2. Br}_2]{\text{1. }^-\text{CH}_2\text{NO}_2} \text{Ph-CH(CH}_2\text{NO}_2\text{)-CHBr-C(Ph)=O} \xrightarrow{\text{KOAc}} \underline{16} + \underline{17} \qquad (4.34)$$

$$\underline{15}$$

$$\ce{CH2=CH2} + \ce{Hg(C(NO2)3)2} \rightarrow \ce{(O2N)3C-CH2-CH2-HgC(NO2)3} \xrightarrow{\text{HCl}} \ce{(O2N)3C-CH2-CH2-HgCl}$$

$$\underline{18}$$

$$\downarrow \ce{CH2=CH2}$$

$$\ce{(O2N)3C-CH2-CH2-Hg-CH2-CH2-C(NO2)3} \qquad (4.35)$$

Alkylation of Nitro Compounds. Synthesis of Alkyl Nitronates

$$\underline{18} \xrightarrow{\text{NH}_2\text{OH}}_{\text{KOH}} \quad {}^{-}\text{C(NO}_2)_2 \quad \xrightarrow{\text{HgCl}} \quad \xrightarrow{\text{HCl}} \quad \text{CH(NO}_2)_2 \xrightarrow{\text{HgCl}} \quad \underline{19} \quad (4.36)$$

$$\underline{19} \xrightarrow{\text{Br}_2} \text{CH(NO}_2)_2\text{-Br} \xrightarrow[25\,°C/24\,h]{\text{KOAc}} \underline{20} \quad (4.37)$$

(isoxazoline N-oxide **20**)

$$\not\to \underline{21}$$ (1,1-dinitrocyclopropane)

nitromethanide occurs on the carbon atom in analogy to reaction (4.17). This is not always the case. The addition of tetranitromethane to cyclohexene gives the nitronate ester, which then cycloadds to a second molecule of cyclohexene to give the isoxazolidine **22**, eq. (4.38).

$$\text{(cyclohexene + } {}^{+}\text{NO}_2 \cdots {}^{-}\text{OC(NO}_2)_3) \rightarrow \text{(nitronate ester)} \xrightarrow{\text{cyclohexene}} \underline{22} \quad (4.38)$$

With few exceptions acylation of nitronate salts occurs on oxygen,[1,13,119,141–150] but the yield of the isolated product **23** is poor, eq. (4.39). For $R^2 = H$ compound **23** rearranges into the hydroxamic acid derivatives **24** and **25** or ultimately gives the free acids, eq. (4.40) route a. The intermediacy of **23** was demonstrated by trapping with a dipolarophile, eq. (4.41).[144] It is suggested that **23** is the active dipolarophile in reaction (4.41) and not the nitrile oxide, which conceivably could be formed by eliminating acid, eq. (4.42), because no nitrile oxide dimer was observed, as it was for the closely related acylation of primary nitro compounds with isocyanate, eq. (4.43).[151]

$$\underset{R^2}{\overset{R^1}{>}}\!\!=\!\!\underset{O}{\overset{O^-}{N^+}} \;+\; \text{RCOX} \;\longrightarrow\; \underset{R^2}{\overset{R^1}{>}}\!\!=\!\!\underset{O^-}{\overset{OCOR}{N^+}} \;+\; X^- \quad (4.39)$$

$$\underline{23}$$

X = RCOO, Cl, Br or Ketene

$$23 \xrightarrow[a]{\Delta} \underset{\underset{OCOR}{|}}{R^1C=NOH} \longrightarrow \underset{\underset{O}{\|}}{R^1C-NHOCOR} \quad (4.40)$$

$$23 \downarrow b$$

$$\underset{26}{\underset{R^2}{R^1}\diagdown\diagup_{OCOR}^{NO}} \quad 24 \qquad 25$$

$$R^1CH_2NO_2 \xrightarrow[NaOAc]{Ac_2O} 23 \xrightarrow{CH_3OOC\equiv COOCH_3} \text{(isoxazole with } CH_3OOC, R^1, CH_3OOC\text{)} \quad (4.41)$$

R^1 = Alkyl, Aryl; yields 30–82%

$$\underset{H}{\overset{R^1}{\diagdown}}N\underset{O}{\diagup^{OCOR}} \longrightarrow |R^1C\equiv NO| \quad (4.42)$$

$$\underset{H}{\overset{R^1}{\diagdown}}C=N\underset{O}{\diagup^{O^-}} \xrightarrow[Et_3N]{PhNCO} \underset{H}{\overset{R^1}{\diagdown}}C=N\underset{O}{\diagup} \text{(PhNH-CO-O)} \longrightarrow |R^1C\equiv NO| \quad (4.43)$$

Secondary nitro compounds give more stable acyl derivatives, which occasionally can be isolated.[141,143] They rearrange into nitroso-acyloxy compounds (**26**), eq. (4.40) route b.[143,149]

C-Acylation can be accomplished with methoxymagnesium methyl carbonate, Stiles' reagent, eq. (4.44), route a,[152,153] acylimidazole, eq. (4.44) route b,[154,155] and acylcyanide, eq. (4.44) route c.[156]

$$RCH=NO_2^- \begin{cases} \xrightarrow{a\ Mg(OCH_3)(OCO_2CH_3)} & R\underset{COOCH_3}{\overset{|}{C}}HNO_2 \\ \xrightarrow{b\ \text{imidazole-}COR^1} & R\underset{COR^1}{\overset{|}{C}}HNO_2 \\ \xrightarrow{c\ R^1COCN} & R\underset{COR^1}{\overset{|}{C}}HNO_2 \end{cases} \quad (4.44)$$

In contrast to simple nitronates the double deprotonated nitroalkanes and nitroalkenes give good yields of *C*-alkylated or *C*-acylated products, eq. (4.45)–(4.48).[83–87]

$$RCH_2NO_2 \xrightarrow[\text{THF/HMPT}]{\text{LiBu, -90 °C}} \underset{\underline{27}}{\overset{R}{\underset{-}{>}}=N\overset{O^-}{\underset{O}{<}}} \xrightarrow{\text{Electrophile}} \overset{R}{\underset{E}{>}}=N\overset{O^-}{\underset{O}{<}} \qquad (4.45)$$

Electrophile = RBr(I), RCOOCH$_3$, (RCO)$_2$O, RCHO, R$_2$CO, $\overset{\diagup}{\diagdown}$O

$$\underset{R^1}{\overset{RCH_2}{>}}CHNO_2 \rightarrow \underset{\underline{28}}{\overset{R\bar{C}H}{\underset{R^1}{>}}=N\overset{O^-}{\underset{O}{<}}} \xrightarrow{\text{Electrophile}} \underset{R^1}{\overset{R\overset{E}{\underset{|}{C}}H}{>}}=N\overset{O^-}{\underset{O}{<}} \qquad (4.46)$$

$$\overset{\diagup\!\!\!=}{}CH_2NO_2 \rightarrow \overset{\diagup\!\!\!=}{}\!\!=N\overset{O^-}{\underset{O}{<}} \xrightarrow{E^+} \overset{\diagup\!\!\!=}{\underset{E}{}}\!\!=N\overset{O^-}{\underset{O}{<}} \qquad (4.47)$$

$$\overset{\diagup\!\!\!=}{}CH_2NO_2 \rightarrow \overset{\diagup\!\!\!=}{}\!\!=N\overset{O^-}{\underset{O}{<}} \xrightarrow{E^+} \overset{\diagup\!\!\!=}{\underset{E}{}}\!\!=N\overset{O^-}{\underset{O}{<}} + \overset{E\diagup\!\!\!=}{}\!\!=N\overset{O^-}{\underset{O}{<}} \qquad (4.48)$$

$$\underline{29} \qquad\qquad\qquad \underline{30} \qquad\qquad \underline{31}$$

The C_1-protons of primary nitronate salts are comparably easy to abstract with a strong base such as butyllithium, thus giving the doubly negatively charged anion **27**, eq. (4.45), which selectively is attacked by the electrophile on C_1. A secondary nitronate salt gives the doubly negatively charged anion **28**, eq. (4.46), which allows an electrophilic attack on the β-carbon atom. The C_2-proton is abstracted in preference to the C_1-proton, $R^1 = H$ in **28**, if R is an activating group as in **29**, eq. (4.48). A mixture of products **30** and **31** is obtained.

4.3 PHYSICOCHEMICAL PROPERTIES OF ALKYL NITRONATES. REACTIONS OF ALKYL NITRONATES

The instability of alkyl nitronates has already been pointed out. Their lifetime can be from minutes to several weeks at room temperature. The major products from spontaneous decomposition or from refluxing the alkyl nitronates in alcohol[114] or water[157] are oximes and carbonyl compounds, as shown in eq. (4.32). This reaction is useful for the preparation of oximes from nitroalkanes. Treating alkyl nitronates under Nef conditions produces aldehydes and ketones,[95] eq. (4.2a). Hydrolysis of the alkyl nitronate to the parent nitro compound does not occur. In this respect the alkyl nitronates differ sharply from the silyl nitronates, which undergo rapid hydrolysis to the nitro compound and silanol. Concentrated sulfuric acid transforms alkyl nitronates into hydroxamic acids; hydrogen chloride gives hydroximic acid chlorides. Nitronic esters derived from primary α-nitro ketones, ethyl nitroacetate, and phenylsulfonylnitromethane eliminate alcohol in the presence of TsOH to give nitrile oxides that cycloadd to olefins or acetylenes.[113]

Alkyl nitronates are reduced to oximes by hydrogen iodide.[98] Catalytic reduction over Pt gives the amine[141]; cf. the hydrogenation of silyl nitronates over Raney-Ni, which also gives complete reduction to the amine.[158]

Infrared and UV absorption spectra have been recorded for a number of nitronate esters. They are similar to the spectra of oximes and nitronic acids. In the IR region an intense C=N absorption is located at 1610–1650 cm^{-1}. No UV spectrum of a simple aliphatic nitronate ester has been recorded, but they should probably have a π-π* transition near 220 nm, close to the absorption of aliphatic nitronic acids. Data for IR and UV absorption are collected in Table 4.5. Relevant ^{1}H NMR data are collected in Table 4.4.

The most important reaction of alkyl nitronates is their propensity to cycloadd in a 1,3-dipolar fashion to alkenes, alkynes, and heterodipolarophiles.[1,101-113,118,139,159-163] The N-alkoxyisoxazolidines eliminate alcohol in an acid-catalyzed reaction and give 2-isoxazolines, eq. (4.49). The

$$\underset{R^1}{\overset{R^2}{>}}C=N\underset{OR}{\overset{O}{<}} + \overset{X}{=} \longrightarrow \begin{array}{c} R^1 \\ R^2 \\ X \overset{}{\underset{O}{\diagup}} N\underset{OR}{} \end{array} \overset{H^+}{\longrightarrow} \begin{array}{c} R^1 \\ X \overset{}{\underset{O}{\diagup}} N \\ R^2 = H \end{array} \quad (4.49)$$

reaction, which proceeds in good yields at room temperature in methylene chloride, was developed by Tartakovskii and his group (Ref. 101 and later papers). The dinitro derivatives, e.g., are reactive species, and the reactivity decreases in the series

$$\underset{O_2N}{\overset{O_2N}{>}}C=N\underset{OCH_3}{\overset{O}{<}} \quad > \quad \underset{Alk}{\overset{(H)Alk}{>}}C=N\underset{OCH_3}{\overset{O}{<}} \quad \gg \quad \underset{Alk}{\overset{(H)Alk}{>}}C=N\underset{OSi(CH_3)_3}{\overset{O}{<}}$$

32   **33**   **34**

At room temperature dinitro nitronates add to cyclohexene,[139] which cannot be forced to react with **33** or **34**. Electron withdrawing groups on the α-carbon atom increases thus the reactivity of the nitronates. The same trend has been noted in the nitrone series in that C-acyl groups increase[164] the reactivity, whereas C-O-alkyl groups are reported to decrease[165] the reactivity of nitrones.

Some N,O-dialkyl-N-nitronates,

$$Alk-N=N\underset{O}{\overset{OAlk}{<}}$$

have also been prepared by alkylation of the N-nitronate salts with alkyl halides.[168,170,171] They are relatively stable compounds, and the lower-molecular-weight members of the series can be distilled. Geometrical isomerism

Table 4.5. IR and UV Data for Some Alkyl Nitronates

Compound	IR (cm^{-1}) C=N	UV λ_{max} (ε)	Refs.
4-Br-C$_6$H$_4$-CH=N(OCH$_3$)O	1,610–1,620 (s, Nujol)	288 (32,700), 299 sh (27,300) EtOH	95
(Ar)Alk-CH=N(OAlk)O	1,610–1,650 (s, CCl$_4$, CHCl$_3$)		95
3-NO$_2$-isoxazoline N-oxide	1,620, 1,640 (s)	320 (8,366) H$_2$O	166
camphor-derived nitronate, OCPh$_3$	1,727 (s) C=O	288 (13,900) Et$_2$O	167
pyrazoline N-oxide ^{a}	1,620 (N=N)	220 (log ε 3.44)	168
CH$_3$-N=N(OCH$_3$)O	1,587 (vs) 1,577 (vs) (N=N) 1,264 (s) (NO$_2$) 994 (s) (NO$_2$)		169
CH$_3$-N=N(OCH$_2$CH$_3$)O ^{b}		205 (log ε 3.80)	170

^{a}^{1}H NMR: δ 4.4 (m).
^{b}^{1}H NMR (CDCl$_3$): δ 1.32 (3H, t, $J = 7$ Hz), 3.22 (3H, s), 4.40 (2H, q, $J = 7$ Hz).[172]

has been observed.[170-172] The chemistry of aliphatic nitramines has been reviewed.[173] Some spectral data are collected in Table 4.5.

In contrast to the C-nitronates, N-isopropyl-O-methyl-N-nitronate could not be forced to add to methyl acrylate or ethyl vinyl ether in a dipolar reaction, eq. (4.50).[174] The resistance of the analogous aromatic azoxy system in 1,3-dipolar cycloaddition has been noted earlier.[175] It can be overcome by using the strained trans-cyclooctane, eq. (4.51), which produced compounds **35** and **36** in subsequent reactions.[176,177] Calculating the electron structure of N-methyl-O-methyl-N-nitronate (optimized structure by CNDO/2) using a 4-31 G basis set explains the unreactive N-nitronates in 1,3-dipolar cycloadditions. Figure 4.1 shows the highest occupied molecular orbital (HOMO) and the lowest unoccupied molecular orbital (LUMO) for the C-nitronate to the left and methyl

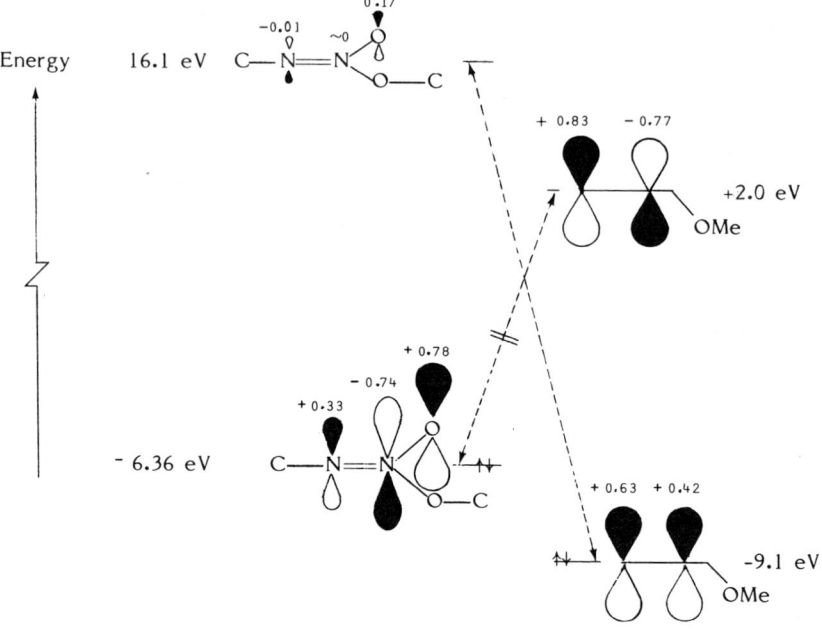

Figure 4.1. Representation of the frontier orbitals in *N*-methyl-*O*-methyl-*N*-nitronate calculated by 4-31 G basis set and the frontier orbitals in methyl vinyl ether.

vinyl ether to the right. It appears that an interaction between the HOMO of the N-nitronate and the LUMO of methyl vinyl ether is a symmetry forbidden reaction, and that an interaction between LUMO of the nitronate and HOMO of the methyl vinyl ether is rather unfavorable because of the small orbital coefficients at the reactive sites and the large HOMO-LUMO energy gap (Section 1.7).[174,178]

4.4 SYNTHESIS OF SILYL NITRONATES

Primary and secondary nitro compounds, **37**,[179,180] are O-silylated by a variety of silylating agents, **39–46**, eq. (4.52). Compound **39** is a weak silylating agent that only attacks acidic C^1-protons as in NO_2CH_2COOEt. It does not attack nitromethane.[185] Primary and secondary nitro compounds require heating at 60–80°C with bis-trimethylsilylacetamide (BSA) **40** in benzene or still more efficiently heating with bis-trimethylsilyltrifluoroacetamide **41**. The trimethylchlorosilane-triethylamine reagent **42** transfers the lower homologues of primary nitro compounds into silyl nitronates in benzene as solvent at 25–60°C, but leaves the secondary nitro compounds practically untouched.[158,186] However, if the silylation is carried out in acetonitrile as solvent or in benzene: acetonitrile 2:1, silylation of primary as well as secondary nitro compounds

$$\underset{\underset{37}{}}{\overset{R^1}{\underset{R^2}{\diagdown}}\underset{H}{\overset{|}{C}}-NO_2} + (CH_3)_3SiX \longrightarrow \underset{\underset{38}{}}{\overset{R^1}{\underset{R^2}{\diagdown}}C=N\overset{O}{\underset{O-Si(CH_3)_3}{\diagup}}} + HX \quad (4.52)$$

φNHCONφ
|
Si(CH₃)₃

39[181-184]

$CH_3C\overset{OSi(CH_3)_3}{\underset{N-Si(CH_3)_3}{\diagdown}}$

BSA, **40** [185,186]

$CF_3C\overset{OSi(CH_3)_3}{\underset{N-Si(CH_3)_3}{\diagdown}}$

BSTFA, **41**[186]

$ClSi(CH_3)_3$, $(C_2H_5)_3N$

42 [158,186-193]

$CF_3SO_3Si(CH_3)_3$

Trimethylsilyl triflate

43[194,195]

$ClSi(CH_3)_3$, $LiN(C_2H_5)_2$

44 [196-201]

$ClSi(CH_3)_3$, Li_2S

45[202,203]

$ClSi(CH_3)_3$ + Ag^+ or Hg^{2+}

46 [182,204]

occurs smoothly at 25–50°C within 24 h.[187–193] This procedure is experimentally simple and inexpensive.

Trifluorosulfonotrimethylsilane (trimethylsilyl triflate) (**43**) is a very strong silylating agent that, combined with trimethylamine, can introduce two silyl groups into primary nitro compounds, eq. (4.53).[194,195] The trimethylsilyl ester of an α-hydroxylated aldoxime (**47**) is the final product of this reaction.

$$RCH_2CH_2NO_2 \xrightarrow[Et_3N]{43} RCHCH=N(H)(OSi(CH_3)_3) \longrightarrow$$

$$[RCH=CHNO \cdot OSi(CH_3)_3] \xrightarrow[Et_3N]{43} \underset{OSi(CH_3)_3}{RCHCH=NOSi(CH_3)_3}$$

(4.53)

47

Trimethylchlorosilane combined with the strong base lithium diethylamide, **44**, silylates efficiently primary and secondary nitro compounds at −70°C in ether.[196–201] It has been reported that the trimethylchlorosilane–lithium sulfide reagent **45** is a mild reagent and suitable for silylation of secondary nitro compounds.[202,203]

The silyl nitronates are sensitive to humidity. Silylations should therefore be carried out in dry aprotic solvents like benzene, methylene chloride, chloroform, acetonitrile, or ether, and the product should be protected from humid air. They should be stored in the refrigerator, or what is best, used directly in further reactions.

4.5 PHYSICOCHEMICAL PROPERTIES OF SILYL NITRONATES

The silyl nitronates are more stable than the alkyl nitronates. The trimethylsilyl ester of *aci*-nitroproane (**48**) distills at 58°C/12 mmHg.[158] Compounds **49**[183] and **50**[197] distill without decomposition at 50–51°C/0.65 mmHg and 90–100°C/0.01 mmHg (Kugelrohr), respectively. Tables 4.6 and 4.7 contain physicochemical data of some nitronate esters. The stability of the nitronate esters increases by use of the *t*-butyldimethylsilyl group as, e.g., in compound **50**.

For **50** the resistance toward hydrolysis is simultaneously increased by the bulky group. The parent nitro compounds (**37**) are rapidly formed by treating silyl nitronates (**38**) with water or alcohol, eq. (4.54). The silyl nitronates (**38**) are stable toward the base, but they decompose when treated with catalytic amounts

Table 4.6. Physicochemical Properties of Silyl Nitronates, IR, and UV Data

Compound	Mp (°C) Bp (°C/mmHg) Yield (%)	UV λ_{max} (nm) (ε), CH_2Cl_2	IR (cm^{-1}) CH_2Cl_2	Refs.
CH_3, H, =N, $OSi(CH_3)_3$	64/25 (64) 65–67/17 (~100)	240 (5,100)	1,622, C=N	158, 185
CH_3CH_2, H, =N, $OSi(CH_3)_3$	58/12 (79) 67/17 (~100)	240 (6,050)	1,617	158, 185
CH_3, CH_3, =N, $OSi(CH_3)_3$	72–77/18 (~100)	240 (7,000)	1,626	185
Ph, H, =N, $OSi(CH_3)_3$	85–87/0.5 (~100)	280 (7,000)	1,590	185
C_5H_{11}, H, =N, $OSi(CH_3)_2$ +	80–90/0.02 (78)		3,100, HC=N 1,615, C=N 1,250, Si–CH_3 1,110, Si–O 835, Si–C 790, Si–C	197
cyclopentyl =N, $OSi(CH_3)_2$ +	−20 90–100/0.01 (62)	233 (11,900)	1,645 1,250 835 790	197
C_6H_{13}, CH_3, =N, $OSi(CH_3)_2$ +	90–100/0.01 (~95)		1,615, 835 1,250, 790 1,100	197
CH_3OOC, H, =N, $OSi(CH_3)_3$	50–51/0.65 (~100)	265 (10,000)	1,591, C=N 1,735, C=O	183
O_2N, H, =N, $OSi(CH_3)_3$	(~100)	303 (7,000)	1,600, C=N 1,575, NO_2 1,330, NO_2	183
H_3C, N=N, $OSi(CH_3)_3$	59/17 (96)		1,500–1,510,[a] 1550[b] 1,290–1,300, 1,255	206
CH_3–C$_6$H$_4$–SO_2, N=N, $OSi(CH_3)_3$	oil (~95)	1,510	1,530 1,262	207

[a] N-silyl derivative.
[b] O-silyl derivative.

Table 4.7. NMR Data of Silyl Nitronates

Compound Mp (°C), Yield ( ) Bp (°C/mmHg)	^{1}H NMR, δ, J Hz CCl$_4$ (TMS)^a	^{13}C NMR, δ	Refs.
H$_2$C=N(O)OSi(CH$_3$)$_3$	5.55		205
CH$_3$CH=N(O)OSi(CH$_3$)$_3$	1.84 (d, $J = 6$) 6.05 (q, $J = 6$)	−0.9, SiCH$_3$ 109.8, C$^\alpha$ 11.4, C$^\beta$	205
(CH$_3$)$_2$C=N(O)OSi(CH$_3$)$_3$	0.18 (s, SiCH$_3$) 1.90 (s)	0.2, SiCH$_3$ 118.3, C$^\alpha$ 18.0, C$^\beta$	205
PhCH=N(O)OSi(CH$_3$)$_3$	0.18 (s, SiCH$_3$) 6.74 (s), 7.05 (m), 7.66 (m)		205
(CH$_3$OOC)$_2$C=N(O)OSi(CH$_3$)$_3$ (~97)	0.30 (s, SiCH$_3$) 3.80	−1.7, SiCH$_3$ 114.2, C$^\alpha$ 158.8, C=O 51.9, OCH$_3$	205
O$_2$N−CH=N(O)OSi(CH$_3$)$_3$	0.33, 8.25	−0.7, SiCH$_3$ broad, C$^\alpha$	205
CH$_3$OOCCH$_2$CH$_2$C(H)=N(O)OSi(CH$_3$)$_3$ 92-94/1 (91)	0.30 (s), 2.46 (m, 4H), 3.69 (s), 6.2 (1H, m) (CDCl$_3$)		186
(CH$_3$)$_3$SiOCH$_2$CH=N(O)OSi(CH$_3$)$_3$ (~100)	0.33 (s), 4.38 (d, $J = 5.6$), 6.29 (t, $J = 5.6$) (CDCl$_3$)		186
Cyclohexyl−C[OSi(CH$_3$)$_3$](H)=N(O)OSi(CH$_3$)$_3$ 100/0.5 (69)	0.13 (9H, s), 0.30 (9H, s), 1.2−2.0 (10H, m), 6.03 (s) (CDCl$_3$)		186
Cyclopentanone N-silyl nitronate OSi(CH$_3$)$_2$ −20, 90-100/0.01 (62)	0.25 (6H, s), 0.88 (9H, s), 1.8 (4H, m), 2.45 (4H, m)	−3.95, SiCH$_3$, 17.78 C−, 25.57, C^{3,4}, 26.03, C−CH$_3$, 29.85, C^{2,5}, 132.15, C$^\alpha$	197

Table 4.7. (continued)

Compound Mp (°C), Yield ( ) Bp (°C/mmHg)	^{1}H NMR, δ, J Hz CCl$_4$ (TMS)^a	^{13}C NMR, δ	Refs.
C$_5$H$_{11}$, CH$_3$ >C=N–O–OSi(CH$_3$)$_3$ (~95)	0.25 (6H, s), 0.9 (9H, m), 0.8–1.65 (9H, m) 1.90 (3H, s), 2.25 (2H, m)		197
CH$_3$, CH$_3$ >C=N–O–OSi(CH$_3$)$_3$ 120–140/10 (30)	0.3 (6H, s), 0.95 (9H, s), 2.0 (6H, s)	–4.5, SiCH$_3$, 17.1, C^{tert} – 17.6, C$^\beta$, 25.4, C CH$_3$, 120.1, C$^\alpha$	197
CH$_3$–C$_6$H$_4$–SO$_2$–N=N–OSi(CH$_3$)$_3$	0.33 (9H, s), 2.38 (3H, s), 7.25 (2H, d, $J = 9$), 7.7 (2H, d, $J = 9$)	–1.14, SiCH$_3$, 21.53, CH$_3$, 146.42, C^1, 129.85, 129.64, C^2, C^3, 132.77, C^4	207

^{a}TMS = tetramethylsilane.

$$\underset{\mathbf{38}}{R^1R^2C=N(O)OSi(CH_3)_3} \xrightarrow{H_2O} \underset{\mathbf{37}}{R^1R^2CH-NO_2} + (CH_3)_3SiOH \quad (4.54)$$

$$\searrow \text{EtOH}$$

$$\mathbf{37} + (CH_3)_3SiOEt$$

of acid or on standing long at room temperature. Thus, adding a small amount of p-TsOH or borontrifluoride etherate to compound **48** causes a rapid decomposition, whereas adding triethylamine stabilizes the compound. Bis-trimethylsilyl ether (**51**), trimethylsilanol (**52**), propanal oxime trimethylsilyl ether (**53**), 3,4-diethylfuroxan (**54**), nitropropane (**55**), and an intractable polymeric material, probably formed by polymerization of an intermediate nitrile oxide, have been identified in the decomposition product, eq. (4.55).

$$\mathbf{48} \xrightarrow{H^+} \underset{\mathbf{51}}{(CH_3)_3SiOSi(CH_3)_3} + \underset{\mathbf{52} \text{ (major)}}{(CH_3)_3SiOH} + \underset{\mathbf{53}}{C_2H_5CH=N-OSi(CH_3)_3} +$$

$$+ \underset{\mathbf{54} \text{ (minor)}}{\text{3,4-diethylfuroxan}} + \underset{\mathbf{55}}{CH_3CH_2CH_2NO_2} + \underset{\text{(major)}}{\text{polymer}} \quad (4.55)$$

When primary *N*-nitramines are silylated with BSA (**40**), eq. (4.56), a rapidly equibrating mixture of *N*- and *O*-silylated products is obtained.[206-210] These products are very moisture labile. The amount of the *O*-silylated product **57** increases at higher temperature and for compounds with bulkier silyl groups.

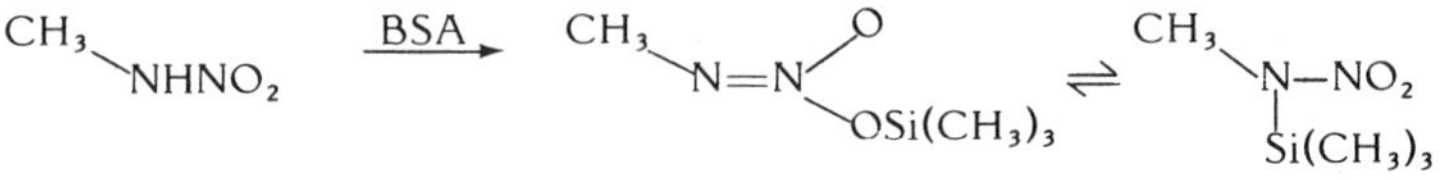

$$\underset{\mathbf{56}}{\text{CH}_3\text{NHNO}_2} \xrightarrow{\text{BSA}} \underset{\mathbf{57}}{\text{CH}_3\text{N}=\text{N}(\text{O})\text{OSi}(\text{CH}_3)_3} \rightleftharpoons \underset{\mathbf{58}}{\text{CH}_3\text{N}(\text{Si}(\text{CH}_3)_3)-\text{NO}_2} \quad (4.56)$$

The tosylate **59** occurs only in the *O*-form. At low temperature the two forms **57** and **58** are frozen out and ^{29}Si magnetic resonance shows two peaks that collapse at room temperature. The reaction **57** ⇌ **58** is an intramolecular process, eq. (4.57), because no dilution and solvent effects are observed, which would influence a dissociative mechanism.

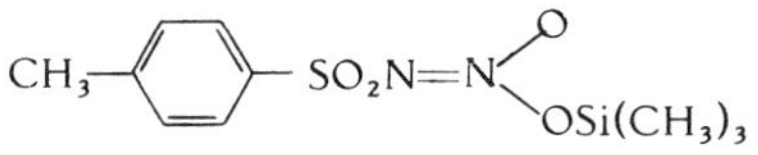

$$\text{CH}_3\text{-C}_6\text{H}_4\text{-SO}_2\text{N}=\text{N}(\text{O})\text{OSi}(\text{CH}_3)_3$$

59

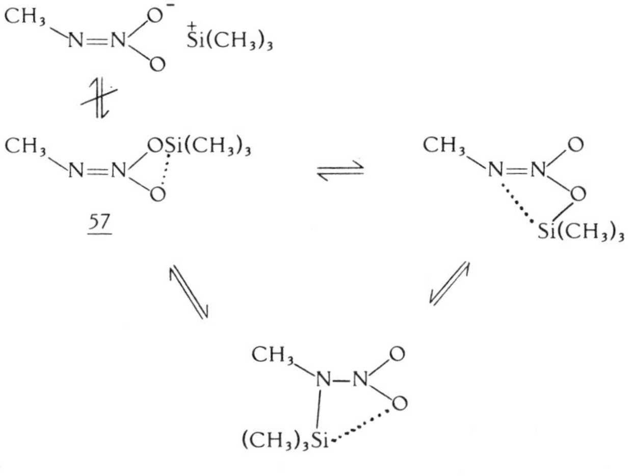

(4.57)

The NMR spectra of silyl nitronates (**38**) show one single compound, even though two tautomers (**38a,b**), eq. (4.58), are possible. This contrasts with the alkyl nitronates (Section 4.3), which show the existence of two isomers by ^{1}H NMR. This is accounted for by rapid intramolecular jumping of the silyl group between the oxygens, eq. (4.58).[197,205] A five-coordinate symmetrical structure

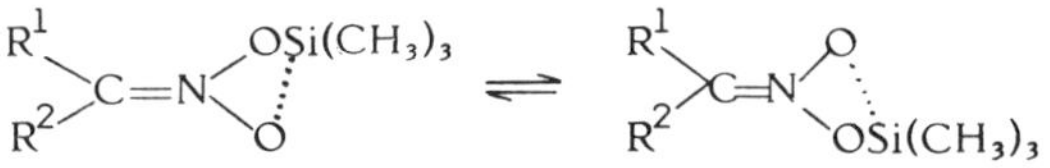

$$\underset{\mathbf{38a}}{\overset{R^1}{\underset{R^2}{>}}\text{C}=\text{N}(\text{O})\text{OSi}(\text{CH}_3)_3} \rightleftharpoons \underset{\mathbf{38b}}{\overset{R^1}{\underset{R^2}{>}}\text{C}=\text{N}(\text{O})\text{OSi}(\text{CH}_3)_3} \quad (4.58)$$

60 is not supported by an x-ray investigation.[197] Compound **61** does not show any discernible splitting of the CH_3 group even at $-125°C$, but the cyclopentyl derivative (**50**) shows separated 2,5-methylene signals at $-75°C$ coalescing at ca. $-65°C$ in $CHClF_2$ as solvent, corresponding to an activation energy of $E_a \sim 10$ kcal mol^{-1}.

60   **61**

A few other inorganic nitronate esters[179] have also been described, such as O-diethylboryl nitronates (**62**),[207,211-213] organotin and organolead nitronates (**63**),[214] and *aci*-nitrophosphites (**64**).[215] These nitronates are less stable than their silicon analogues.

62   **63** M = Sn, Pb   **64**

4.6 . REACTIONS OF SILYL NITRONATES

The silyl nitronates are characterized by being easily cleaved by water and alcohol, eq. (4.54). They show promising synthetic applicabilities as reagents in the nitro-aldol reaction (Henry's reaction),[26] eq. (4.59), and in 1,3-dipolar reactions, eq. (4.62). The nitro-aldol reaction is reversible, and eliminating water leads to unstable nitroolefins, which lower the yields. The reduction to amino alcohols is frequently a problem. It was found that heating the trimethylsilyl ester of *aci*-nitroethane with benzaldehyde in the presence of triethylamine gave 1-phenyl-1-trimethylsilyloxy-2-nitropropane.[158]

(4.59)

The reaction is catalyzed by fluoride ions and proceeds smoothly in good yields even at $-78°C$ in tetrahydrofuran (THF), eq. (4.60).[196,198] The transfer of the silyl group to the hydroxy group stabilizes the nitro alcohol, and

$$R^1R^2C=N(O)OSi(CH_3)_3 + R^3CHO \xrightarrow[-78\ ^\circ C]{F^-,\ THF} \underset{\mathbf{65}}{R^1R^2C(NO_2)-C(OSi(CH_3)_3)R^3} \xrightarrow{LiAlH_4} \underset{\mathbf{66}}{R^1R^2C(NH_2)-C(OSi(CH_3)_3)R^3} \quad (4.60)$$

R^1, R^2 = alkyl, H, aryl
R^3 = alkyl, aryl

the protected derivative **65** can directly be reduced by lithium aluminum hydride to the amino alcohol, which is obtained as a mixture of isomers. The diastereomeric preference is small for the formation of the vicinal trimethylsilyloxy–nitro compound and absent in the amino alcohol. Diastereomerically enriched nitro aldols can be prepared from the reaction of doubly deprotonated nitro compounds with aldehydes. Subsequent protonation with acetic acid at $-100\ ^\circ C$ in the presence of HMPT gives predominantly the threo-product **68**, eq. (4.61). Essentially the same enrichment of **68** is obtained if one starts from about a 1:1 mixture of threo and erythro forms **68** (the Henry reaction products), which are doubly deprotonated to the dianion **67** and then acidified in the presence of HMPT. The threo/erythro ratio lies characteristically at ca. 3–5:1. When the *t*-butyldimethylsilyl protected nitro alcohol is deprotonated with lithium diisopropylamide in THF and reprotonated at $-100\ ^\circ C$, the erythro form is produced with high stereoselectivity, eq. (4.61), and a threo/erythro ratio ca. 5/95.[199,216]

The silyl nitronates are versatile reagents for the preparation of 2-isoxazolines, which subsequently can be transformed into a variety of heterocycles, cyclopentanones, and oxygenated hydrocarbons, etc.[158,188,193] They are much less reactive than nitrile oxides. Since they are thermolabile, it is advisable to run the addition below ca. $80\ ^\circ C$. They cycloadd to activated dipolarophiles, such as conjugated dienes, styrene, and α,β-unsaturated carbonyl compounds. They react slower with monosubstituted olefins, such as allyl acetate, and they are unreactive towards 1,2-disubstituted olefins and vinyl chloride. The primary cycloadduct is the *N*-silyloxyisoxazolidine, which easily splits off silanol on treatment with acid with formation of 2-isoxazolidines, eq. (4.62).

Organolithium reagents nucleophilically attack silyl nitronates (from primary nitro compounds), giving oximes in modest yields, eq. (4.63).[217] With secondary silyl nitronates the reaction is more complicated. An α-proton is abstracted, and presumably via an intermediary nitroso compound an α-alkylated oxime is obtained, eq. (4.64). Grignard reagents attack the silicon atom, forming magnesium nitronates, eq. (4.65).

The yield of thiohydroximates can be improved by reacting silyl nitronates with thiols in the presence of triethylamine as demonstrated in the synthesis of

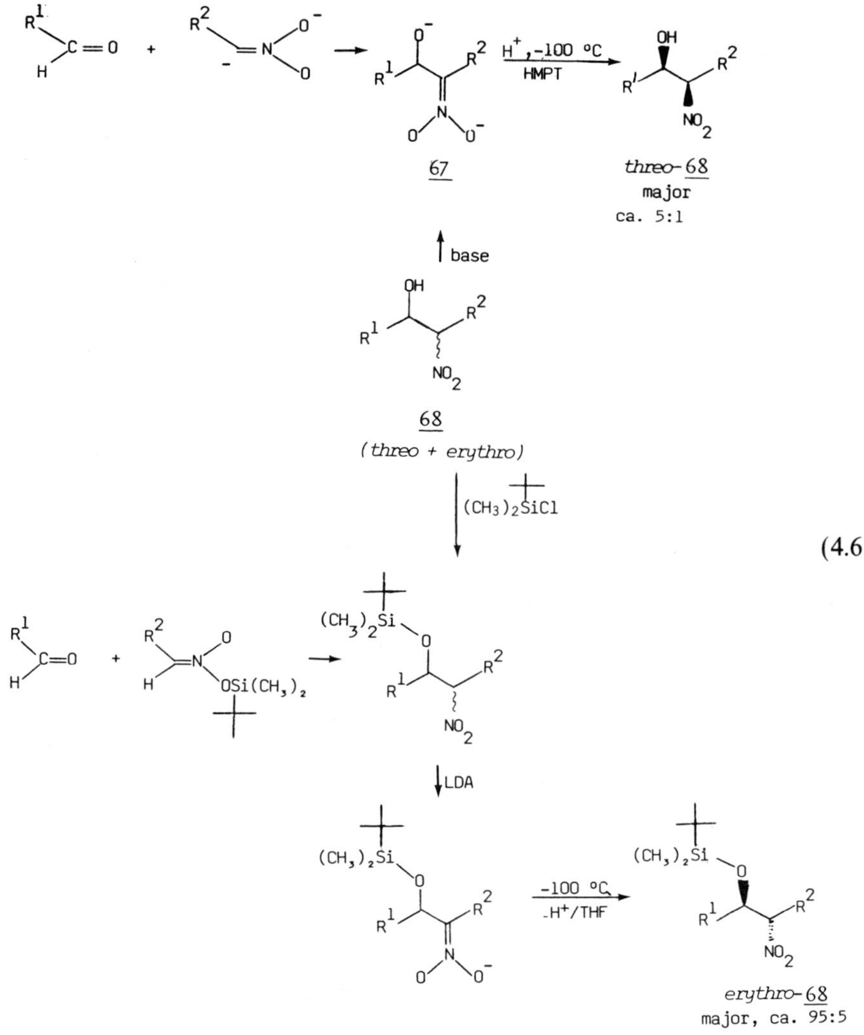

Reactions of Silyl Nitronates

(4.61)

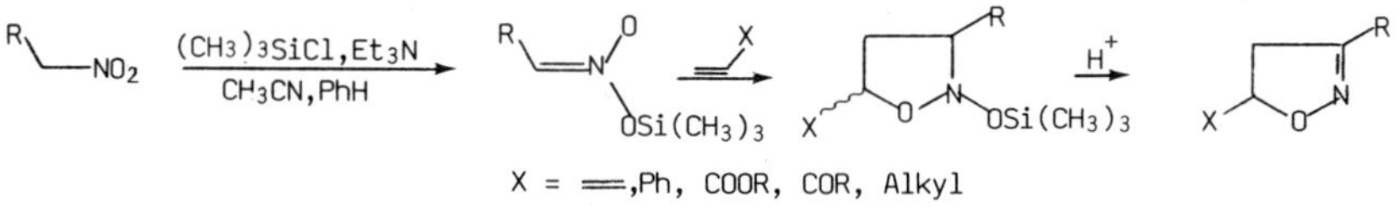

(4.62)

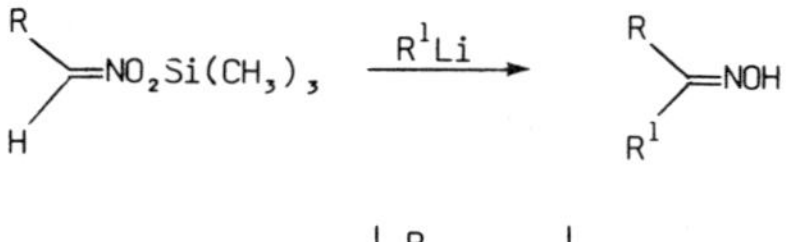

(4.63)

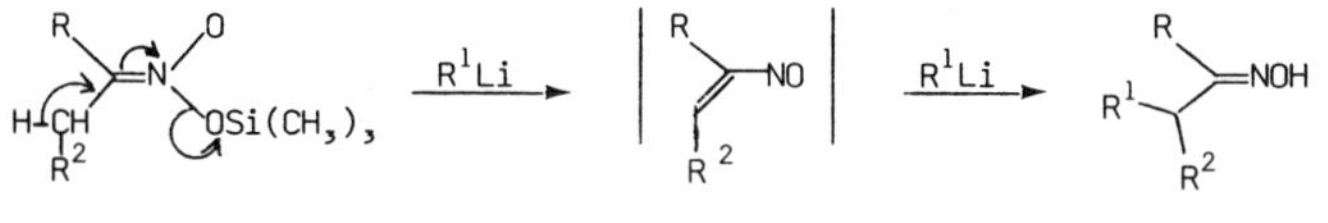

(4.64)

glucosinolates (4.66).[201] The *t*-butyldimethylsilyl ester of *aci*-nitroethane was used to trap the elusive thioaldehyde, eq. (4.67).[200]

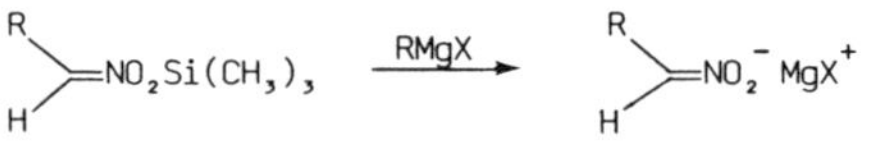
R
H
=NO₂Si(CH₃)₃
RMgX
=NO₂⁻ MgX⁺

(4.65)

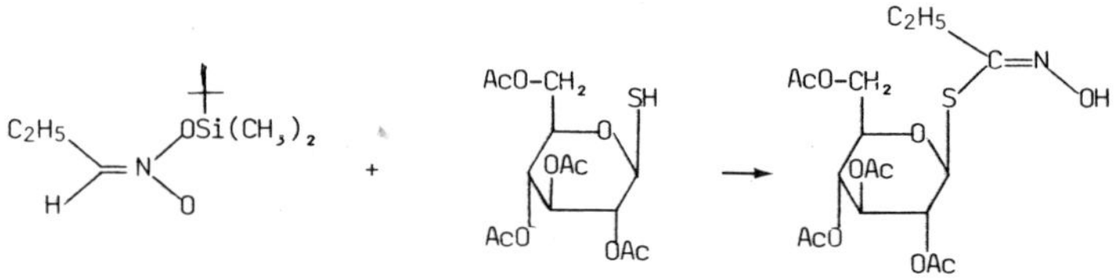
C2H5
OSi(CH₃)₂
AcO-CH₂
SH
OAc
AcO
C=N
OH
S

(4.66)

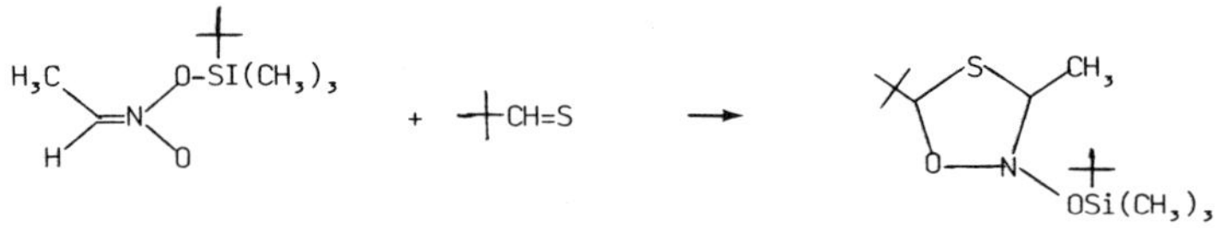
H₃C
O-SI(CH₃)₃
CH=S
S
CH₃
OSi(CH₃)₃

(4.67)

Silyl nitronates are reported to be reduced to *O*-silyl oximes by trimethyl phosphite.[218] They are transformed to carbonyl compounds by ceric salts[203] or by treatment with hydrogen chloride in pentane, eq. (4.68).[218]

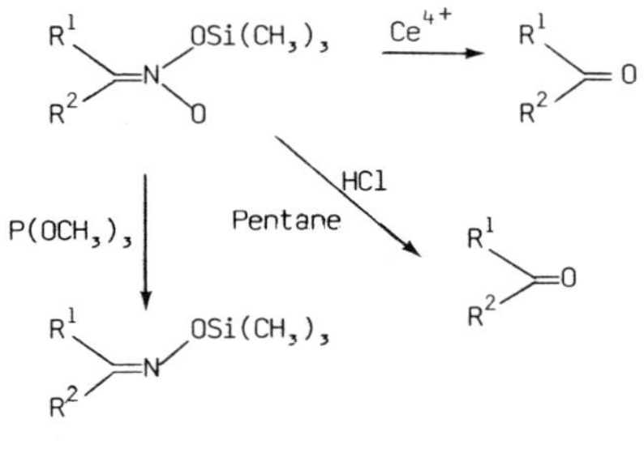
OSi(CH₃)₃
Ce⁴⁺
HCl
Pentane
P(OCH₃)₃
OSi(CH₃)₃

(4.68)

REFERENCES

1. A. T. Nielsen In "The Chemistry of the Nitro and Nitroso Groups" (H. Feuer, Ed.), Wiley, New York, 1969, p. 349.
2. E. Breuer In "The Chemistry of Amino, Nitroso and Nitro Compounds and their Derivatives (S. Patai, Ed.), Wiley, New York, 1982, p. 459.
3. A. F. Holleman, *Rec. Trav. Chim. 4* (1895), 121.
4. A. Hantzsch; O. W. Schultze, *Chem. Ber. 29* (1896), 699; M. I. Konowalow, *Chem. Ber. 29* (1896), 2193.
5. J. U. Nef, *Liebigs Ann. 280* (1894), 264.
6. E. Bamberger, *Chem. Ber. 35* (1902), 54.
7. R. Junell, *Diss. Uppsala, Sweden* 1935; D. Turnbull; S. H. Maron, *J. Am. Chem. Soc. 65* (1943), 212.
8. S. G. Mairanovskii; V. M. Belokov; E. B. Korchemnaya; V. A. Klimova; S. S. Novikov, *Izv. Akad. Nauk SSSR, Ser. Khim. Nauk 1960*, 1787.
9. Ya. J. Turyan; Yu. M. Tyurin; P. M. Zaitsev, *Dokl. Akad. Nauk SSSR 134* (1960), 850.
10. S. S. Novikov; V. I. Slovetskii; S. A. Shevelev; A. A. Fainzil'berg, *Izv. Akad. Nauk SSSR, Ser. Chim. Nauk 1962*, 598.

References

11. J. S. Belev; L. G. Hepler, *J. Am. Chem. Soc. 78* (1956), 4005.
12. V. J. Slovetskii; S. A. Shevelev; A. A. Fainzil'berg; S. S. Novikov, *Zh. Vses. Khim. Obshch. im D. J. Mendeleeva 6* (1961), 599, 707.
13. V. J. Slovetskii; S. A. Shevelev; V. J. Erashko; L. J. Berynkova; A. A. Fainzil'berg; S. S. Novikov, *Izv. Akad. Nauk SSSR, Ser. Khim. Nauk 1966*, 655.
14. H. G. Adolph; M. J. Kamlet, *J. Am. Chem. Soc. 88* (1966), 4761.
15. A. Hantzsch; A. Veit, *Chem. Ber. 32* (1899), 607.
16. E. Bamberger; R. Seligman, *Chem. Ber. 35* (1902), 3884; *36* (1903), 701.
17. W. Wislicenus; A. Endres, *Chem. Ber. 41* (1908), 4121.
18. C. D. Nenitzescu; D. A. Isacescu, *Chem. Ber. 63* (1930), 2484.
19. E. B. Hodge, *J. Am. Chem. Soc. 73* (1951), 2341.
20. W. E. Noland, *Chem. Rev. 55* (1955), 137.
21. W. Steinkopf; B. Jürgens, *J. Prakt. Chem. 84* (1911), 686.
22. V. Meyer; C. Wurster, *Chem. Ber. 6* (1873), 1168.
23. N. Kornblum; R. A. Brown, *J. Am. Chem. Soc. 87* (1965), 1742.
24. W. E. Noland; J. H. Cooley; P. H. McVeigh, *J. Am. Chem. Soc. 81* (1959), 1209.
25. J. Armand, *Bull. Soc. Chim. France 1965*, 3246.
26. (a) O. von Schickh; G. Apel; H. G. Padeken; H. H. Schwarz; A. Segnitz In Houben-Weil, "Methoden der Organischen Chemie", *10:1* (1971) 1. (E. Müller, Ed.) G. Thieme Verlag, Stuttgart.
    (b) E. D. Bergmann; D. Ginsburg; R. Pappo, *Org. React. 10* (1959), 179.
27. V. Meyer, *Chem. Ber. 6* (1873), 1492.
28. H. Metzger In Houben-Weyl, "Methoden der Organischen Chemie", *10:4* (1968) 1. (E. Müller, Ed.) G. Thieme Verlag, Stuttgart.
29. M. O. Forster, *J. Chem. Soc. 77* (1900), 251.
30. H. Schechter; F. T. Williams, Jr. *J. Org. Chem. 27* (1962), 3699.
31. C. Lagercrantz; S. Forshult; T. Nilsson; K. Torssell, *Acta Chem. Scand. 24* (1970), 550.
32. F. Kienzle; G. W. Holland; J. L. Jernow; S. Kwoh; P. Rosen, *J. Org. Chem. 19* (1973), 3440.
33. S. Nametkin; A. Zabrodena, *Chem. Ber. 69* (1936), 1789.
34. J. H. Clark; D. G. Stork, *J. Chem. Soc., Chem. Commun. 1982*, 635.
35. F. S. Alvarez; D. Wren, *Tetrahedron Lett. 1973*, 569.
36. V. A. Klimova; K. S. Zabrodina, *Izv. Akad. Nauk SSSR, Ser. Khim. Nauk 1961*, 176.
37. J. E. McMurry, *Accts. Chem. Res. 7* (1974), 281.
38. T. L. Ho; C. M. Wong, *Synthesis 1974*, 196.
39. R. Kirchhoff, *Tetrahdron Lett. 1976*, 2533.
40. G. A. Olah, G. K. S. Parkash; T. L. Ho, *Synthesis 1976*, 810.
41. Y. Akita; M. Inaba; H. Uchida; A. Ohta, *Synthesis 1977*, 792.
42. G. A. Russell; A. J. Moye; E. G. Janzen; S. Mak; E. R. Talaty, *J. Org. Chem. 32* (1967), 137.
43. H. Feuer; P. M. Pivawer, *J. Org. Chem. 31* (1966), 3152.
44. T. Simmons; R. F. Love; K. L. Kreuz, *J. Org. Chem. 31* (1966), 2400.
45. E. B. Hodge; R. Abbott, *J. Org. Chem. 27* (1962), 2254.
46. A. T. Nielsen; T. G. Archibald, *J. Org. Chem. 34* (1969), 1470.
47. F. Kienzle; J.-Y. Fellmann; J. Stadlwieser, *Helv. Chim. Acta 67* (1984), 789.
48. M. Miyashita; T. Yanami; A. Yoshikoshi, *J. Am. Chem. Soc. 98* (1976), 4679; *Org. Synth. 60* (1981), 117.
49. M. Miyashita; T. Kumazawa; A. Yoshikoshi, *Chem. Lett. 1980*, 1043.
50. M. Miyashita; T. Yanami; T. Kumazawa; A. Yoshikoshi, *J. Am. Chem. Soc. 106* (1984), 2149.
51. (a) D. Seebach; M. A. Brook, *Helv. Chim. Acta 68* (1985), 319.
    (b) M. Ochiai; M. Arimoto; E. Fujita, *Tetrahedron Lett. 1981*, 1115.
52. L. W. Herman; J. W. ApSimon, *Tetrahedron Lett. 1985*, 1423.
53. L. A. Kaplan, *J. Org. Chem. 29* (1964), 2256.
54. B. Klewe, *Acta Chem. Scand. 26* (1972), 1049.
55. D. J. Sutor; F. J. Llewellyn; H. S. Maslen, *Acta Cryst. 7* (1954), 145.
56. C. N. R. Rao In "The Chemistry of the Nitro and Nitroso Groups", Part 1. (H. Feuer, Ed.), Wiley, New York, 1969, p. 79.
57. P. Noble, Jr.; E. G. Borgardt; W. L. Reed, *Chem. Rev. 64* (1964), 19.
58. F. T. Williams, Jr.; P. W. K. Flanagan; W. J. Taylor; H. Schechter, *J. Org. Chem. 30* (1965), 2674.
59. M. J. Kamlet; D. J. Glower, *J. Org. Chem. 27* (1962), 537.
60. W. Kemula; W. Turnovska-Rubaszewska, *Rocz. Chem. 35* (1961), 1169; *37* (1963), 1597.
61. S. S. Novikov; V. J. Slovetskii; A. A. Fainzil'berg; S. A. Shevelev; V. A. Shlyapochnikov; A. I. Ivanov; V. A. Tartakovskii, *Tetrahedron 20*, Suppl. 1 (1964), 119.

62. Z. Buczkowski; T. Urbanski, *Spectrochim. Acta 18* (1962), 1187.
63. H. Feuer; C. Savides; C. N. R. Rao, *Spectrochim. Acta 19* (1963), 431.
64. V. I. Erashko; S. A. Shevelev; A. A. Fainzil'berg, *Russian Chem. Rev. 35* (1966), 719 (Uspekhi Khim., Engl. Transl.).
65. H. B. Hass; M. L. Bender, *J. Am. Chem. Soc. 71* (1949), 1767, 3482.
66. L. Weisler; R. W. Helmkamp, *J. Am. Chem. Soc. 67* (1945), 1167.
67. A. Hantzsch; A. Rinckenberger, *Chem. Ber. 32* (1899), 628.
68. A. Hantzsch; R. Caldwell, *Chem. Ber. 39* (1906), 2472.
69. W. Reich; G. Rose; W. Wilson, *J. Chem. Soc.* (1947), 1234.
70. G. Hammond; W. D. Emmons; C. Parker; B. Graybill; J. Waters; M. Hawthorne, *Tetrahedron 19*, Suppl. 1 (1963), 177.
71. K. P. Park; L. B. Clapp, *J. Org. Chem. 29* (1964), 2108.
72. P. Duden, *Chem. Ber. 26* (1893), 3003.
73. N. Kornblum; H. J. Taylor, *J. Org. Chem. 28* (1963), 1424.
74. A. R. Katritzky; G. de Ville; R. C. Patel, *J. Chem. Soc., Chem. Commun 1979*, 602; *Tetrahedron 37*, Suppl. *1* (1981), 25.
75. A. R. Katritzky; G. Musumarra, *Chem. Soc. Rev. 13* (1984), 47.
76. A. R. Katritzky; M. A. Kashmiri; D. K. Wittmann, *Tetrahedron 40* (1984), 1510.
77. G. A. Russell; J. Hershberger; K. Owens, *J. Am. Chem. Soc. 101* (1979), 1312.
78. G. A. Russell; M. Jawdosiuk; F. Ros, *J. Am. Chem. Soc. 101* (1979), 3378.
79. S. Zen; E. Kaji, *Bull. Chem. Soc. Japan 43* (1970), 2277; *46* (1973), 339; *Chem. Pharm. Bull. Tokyo 22* (1974), 477.
80. A. S. Polyanskaya; V. V. Perekalin; L. A. Bocharova; R. I. Ivanova, USSR P. 172 748 (1965).
81. P. A. Wade; H. R. Hinney; N. V. Amin; P. D. Veil; S. D. Morrow; S. A. Hardinger; M. S. Saft, *J. Org. Chem. 46* (1981), 765.
82. P. A. Wade; S. D. Morrow; S. A. Hardinger, *J. Org. Chem. 47* (1982), 365; cf. A. M. Khawaga; M. T. Ismael; A.-M. A. Abdel-Wahab, *Gazz. Chim. Ital. 112* (1982), 235.
83. D. Seebach; F. Lehr, *Angew. Chem. 88* (1976), 540.
84. D. Seebach; R. Henning; F. Lehr; J. Gonnermann, *Tetrahedron Lett. 1977*, 1161; *Chem. Ber. 112* (1979), 234.
85. R. Henning; F. Lehr; D. Seebach, *Helv. Chim. Acta 59* (1976), 2213.
86. (a) D. Seebach; R. Henning; T. Mukhopadhyay, *Chem. Ber. 115* (1982), 1705.
    (b) M. Eyer; D. Seebach, *J. Am. Chem. Soc. 197* (1985), 3601.
87. D. Seebach; R. Henning; F. Lehr, *Angew. Chem. 90* (1978), 479.
88. N. Kornblum; R. E. Michel; R. C. Kerber, *J. Am. Chem. Soc. 88* (1966), 5660, 5662.
89. G. A. Russell; W. C. Danen, *J. Am. Chem. Soc. 88* (1966), 5663.
90. N. Kornblum, *Angew. Chem. 87* (1975), 797.
91. N. Kornblum In "The Chemistry of Amino, Nitroso, Nitro Compounds and their Derivatives", Suppl. F (S. Patai, Ed.), Wiley, Chichester, 1982, p. 361.
92. R. C. Kerber; G. W. Urry; N. Kornblum, *J. Am. Chem. Soc. 87* (1965), 4520.
93. N. Kornblum, P. Pink, *Tetrahedron 19*, Suppl. 1 (1963), 17.
94. R. C. Kerber; A. Porter, *J. Am. Chem. Soc. 91* (1969), 366.
95. N. Kornblum; R. A. Brown, *J. Am. Chem. Soc. 85* (1963), 1359; *86* (1964), 2681.
96. H. Sato; T. Kusumi; K. Imaye; H. Kakisawa, *Bull. Chem. Soc. Japan 49* (1976), 2815.
97. L. G. Donaruma, *J. Org. Chem. 22* (1957), 1024.
98. F. Arndt; J. D. Rose, *J. Chem. Soc. 1935*, 1.
99. M. K. Shakhova; M. I. Budagyants; G. I. Samokhvalov; N. A. Preobashchenskii, *Zh. Obshch. Khim. USSR 32* (1962), 2832.
100. P. E. Fanta; R. M. W. Rickett; D. S. Jones, *J. Org. Chem. 26* (1961), 938.
101. V. A. Tartakovskii; I. E. Chlenov; S. S. Smagin; S. S. Novikov, *Izv. Akad. Nauk SSSR, Ser. Khim. Nauk 1964*, 583; *1965*, 552.
102. V. A. Tartakovskii; I. E. Chlenov; S. L. Joffe; G. V. Lagodzinskaya; S. S. Novikov, *Zh. Org. Khim. 2* (1966), 1593.
103. V. A. Tartakovskii; O. A. Lukyanov; N. I. Shlykova; S. S. Novikov, *Zh. Org. Khim. 3* (1967), 980; *4* (1968), 231.
104. V. A. Tartakovskii; I. E. Savostyanova; S. S. Novikov, *Zh. Org. Khim. 4* (1968), 240.
105. V. A. Tartakovskii; I. E. Chlenov; N. S. Morozova; S. S. Novikov, *Izv. Akad. Nauk SSSR, Ser. Khim. Nauk 1966*, 370.
106. V. A. Tartakovskii; I. E. Chlenov; G. V. Lagodzinskaya; S. S. Novikov, *Dokl. Akad. Nauk SSSR 161* (1965), 136.

107. V. A. Tartakovskii; Z. Ya. Lapshina; I. A. Savostyanova; S. S. Novikov, *Zh. Org. Khim. 4* (1968), 236.
108. G. A. Shvekhgeimer; N. I. Sobtsova; A. Baranski, *Rocz. Chem. 46* (1972), 1543, 1735, 1741.
109. A. A. Onishchenko; I. E. Chlenov; L. M. Makarenkova; V. A. Tartakovskii, *Izv. Akad. Nauk SSSR, Ser. Khim. Nauk 1971*, 1560.
110. R. Grée; R. Carrié, *Bull. Soc. Chim. France 1975*, 1314, 1319.
111. R. Grée; F. Tonnard; R. Carrié, *Tetrahedron 32* (1976), 675.
112. R. Grée; R. Carrié, *Tetrahedron Lett. 1971*, 4117.
113. P. A. Wade; N. V. Amin; H.-K. Yen; D. T. Price; G. F. Huhn, *J. Org. Chem. 49* (1984), 4595.
114. T. Severin; B. Brück, *Chem. Ber. 98* (1965), 3847; T. Severin; B. Brück; P. Adhikary, *Chem. Ber. 99* (1966), 3097.
115. L. A. Cohen; W. M. Jones, *J. Am. Chem. Soc. 85* (1963), 3397.
116. E. Bamberger, *Chem. Ber. 35* (1902), 54; E. Bamberger; J. Grob, *Chem. Ber. 35* (1902), 67; E. Bamberger; J. Frey, *Chem. Ber. 35* (1902), 82.
117. H. J. Backer, *Rec. Trav. Chim. Pays Bas 69* (1950), 610.
118. C. Bellandi; M. DeAmici; C. DeMicheli; R. Gandolfi, *Heterocycles 22* (1984), 2187.
119. C. D. Nenitzescu; D. A. Isacescu, *Bull. Soc. Chim. Romania 14* (1932), 53; *Chem. Ber. 63* (1930), 2484.
120. R. E. McCoy; R. E. Gohlke, *J. Org. Chem. 22* (1957), 286.
121. H. B. Hass; M. L. Bender In *Organic Synth.* Coll. Vol. IV *1963*, 932.
122. S. V. Lieberman, *J. Am. Chem. Soc. 77* (1955), 1114.
123. M. Montavon; H. Lindlar; R. Marbet; R. Rüegg; G. Ryser; S. Saucy; P. Zeller; O. Isler, *Helv. Chim. Acta 40* (1957), 1250.
124. (a) J. Kimura; A. Kawashima; M. Sugizaki; N. Nemoto; O. Mitsuňobu, *J. Chem. Soc., Chem. Commun. 1979*, 303; O. Mitsunobu; J. Kimura; T. Shimizu; A. Kawashima, *Chem. Lett. 1980*, 927.
    (b) J. S. Meek; J. S. Fowler; P. A. Monroe; T. J. Clark, *J. Org. Chem. 33* (1968), 223; J. S. Meek; J. S. Fowler, *J. Org. Chem. 33* (1978), 226.
125. E. P. Kohler, *J. Am. Chem. Soc. 46* (1924), 503, 1733.
126. E. P. Kohler; G. R. Barrett, *J. Am. Chem. Soc. 46* (1924), 2105; *48* (1926), 1770.
127. L. I. Smith, *Chem. Rev. 23* (1938), 255.
128. A. T. Nielsen; T. G. Archibald, *Tetrahedron Lett. 1968*, 3375.
129. Zh. A. Krasnaya; T. S. Stytsenko; E. P. Prokofiev; J. P. Yakovlev; V. F. Kucherov, *Izv. Akad. Nauk, Ser. Khim. Nauk SSSR 1974*, 845.
130. E. Kaji; H. Ishikawa; S. Zen, *Bull. Soc. Chim. Japan 52* (1979), 2928.
131. E. Kaji; S. Zen, *Heterocycles 13* (1979), 187; *Chem. Pharm. Bull. Japan 28* (1980), 479.
132. K. Takahashi; E. Kaji; S. Zen, *Nippon Kagaku Kaishi 1983*, 1678; *Synthetic Commun. 14* (1984), 139.
133. S. Zen; K. Takahashi; E. Kaji; H. Nakamura; Y. Iitaka, *Chem. Pharm. Bull. 31* (1983), 1814.
134. V. A. Tartakovskii; S. S. Novikov; T. I. Godovikova, *Izv. Akad. Nauk, Ser. Khim. Nauk SSSR 1961*, 1042.
135. V. A. Tartakovskii; I. A. Savostyanova; B. G. Gribov; S. S. Novikov, *Izv. Akad. Nauk, Ser. Khim. Nauk SSSR 1963*, 1328.
136. V. A. Tartakovskii; B. G. Gribov; S. S. Novikov, *Izv. Akad. Nauk, Ser. Khim. Nauk SSSR 1965*, 1074.
137. V. A. Tartakovskii; B. G. Gribov; I. A. Savostyanova; S. S. Novikov, *Izv. Akad. Nauk, Ser. Khim. Nauk SSSR 1965*, 1644.
138. V. A. Tartakovskii; A. A. Onishchenko; G. V. Lagodzinskaya; S. S. Novikov, *Zh. Org. Khim. 3* (1967), 165.
139. K. Torssell, *Acta. Chem. Scand. 21* (1967), 1392.
140. V. A. Tartakovskii; L. A. Nikonova; S. S. Novikov, *Izv. Akad. Nauk, Ser. Khim. Nauk SSSR 1966*, 1290.
141. J. T. Thurston; R. L. Shriner, *J. Org. Chem. 2* (1937), 183.
142. E. P. Stefl; M. F. Dull, *J. Am. Chem. Soc. 69* (1947), 3037.
143. T. Urbański; W. Gurzynska, *Rocz. Chem. 25* (1951), 213.
144. A. McKillop; R. J. Kobylecki, *Tetrahedron 30* (1974), 1365.
145. T. Urbański, *J. Chem. Soc. 1949*, 3374.
146. W. Steinkopf; B. Jürgens, *J. Prakt. Chem. 84* (1911), 686.
147. H. Wieland; Z. Kitasato, *Chem. Ber. 62* (1929), 1250.
148. T. Fujizawa; Y. Kurita; T. Sato, *Chem. Lett. 1983*, 1537.

149. E. H. White; W. J. Considine, *J. Am. Chem. Soc. 80* (1958), 626.
150. E. Kaji; K. Harada; S. Zen, *Chem. Pharm. Bull. 26* (1978), 3254.
151. T. Mukaiyama; T. Hoshino, *J. Am. Chem. Soc. 82* (1960), 5339.
152. H. L. Finkbeiner; M. Stiles, *J. Am. Chem. Soc. 85* (1963), 616.
153. H. L. Finkbeiner; G. W. Wagner, *J. Org. Chem. 28* (1963), 215.
154. D. C. Baker; S. R. Putt, *Synthesis 1978*, 478.
155. R. L. Crumbie; J. S. Nimitz; H. S. Mosher, *J. Org. Chem. 47* (1982), 4040.
156. G. B. Bachman; T. Hokama, *J. Am. Chem. Soc. 81* (1959), 4882.
157. E. Bamberger; J. Grob, *Chem. Ber. 35* (1902), 67.
158. K. Torssell; O. Zeuthen, *Acta Chem. Scand. B32* (1978), 118.
159. V. A. Tartakovskii; A. A. Onishchenko; I. E. Chlenov; S. S. Novikov, *Dokl. Akad. Nauk USSR 167* (1966), 844.
160. I. E. Chlenov; N. S. Morosova; V. A. Tartakovskii, *Izv. Akad. Nauk, Ser. Khim. Nauk SSSR 32* (1983), 1889.
161. E. Vedejs; R. A. Buchanan, *J. Org. Chem. 49* (1984), 1840.
162. E. Vedejs; D. A. Perry, *J. Am. Chem. Soc. 105* (1983), 1683.
163. V. A. Tartakovskii; O. A. Lukyanov; S. S. Novikov, *Dokl. Akad. Nauk SSSR 178* (1968), 123.
164. R. Huisgen; H. Seidl; I. Brüning, *Chem. Ber. 102* (1969), 1102.
165. J. B. Hendrickson; S. A. Pearson, *Tetrahedron Lett. 1983*, 4657.
166. A. I. Ivanov; I. E. Chlenov; V. A. Tartakovskii; V. I. Slovetskii; S. S. Novikov, *Izv. Akad. Nauk SSSR, Ser. Khim. Nauk 1965*, 1491.
167. A. Young; O. Levand; W. K. H. Luke; H. O. Larson, *Chem. Commun. 1966*, 230.
168. O. A. Lukyanov; A. A. Onishchenko; V. P. Gorelik; V. A. Tartakovskii, *Izv. Akad. Nauk SSSR, Ser. Khim. Nauk 1973*, 1294.
169. N. Jonathan, *J. Mol. Spectrosc. 5* (1960), 101.
170. A. H. Lamberton; G. Newton, *J. Chem. Soc. 1961*, 1797.
171. P. Bruck; A. H. Lamberton, *J. Chem. Soc. 1955*, 3997; *1957*, 4198.
172. A. H. Lamberton; H. M. Yusuf, *J. Chem. Soc. (C) 1969*, 397.
173. A. L. Fridman; V. P. Ivshin; S. S. Novikov, *Uspekhi Khim. 38* (1969), 640 (Engl. Trans.).
174. K. A. Jørgensen; K. Torssell, unpublished results.
175. S. R. Challand; C. W. Rees; R. C. Storr, *J. Chem. Soc., Chem. Commun. 1973*, 837.
176. R. Huisgen; R. P. Gambra, *Tetrahedron Lett. 1982*, 55; *Chem. Ber. 115* (1982), 2242.
177. R. C. Storr In "1,3-Dipolar Cycloaddition Chemistry" (A. Padwa, Ed.), Vol. 2, Wiley, New York, 1984, p. 169.
178. G. L. Salem, "Electrons in Chemical Reactions", Wiley, New York, 1982, p. 171.
179. S. L. Joffe; L. M. Leonteva; V. A. Tartakovskii, *Uspekhi Khim. 46* (1977), 1658.
180. W. P. Weber, "Silicon Reagents for Organic Synthesis", Springer-Verlag, Berlin, 1983; V. A. Tartakovskii, *Izv. Akad. Nauk SSSR, 1984*, 165.
181. S. L. Joffe; M. V. Kashutina; V. M. Shitkin; A. Z. Yankelevich; A. A. Levin; V. A. Tartakovskii, *Izv. Akad. Nauk SSSR, Ser. Khim. Nauk 1972*, 1341.
182. S. L. Joffe; M. V. Kashutina; V. M. Shitkin; A. A. Levin; V. A. Tartakovskii, *Zh. Org. Khim. 9* (1973), 896.
183. M. V. Kashutina; S. L. Joffe; V. M. Shitkin; N. O. Cherskaya; V. A. Korenevskii; V. A. Tartakovskii, *Zh. Obshch. Khim. 43* (1973), 1715.
184. S. L. Joffe; L. M. Makarenkova; V. A. Tartakovskii, *Izv. Akad. Nauk SSSR, Ser. Khim. Nauk 1974*, 463.
185. M. V. Kashutina; S. L. Joffe; V. A. Tartakovskii, *Dokl. Akad. Nauk SSSR 218* (1974), 109.
186. S. C. Sharma; K. Torssell, *Acta Chem. Scand. B33* (1979), 379.
187. S. K. Mukerji; K. Torssell, *Acta Chem. Scand. B35* (1981), 643.
188. S. H. Andersen; N. B. Das; R. D. Jørgensen; G. Kjeldsen; J. S. Knudsen; S. C. Sharma; K. B. G. Torssell, *Acta Chem. Scand. B36* (1982), 1.
189. N. B. Das; K. B. G. Torssell, *Tetrahedron 39* (1983), 2227, 2247.
190. S. K. Mukerji; K. K. Sharma; K. B. G. Torssell, *Tetrahedron 39* (1983), 2231.
191. M. Asaoka; T. Mukuta; H. Takei, *Tetrahedron Lett. 1981*, 735.
192. M. Asaoka; M. Abe; T. Mukuta; H. Takei, *Chem. Lett. 1982*, 215.
193. K. B. G. Torssell; A. C. Hazell; R. G. Hazell, *Tetrahedron 41* (1985), 5569.
194. H. Feger; G. Simchen, *Synthesis 1981*, 378.
195. H. Emde; D. Domsch; H. Feger; U. Frick; A. Götz; H. H. Herrgott; K. Hofmann; W. Kober; K. Krägeloh; T. Oesterloe; W. Steppan; W. West; G. Simchen, *Synthesis 1982*, 1.
196. E. W. Colvin; D. Seebach, *J. Chem. Soc., Chem. Commun. 1978*, 689.

197. E. W. Colvin; A. K. Beck; B. Bastani; D. Seebach; Y. Kai; J. D. Dunitz, *Helv. Chim. Acta 63* (1980), 697.
198. E. W. Colvin; A. K. Beck; D. Seebach, *Helv. Chim. Acta 64* (1981), 2264.
199. D. Seebach; A. K. Beck; T. Mukhopadhyay; E. Thomas, *Helv. Chim. Acta 65* (1982), 1101.
200. E. Vedejs; D. A. Perry, *J. Am. Chem. Soc. 105* (1983), 1683.
201. T. H. Keller; L. J. Yelland; M. H. Benn, *Can. J. Chem. 62* (1984), 437.
202. G. A. Olah; B. G. B. Gupta; S. C. Narang; R. Malhotra, *J. Org. Chem. 44* (1979), 4272.
203. G. A. Olah; B. G. B. Gupta, *Synthesis 1980*, 44.
204. S. L. Joffe; L. M. Leonteva; L. M. Makarenkova; V. F. Pyaterikov; V. A. Tartakovskii, *Izv. Akad. Nauk SSSR, Ser. Khim. Nauk 1975*, 1146.
205. S. L. Joffe; V. M. Shitkin; B. N. Khasapov; M. V. Kashutina; V. A. Tartakovskii; M. Ya. Myagi; E. T. Lippmaa, *Izv. Akad. Nauk SSSR, Ser. Khim. Nauk 1973*, 2146.
206. S. L. Joffe; L. M. Makarenkova; V. A. Tartakovski, *Izv. Akad. Nauk SSSR, Ser. Khim. Nauk 1974*, 463.
207. S. L. Joffe; L. M. Leont'eva; L. M. Makarenkova; A. L. Blyumenfel'd; V. F. Pyaterikov; V. A. Tartakovskii, *Izv. Akad. Nauk SSSR, Ser. Khim. Nauk 1975*, 1146.
208. M. Ya. Myagi; E. T. Lippmaa; S. L. Joffe; V. A. Tartakovskii; A. S. Shashkov; B. N. Khasapov; L. M. Makarenkova, *Izv. Akad. Nauk SSSR, Ser. Khim. Nauk 1973*, 1431.
209. S. L. Joffe; A. L. Blyumenfel'd; A. S. Shashkov; L. M. Makarenkova; V. A. Tartakovskii, *Izv. Akad. Nauk SSSR, Ser. Khim. Nauk 1976*, 2320.
210. S. L. Joffe; L. M. Makarenkova; A. L. Blyumenfel'd; I. A. Maslina; V. A. Tartakovskii, *Isv. Akad. Nauk SSSR, Ser. Khim. Nauk 1976*, 2326.
211. O. P. Shitov; S. L. Joffe; L. M. Leont'eva; V. A. Tartakovskii, *Zh. Obshch. Khim. 43* (1973), 1266.
212. O. P. Shitov; L. M. Leont'eva; S. L. Joffe; B. N. Khasapov; V. M. Novikov; A. U. Stepanyants; V. A. Tartakovskii, *Izv. Akad. Nauk SSSR, Ser. Khim. Nauk 1974*, 2782.
213. S. L. Joffe; L. M. Leont'eva; A. L. Blyumenfel'd; O. P. Shitov; V. A. Tartakovskii, *Izv. Akad. Nauk SSSR, Ser. Khim. Nauk 1974*, 1659.
214. J. Lorbeth; G. Lange, *J. Organomet. Chem. 54* (1973), 165.
215. T. Mukaiyama; H. Nambu, *J. Org. Chem. 27* (1962), 2201.
216. D. Seebach; A. K. Beck; F. Lehr; T. Weller; E. W. Colvin, *Angew. Chem. 93* (1981), 422.
217. E. W. Colvin; A. D. Robertson; D. Seebach; A. K. Beck, *J. Chem. Soc., Chem. Commun. 1981*, 952.
218. D. Seebach; E. W. Colvin; F. Lehr; T. Weller, *Chimia 33* (1979), 1.

Chapter 5

Applications of Nitrile Oxides, Nitrones, Nitronates, and Intermediate Isoxazoles, Isoxazolines, and Isoxazolidines in Synthesis

5.1 2-NITROALCOHOLS AND 2-AMINOALCOHOLS

The nitroaldol reaction or Henry's reaction is a classical carbon-carbon coupling leading to biologically and synthetically important aminoalcohols on reduction, carbonyl compounds under Nef's conditions, and nitroolefins on dehydration. Owing to the reversible character of the nitroaldol reaction, lithium aluminum hydride reduction of the nitroalcohol often gives a low yield of the aminoalcohol. It was observed that the trimethylsilylnitronate of nitroethane reacted with benzaldehyde in refluxing benzene to give a diastereomeric mixture of 1-phenyl-1-trimethylsilyloxy-2-nitropropane (**1**) with fixed 1,2-hydroxy-nitro structure, eq. (5.1).[1] The reaction proceeds readily with a wide range of aromatic and aliphatic aldehydes and primary and secondary nitro compounds at $-78°C$ in the presence of catalytic amounts of fluoride ions,[2-4] eq. (5.2). Ketones do not react under these conditions. Secondary nitro compounds give predominantly the nitroalcohols (**3**), which can be silylated in a second step. The diastereomeric excess in **2** or **3** is low, but by carrying out deprotonation of **2** with lithium diisopropylamide followed by protonation with

$$\underset{\underline{1}}{\text{CH}_3\text{CH}=N(O)\text{OSi(CH}_3)_3} + \text{PhCHO} \xrightarrow[80\ °C]{\text{Benzene}} \text{Ph-CH(OSi(CH}_3)_3)\text{-CH(NO}_2)\text{CH}_3 \quad (5.1)$$

$$R^1R^2C=N(OSi(CH_3)_3)O + R^3CHO \xrightarrow[-78°C]{F^-, THF} \begin{array}{c} a \\ R^1=H \\ b \end{array}$$

(via a): **2** HR²C(OSi(CH₃)₃)−CR³(NO₂) ; then ClSi(CH₃)₃

(via b): **3** R¹R²C(OH)−CR³(NO₂)

$R^1, R^2 = H$, Alkyl, Aryl

$R^3 =$ Alkyl, Aryl

(5.2)

acetic acid at $-90°C$ the *erythro*-isomer becomes the major product, ca. 90%. The diastereomeric ratio is improved under carefully defined conditions by using the dimethyl-*t*-butylsilyl nitronate in eq. (5.2), *erythro:threo* ca. 95:5.[4] It was also found that careful protonation of doubly deprotonated nitroalcohols (**4**) with acetic acid in the presence of HMPT predominantly gives the *threo*-isomer **6**, typically in a ratio of ca. 3–10:1, eq. (5.3); cf. Section 4.6. The highest diastereoselectivity is obtained from nitroalcohols derived from aromatic aldehydes. The protonation occurs from the least sterically hindered side, and the outcome of the reaction is explained by the supposition that the doubly deprotonated species has conformation **7** and the silylated species has conformation **8**.

$$\mathbf{4} \xrightarrow[\text{THF, HMPT}]{\text{HOAc}, -90°C} \mathbf{5} + \mathbf{6} \quad (5.3)$$

4: R¹R²C(−N(O⁻)(O⁻))
5, **6**: R¹R²C(OH)−CH(NO₂) (diastereomers)

$R^1 = CH(CH_3)_3, C(CH_3)_3, C_5H_{11}, C_6H_5, p\text{-}OCH_3\text{-}C_6H_4, p\text{-}F\text{-}C_6H_4$

$R^2 = CH_3, C_2H_5$

7 (threo): H⁺ attack shown on conformation with NO₂⁻, R¹, H, O⁻, R² substituents

8 (erythro): H⁺ attack shown on conformation with R², R¹, H, (CH₃)₂SiO, NO₂⁻ substituents

The diastereomeric enrichment in the nitroaldoles is preserved in controlled reduction over neutral Raney-Ni, eq. (5.4).[4] The silyloxy group is cleaved by lithium aluminum hydride reduction. Direct reduction of the nitrosilyloxy compound with lithium aluminum hydride leads to isomerization.

$$\underset{H_{11}C_5}{\overset{\overset{+}{OSi(CH_3)_2}}{\underset{NO_2}{\bigg|}}}C_2H_5 \xrightarrow{\text{Raney-Ni}} \underset{H_{11}C_5}{\overset{\overset{+}{OSi(CH_3)_2}}{\underset{NH_2}{\bigg|}}}C_2H_5 \xrightarrow{\text{LiAlH}_4} \underset{H_{11}C_5}{\overset{OH}{\underset{NH_2}{\bigg|}}}C_2H_5 \quad (5.4)$$

retention of configuration

5.2 α-CYANOKETONES AND 2-CYANO-ALCOHOLS. CARBOXYHYDROXYLATION

The 3-unsubstituted isoxazoles **9** and 2-isoxazolines are distinguished by their lability toward bases. The C^3—H abstraction by the base is synchronized with the cleavage of the N—O bond to give α-cyanoketones or -alcohols, respectively, in an irreversible E2 process, eqs. (5.5),(5.6).[5-7] The reaction shows a primary isotope effect, indicating that C^3—H fission takes part in the rate-determining step. No base-catalyzed hydrogen exchange at C^3 has been observed.

$$\underset{\mathbf{9}}{\overset{R^2}{\underset{R^1}{\bigg\langle}}\underset{O}{\overset{H\curvearrowleft^-B}{\underset{N}{\bigg\rangle}}}} \longrightarrow \underset{R^1}{\overset{R^2}{\bigg\langle}}\underset{OH}{\overset{CN}{\bigg\rangle}} \rightleftarrows \underset{R^1}{\overset{R^2}{\bigg\langle}}\underset{O}{\overset{CN}{\bigg\rangle}} \quad (5.5)$$

$$\underset{\mathbf{10}}{\overset{H\curvearrowleft^-B}{\underset{R}{\bigg\langle}}\underset{O}{\overset{}{\underset{N}{\bigg\rangle}}}} \longrightarrow \underset{R}{\overset{}{\bigg\langle}}\underset{OH}{\overset{CN}{\bigg\rangle}} \quad (5.6)$$

$$\underset{C_6H_5}{\overset{}{\bigg\langle}}\underset{O}{\overset{}{\underset{N}{\bigg\rangle}}} \xrightarrow[\text{2. H}^+]{\text{1. OH}^-} C_6H_5COCH_2CN \quad (5.7)$$

The reaction, which is quite general, was first observed by Claisen,[8,9] eq. (5.7), and has since then been the object of numerous investigations. Woodward-Olofson's peptide synthesis, which is a special case of this reaction, is discussed separately in Section 5.3. The α-cyanoketones can be isolated, but they are unstable compounds, prone to polymerize, and react readily with suitable substrates, such as phenylhydrazine to give pyrazole (**11**),[9] eq. (5.8a), and methylvinylketone to form the cyclohexenone **12**,[10] eq. (5.8b). With allyl bromide the reaction gives compound **13**, eq. (5.8c); with benzaldehyde the corresponding benzal derivative is obtained.[11] A variety of 4-mono-, 5-mono-, and 4,5-disubstituted isoxazoles **9** have been cleaved with a base (aqueous OH^-, RO^-/ROH, R_3N, BuLi, $NaNH_2$, etc.), eq. (5.5), Table 5.1.

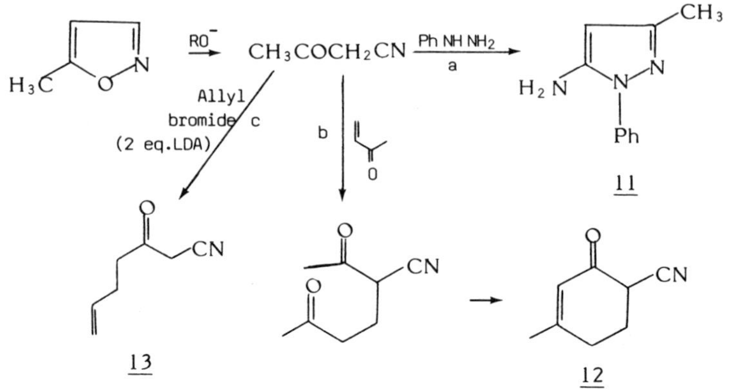

Table 5.1. Ring Opening of 3-Unsubstituted Isoxazoles by Base. Synthesis of α-Cyanoketones, Eq. (5.5)

R^1	R^2	Ref.	R^1	R^2	Ref.
C_6H_5	H	8	CH_3, C_2H_5, C_3H_7	Aryl	21, 22
CH_3	H	9–11, 14		H	23
H	CH_3, Ph	12			
CH_3	CH_3	13			
C_3H_7	H	14	NH_2	COOR, CN	24
C_6H_5	COOH	15	CH_3	Cl, Br	25
COOH	C_6H_5	15	H	Br, NO_2, SO_3H	26, 27
$CH_3\overset{\|}{C}=NOCH_3$	H	16	CH_3	SO_3H	28, 29
COOH, 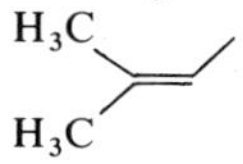 $\underset{H_3C}{\overset{H_3C}{\diagup}}$	H	17	H	H	30
H	COOH	18	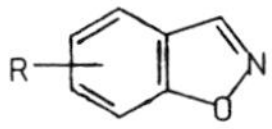		31–33
$HOCH_2$, $HOC(CH_3)_2$	H	19	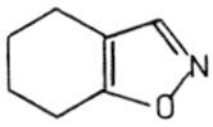		34
CH_3CO	COOR	20			

The preferred route to α-cyanoketones involves formylation of the ketone, oximation to the 3-unsubstituted isoxazole, and finally treatment with a base, eq. (5.9).[34–44] The reaction has been applied in the synthesis of steroids. The same type of cleavage occurs in the thermolysis of 3-carboxy derivatives, e.g. **14**, **15**, eqs. (5.10),(5.11),[11,31–33,45–50] or by treatment of 3-acyl derivatives with a base, eq. (5.12).[48,51]

The functionalization of olefinic bonds is of major synthetic importance. Cyanohydroxylation can be performed by dipolar addition of fulminic acid (**16**) followed by treatment with base. The fulminic acid can either be generated from sodium fulminate and sulfuric acid[53-56] or, better, from formhydroximic acid iodide and triethylamine, eq. (5.13).[52] The fulminic acid has to be generated slowly to avoid dimerization to HON=CH—CNO. Methyl crotonate gives with fulminic acid a mixture of isoxazolines **19** and **20**, 62:38, in a yield of 36%. Isoxazoline **19** undergoes ring fission on treatment with a base to yield the cyanoalcohol **21**.

The trimethylsilylnitronate of nitromethane (**22**) cycloadds to reactive olefins with the formation of N-silyloxyisoxazolidines **23**, which readily eliminate trimethylsilanol on treatment with acid to give the 2-isoxazolines **24**, eq. (5.14).[57-60] This constitutes an alternative route to 2-cyanoalcohols (2-hydroxynitriles).

In analogy with 3-carboxyisoxazoles the 3-carboxyisoxazolines undergo decarboxylative ring opening to give *cis*-cyanoalcohols. Carboethoxyformonitrile oxide (**25**) is cycloadded to the olefin, hydrolyzed in aqueous sodium hydroxide, acidified, and the free acid thermally decarboxylated, eq. (5.15).

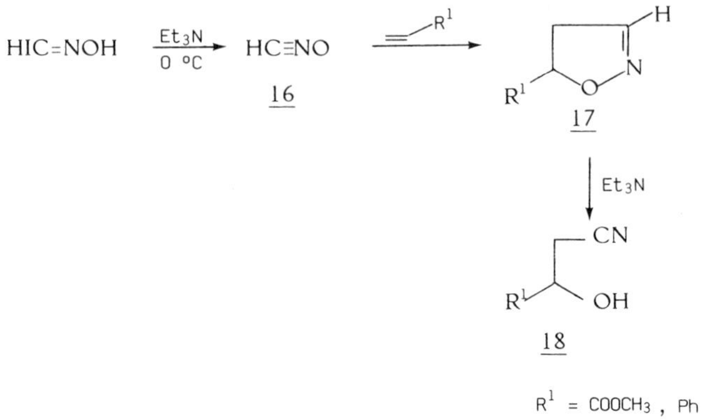

Applications in Synthesis

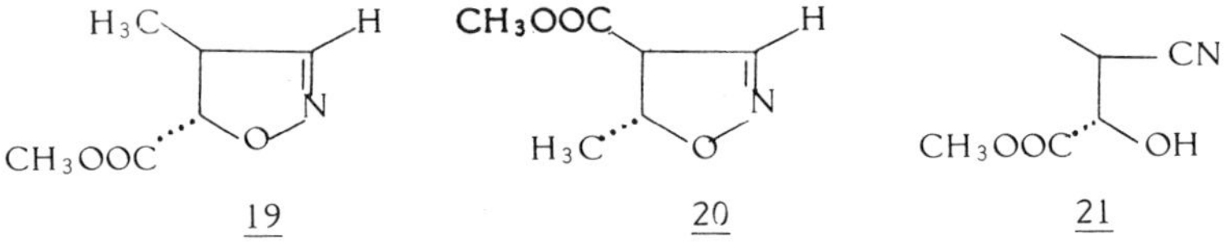

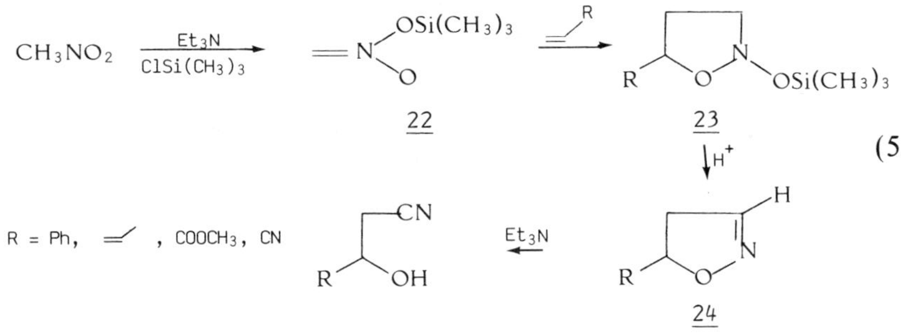

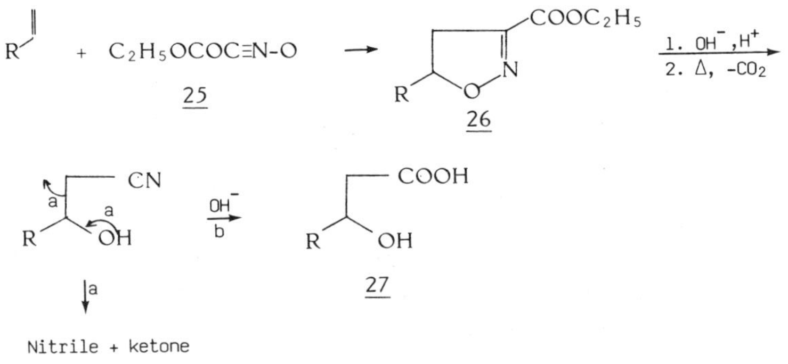

$$\text{HIC=NOH} \xrightarrow[0\ °C]{\text{Et}_3\text{N}} \text{HC≡NO} \quad \underline{16} \quad \xrightarrow{\equiv\!-R^1} \quad \underline{17} \text{ (isoxazoline)} \tag{5.13}$$

Structure **17**: 4,5-dihydroisoxazole with R^1 at C-5, H at C-3.

↓ Et₃N

Structure **18**: $R^1\text{-CH(OH)-CH}_2\text{-CN}$

R^1 = COOCH₃, Ph

19: 3-H, 4-CH₃, 5-CH₃OOC isoxazoline
20: 3-H, 4-CH₃OOC, 5-CH₃ isoxazoline
21: CH₃OOC-CH(OH)-CH(CH₃)-CN... (CH₃OOC–C(–OH)–CN with CH₃)

$$\text{CH}_3\text{NO}_2 \xrightarrow[\text{ClSi(CH}_3)_3]{\text{Et}_3\text{N}} \quad \underset{\underline{22}}{\text{CH}_2\!=\!N(OSi(CH}_3)_3)\text{-O}} \quad \xrightarrow{\equiv\!-R} \quad \underset{\underline{23}}{R\!-\!\text{isoxazolidine-OSi(CH}_3)_3} \tag{5.14}$$

↓ H⁺

R = Ph, =/, COOCH₃, CN

Structure: R-CH(OH)-CH₂-CN ⇌ (Et₃N) ⇌ **24** (isoxazoline with R at 5)

$$R\text{-CH=O} + \text{C}_2\text{H}_5\text{OCOC≡N-O} \longrightarrow \underset{\underline{26}}{R\text{-isoxazoline-COOC}_2\text{H}_5} \xrightarrow[2.\ \Delta,\ -\text{CO}_2]{1.\ \text{OH}^-,\ \text{H}^+} \tag{5.15}$$

$\underline{25}$

R-CH(OH)-CH₂-CN $\xrightarrow[b]{\text{OH}^-}$ R-CH(OH)-CH₂-COOH $\underline{27}$

↓ a

Nitrile + ketone

Prolonged treatment of the cyanoalcohol with base yields directly the β-hydroxy acid **27**. Equation (5.15) was first tested in the steroid series[61–65] and has then found applications in the synthesis of 2-deoxyribose, chiral β-hydroxycarboxylic acids, and (±)-blastomycinone[66–68] cf. Sections 5.21 and 5.23. The fragmentation of isoxazolines prepared from cis- and trans-2-butenes occurs without isomerization. Further fragmentation into nitrile and ketone, eq. (5.15a), has been observed.[64,66] Table 5.2 collects some examples of the cyanohydroxylations, eqs. (5.13)–(5.15) together with results from the hydroxy carboxylation procedures, eqs. (5.16)–(5.18).

Cycloadditions of O-protected 1-nitroalcohols **28**, **29** followed by removal of the protecting group, catalytic reduction, and periodate cleavage give the β-hydroxycarboxylic acid in satisfactory yield, eq. (5.16).[66,67,69] Baeyer-Villiger

oxidation of the intermediate *t*-butyl ketone formed in eq. (5.17) leads to the corresponding hydroxy acid, but the bulky *t*-butyl group lowers the yield of the cleavage. The degradative procedure shown in eq. (5.16) is superior. A third hydroxycarboxylation procedure starting from 3-benzenesulfonyl-substituted 2-isoxazolines[70-75] is outlined in eq. (5.18). The sulfonyl group in **31** is displaced with a methoxy group, and the resulting 3-alkoxyisoxazoline is reduced to the β-hydroxyester **32**.[69] Alternatively, **31** can be reductively cleaved by Na/Hg to the cyanoalcohol **33**.[70,71]

Hydroxycarboxylation has been accomplished by using the methyl or ethyl nitronate of trinitromethane as reagent, which readily cycloadds to alkenes, eq. (5.19).[76] Treatment of the isoxazolidines **34** and **35** with acid gives the

Applications in Synthesis

Table 5.2. Cyano Hydroxylation and Hydroxycarboxylation of Alkenes

Alkene	Product	Method, Eq.	Ref.
Ph-CH=CH$_2$	Ph-CH(OH)-CH$_2$-CN	(13)–(15)	52, 58, 66
CH$_3$OOC-CH=CH$_2$	CH$_3$OOC-CH(OH)-CH$_2$-CN	(13)–(15)	52, 58, 66
norbornene	norbornane-CN, OH	(13), (18)	52, 71
CH$_2$=CH-CH$_2$-CH$_3$	HO-CH(CH=CH$_2$)-CH$_2$-CN	(14)	58
trans-CH$_3$-CH=CH-CH$_3$	H$_3$C-C(OH)(-)-CH(CH$_3$)-CN	(15)	66
cis-CH$_3$-CH=CH-CH$_3$	H$_3$C-C(OH)-CH(CH$_3$)-CN	(15)	66
(cyclopentenyl methyl ketone)	(hydroxy-cyano adduct) + rearr. products	(15)	63
(dioxolane vinyl)	(dioxolane with OH, COOCH$_3$)	(15)	66
cyclopentene	2-hydroxycyclopentane-COOH	(16), (17), (19)	66, 69, 76
C$_3$H$_7$-CH=CH-C$_3$H$_7$	H$_7$C$_3$-CH(OH)-CH(C$_3$H$_7$)-COOH	(16), (17)	69
limonene	limonene with -OH, -CN	(18)	71

hydroxy acids **36** and **37**, respectively. It appears that the synthetic potentialities of trinitromethane cycloaddition are not fully exhausted.

5.3 WOODWARD-OLOFSON'S PEPTIDE SYNTHESIS. CLEAVAGE OF ISOXAZOLIUM SALTS

3-Unsubstituted isoxazolium salts are more susceptible than 3-unsubstituted isoxazoles to base-catalyzed cleavage. Salt **38** is cleaved by potassium benzoate in aqueous medium to N-methyl-N-benzoylacetoacetamide (**39**), eq. (5.20).[9]

Interesting reactions were subsequently reported,[77-80] and mechanistic considerations led to the development of a useful synthesis of peptides.[81,82] The complex reaction in eq. (5.21) starts with an irreversible proton abstraction, synchronized with ring cleavage, to give the benzoyl ketonimine **41**, which subsequently adds acetate to give **42**. The latter rearranges to the enol acetate **43** and ultimately into the isomeric imide **44**. Intermediates of types **41** and **43** have been isolated.[82,83] It was discovered that the intermediate enol acetate **43** is an excellent acylation agent. A suitable nucleophile, e.g., an amino acid with a protected carboxy group becomes acetylated by intercepting **43** before it rearranges to the stable amide **44**, eq. (5.22). Thus, the isoxazolium salt activates the carboxyl group to enter into reaction with a free amino group, as exemplified by the synthesis of carbobenzoxyglycyl-D,L-phenylalanylglycine ethyl ester (**47**) from carbobenzoxyglycyl-D,L-phenylalanine (**45**), glycine ethyl ester and isoxazolium-3′-sulfonate zwitterion (**46**), which is Woodward's Reagent K, as peptide forming reagent,[82] eq. (5.23). Nitromethane and acetonitrile are most often used as solvents, and the reaction proceeds with negligible racemization at 0°C to ambient temperature. A drawback is the high cost of the reagents. The procedure is frequently used in peptide synthesis.[84]

5.4 KETONE ANNELATION. REACTIONS VIA α-DEPROTONATION OF ISOXAZOLES

The masked enaminone functionality of isoxazoles has been used in a variant of Robinson's annelation reaction for constructing six-membered rings. Chloromethylation[85,86] of 3,5-dimethylisoxazole at C^4 to 3,5-dimethyl-4-chloromethylisoxazole (**48**) alkylation with the pyrrolidine enamine of cyclohexanone and catalytic reduction over Pd/C give the enaminoketone **50**, which is cyclized to a mixture of $\Delta^{1,9}$ and $\Delta^{9,10}$ octalones (**51**), eq. (5.24).[87] The cleavage of the isoxazole ring can also be effected by treatment with triethyloxonium tetrafluoroborate followed by refluxing with a base, but the yield is poorer, eq. (5.25). Mechanistic studies reveal that the C^3-methyl is incorporated predominantly into **51**, which indicates that complete hydrolysis, e.g., to the symmetrical compound **52** does not occur but that deacylation ($C^{5'}H_3CO$) occurs to a great extent before the cyclization.[88]

$$(5.24)$$

$$(5.25)$$

The annelation is particularly useful for the synthesis of terpenoids.[89] The tetracyclic ring system of the terpenoid **57** is constructed by alkylation of the sodium enolate of octalin-2,5-dione (**53**), delivering the C/D rings, with the isoxazole **54**, containing the masked A/B ring system, eq. (5.26). The intermediate **55** is reductively cleaved and cyclized to the tricyclic compound **56**, which is converted by reductive methylation and successive treatment with dilute acid and hot aqueous methanolic sodium hydroxide to racemic D-homotestosterone (**57**).

The sesquiterpene ($\pm$)-dehydrofukinone (**62**), eq. (5.27), is prepared by alkylation of Hagemann's ester **58** with the isoxazole **48** followed by removal of the ethoxycarbonyl group and reduction to **59**. Protection of the methylene group at C^6 as the isopropylenol ether and methylation at C^2 give **60**. The protective group is removed, and the isoxazole moiety is cleaved by N-ethylation with triethyloxonium tetrafluoroborate and subsequent treatment with base, which gave the cyclized compound **61**. Preparation of the isopropylenol ether, methylation, and hydrolysis with dilute acid afforded **62**.[90]

The reaction shown in eq. (5.28) demonstrates that selective manipulation of the isoxazole moiety can be done in the presence of very complex functionalities. The isoxazole **63**, which stores a β-ketoamide structure, is used as a building unit in the construction of ring A in the tetracyclines **66**.[91] It is added to **64** in a triethylamine-catalyzed Michael reaction. Deesterification, decarboxylation, and acid-catalyzed aromatization gives **65a**. Sodium hydride–promoted Claisen condensation followed by reductive isoxazole ring cleavage yield the tetracycline derivative **66a**. In a similar sequence the Schiff's base **67** yields **12a–d**.

Alkyl groups adjacent to the isoxazole ring are especially activated in the 5-position. This has been used in a number of cases for base-promoted ($NaNH_2$, BuLi, $LiNH_2$) introduction of alkyl or acyl groups into the side chain, eq. (5.29).[29,92-110] The α-protons become still more labile in the quaternary isoxazolium salts, and weak bases such as piperidine are sufficient as catalysts.

Curcumin has been prepared from N-methyl-3,5-dimethylisoxazolium iodide and vanillin; cf. Chapter 1, eq. (1.22).[111,112] With a sterically small base 3-methyl-5-ethylisoxazole is methylated at the 5-position to the 5-isopropyl derivative.[95] A more space-demanding base such as s-BuLi deprotonates selectively the 3-methyl group of 3-methyl-5-pentylisoxazole.[110]

α-Deprotonation of 3,5-dialkylisoxazoles and subsequent electrophilic substitution is complementary to the doubly deprotonated β-dicarbonyl procedure. β-Ketoesters are converted by N-substituted hydroxylamines to isoxazolin-5-ones, which similarly are deprotonated at the 3α-carbon atom, eq. (5.30). Finally, hydrogenation yields the 4-substituted β-ketoester.[113]

The masked β-dicarbonyl system has been used in a synthesis directed toward the fungal xanthone bikaverin, eq. (5.31). Compound **68** is obtained as the major product via a Wessely-Moser rearrangement.

A variation of the ketone annelation leads to aromatization. The isoxazolium salt **69** cyclizes with pyrrolidine to the tetrahydronaphthalene derivate **70**, whereas sodium hydroxide-catalyzed cyclization gives **71**, eq. (5.32). The latter reaction was applied to the synthesis of ferruginol (**72**), eq. (5.33).[114] Additional aromatizations are discussed in Section 5.6.

5.5 1,3-CARBONYL TRANSPOSITION

1,3-Carbonyl transposition, eq. (5.34), has been realized by using isoxazoles as intermediates.[115] It was of interest to transfer the readily available α- or β-ionones (**73, 74**) to the more precious α- and β-damascones (**75, 76**).

β-Ionone oxime (**74**) is oxidatively cyclized to the isoxazole 77, eq. (5.35). Reduction with sodium in liquid ammonia gives directly the aminoketone **80**,

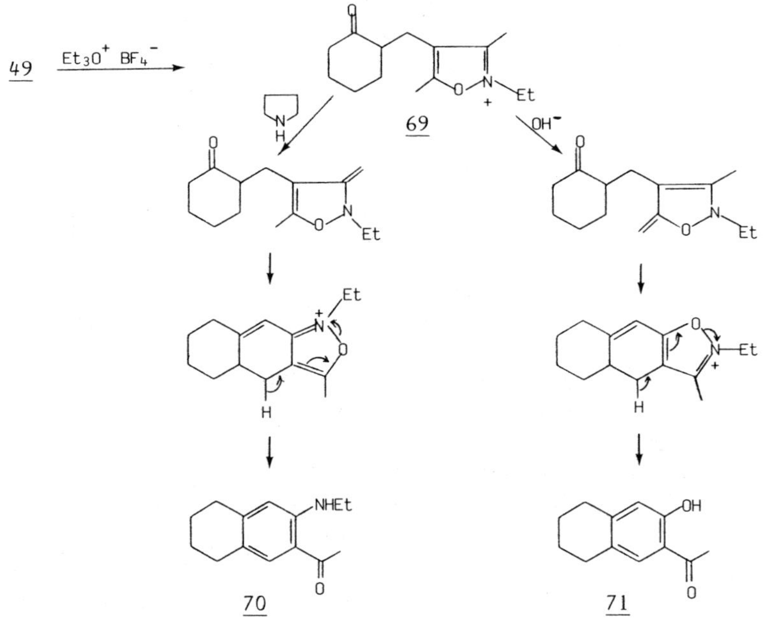

1,3-Carbonyl Transposition

(5.32)

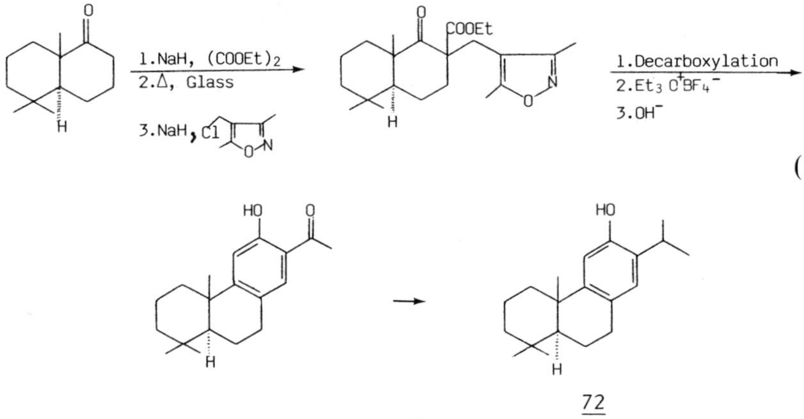

(5.33)

(5.34)

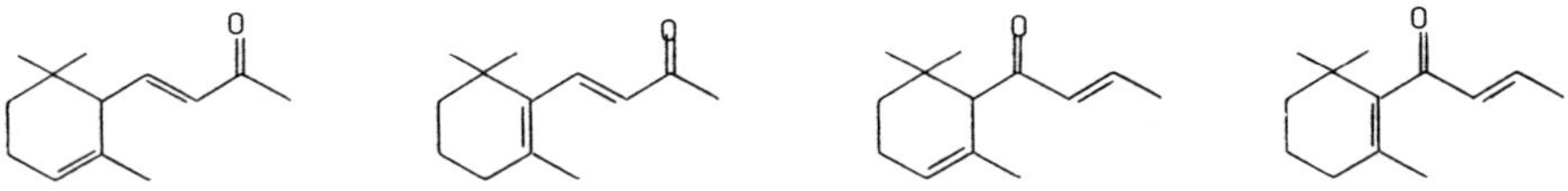

73   74   75   76

$$\text{(5.35)}$$

which readily eliminates ammonia to give β-damascone (**76**). The reduction of **77** to **80** can also be achieved by catalytic hydrogenation to the enaminoketone **78** and subsequent treatment with sodium in liquid ammonia. Acid hydrolysis of **78** gives **79**, demonstrating that the reaction shown in eq. (5.35) is also suitable for conversion of α,β-unsaturated enones to β-diketones. Similarly the simple ketone **81** is transposed to the α,β-unsaturated aldehyde **82**, eq. (5.36). More generally, the sequence shown for eqs. (5.35)–(5.37) can be used for regiospecific generation of enones (**83**) from isoxazoles. If the isomeric enone **84** is desired, the isoxazole is catalytically reduced to the enaminoketone, benzoylated, reduced with sodium borohydride, and hydrolyzed with acid.[97] The cleavages have been applied in the synthesis of dihydrojasmone (**85**),[116] gingerol (**86**),[117,118] eq. (5.38), and prostanoids.[119] The synthesis of **86**, in principle, could now be shortened by cycloaddition of **87** with 1-heptene followed by reduction, eq. (5.39); cf. Section 5.7.

$$\text{(5.36)}$$

$$\text{(5.37)}$$

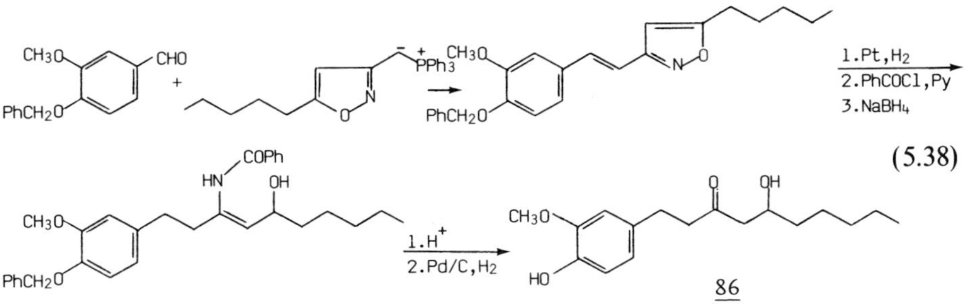

(5.38)

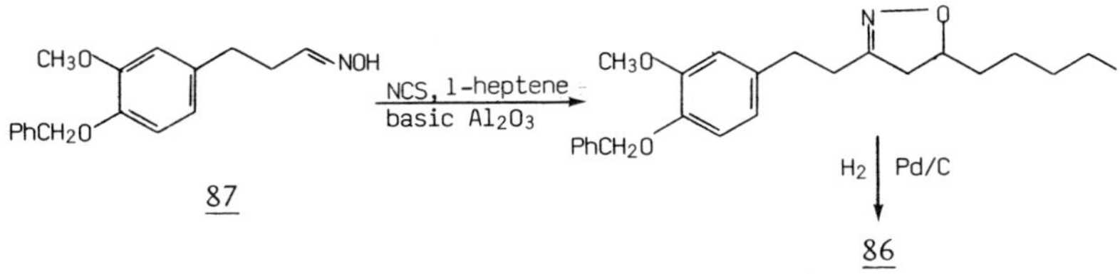

(5.39)

In conclusion, combining the C^α-alkylation, eq. (5.29), with the reductive cleavage, eq. (5.37), constitutes a useful synthesis of α,β-unsaturated ketones from isoxazoles; cf. Section 5.8.

5.6 β-POLYKETONES, PHENOLS, ANILINES

The preparation of β-polyketones is of interest in connection with the biomimetic synthesis of polyketide-based natural products. Reductive cleavage of isoxazoles leads to 1,3-diketones. Acylation of the 5-alkyl side chain (Section 5.4) followed by reductive cleavage and hydrolysis give a β-triketone, eq. (5.40).

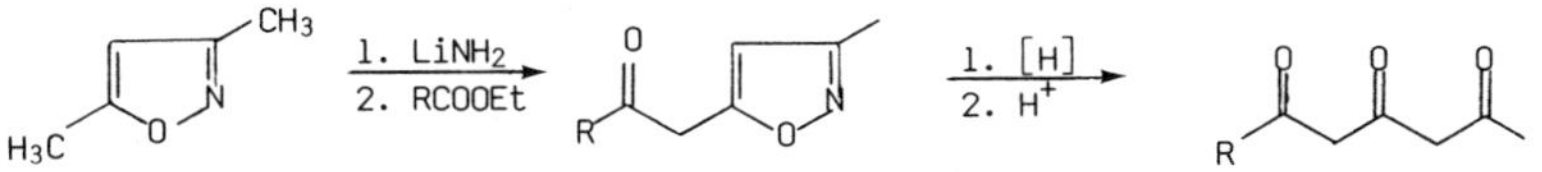

(5.40)

Linear β-tetraketones are available by reduction of methylenediisoxazoles, eqs. (5.41)–(5.43)[102,119–122] or by acylation of 3,5-dimethylisoxazole with ethyl 3,3-ethylenedioxybutanoate and subsequent reductive cleavage to give **96**, eq. (5.44).[104] The equivalent of a linear pentaketide (**97**) is prepared according to eq. (5.45).[103,104] The unmasked polyketide structures **88**, **91**, **94** have been cyclized to the phenol and aniline derivatives **89**, **90**, **92**, **93**, **95**.

The nitrile oxide route to branched C^2-acylated 1,3-dicarbonyl derivatives is a valuable complement to other acylation procedures. The 1,3-dipolar cycloaddition of nitrile oxides to conjugated enaminones, enaminoesters, and corresponding enols is remarkably regiospecific in that the carbonyl group is directed to the C^4-position, eq. (5.46).[86,123–126] Catalytic reduction gives the

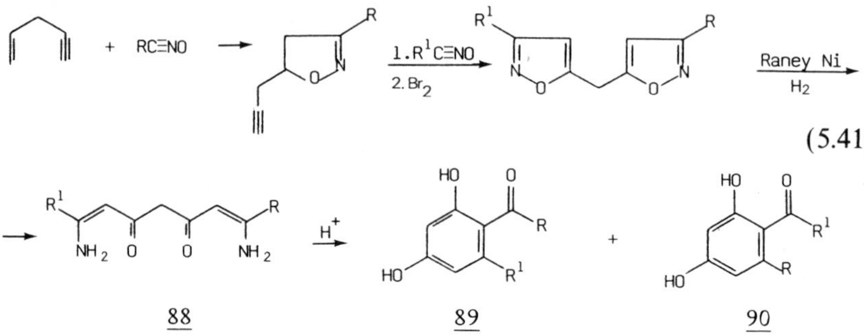

Applications in Synthesis

(5.41)[102]

88 → 89 + 90

R = R¹ = Ph, CH₃
R = Ph, R¹ = CH₃   89 : 90 = 4 : 1

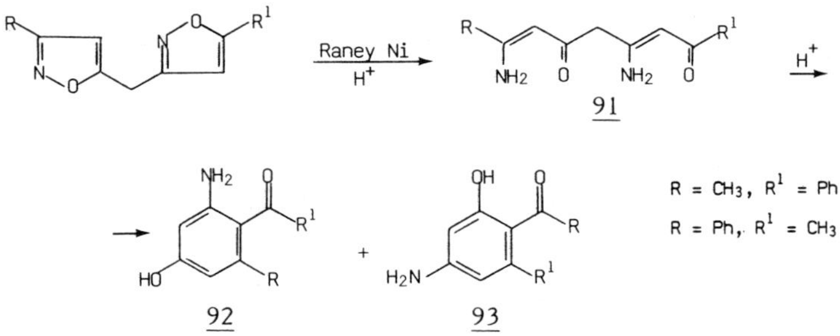

91 → 92 + 93

	92 : 93
R = CH₃, R¹ = Ph	92 : 93 = 96 : 4
R = Ph, R¹ = CH₃	92 ~100

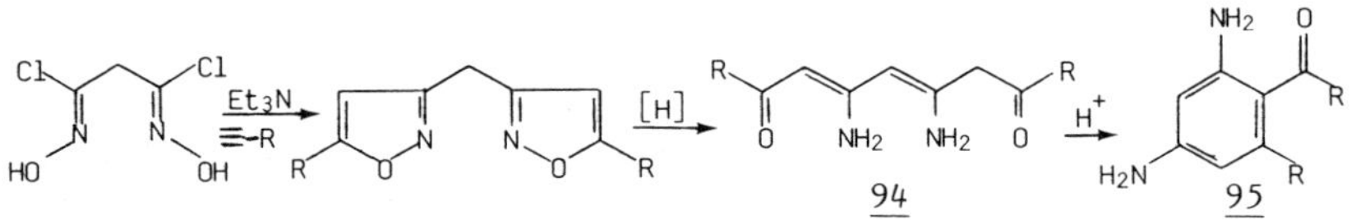

(5.42)[102]

94 → 95

(5.43)[121,122]

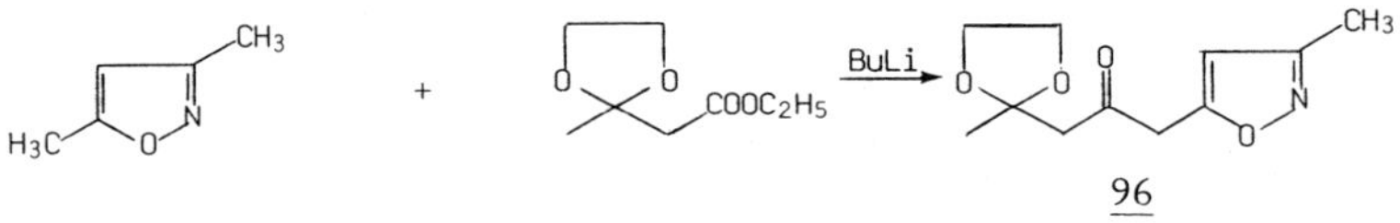

(5.44)

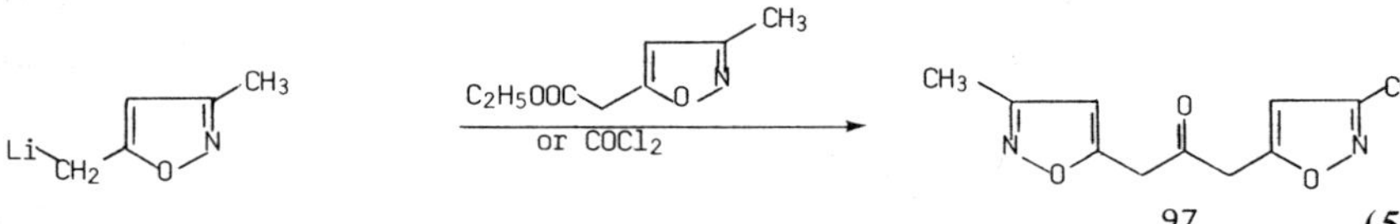

96

(5.45)

97

enaminone **98**, which can be hydrolyzed into the corresponding β-tricarbonyl derivative. The reaction shown in eq. (5.47) demonstrates the versatility of the procedure for synthesizing glutarimide antibiotic (**99**).[127]

5.7 2-ISOXAZOLINES, THE MASKED ALDOLS. SYNTHESIS OF β-HYDROXYKETONES

2-Isoxazolines are masked aldols, and this important fact has been emphasized by several research groups exploring their synthetic potentialities. The usefulness of the isoxazoline route becomes evident when the starting material and the simple experimental conditions of the 1,3-dipolar cycloaddition are considered and when the reaction is compared with another fundamental organic reaction—the Claisen reaction (Section 1.5, Scheme 1.2).[59] The starting materials involved in the isoxazoline route are olefins, aldehydes or primary nitro compounds and, in the Claisen route, carbonyl compounds. Furthermore, the regioselectivity and reactivity of the cycloaddition are predictable (Sections 1.7, 1.8). In short, the isoxazoline route has been developed into a standard procedure for forming carbon-carbon bonds, intermolecularly as well as intramolecularly. This section focuses on principles and preparation of simple representatives of β-hydroxyketones. In subsequent sections further applications are discussed, e.g., the synthesis of heterocycles, carbohydrates, alkaloids, and various natural products. The reductive cleavage,[48,128] as carried out by catalytic hydrogenation[129] or by metals (e.g., Ti^{3+} [130]), has been known for a fairly long time, but isoxazoline methodology has been fully recognized and appreciated only recently.[1,58,59,68,76,131–140,673] The synthesis of 1-phenyl-1-hydroxy-3-butanone [Table 5.3, entries 2, 3; eq. (5.48)] demonstrates some problems

Table 5.3. Synthesis of β-Hydroxyketones by Reductive Cleavage of 2-Isoxazolines

	2-Isoxazoline	Reducing agent	Product (%) yield	Conditions (solv., temp. etc.)	Ref.
1.	(CH₃OOC-substituted 2-isoxazoline)	Pd/C, H₂	β-hydroxyketone with OCH₃ (41)	EtOH/HOAc, 25°C, 3 atm.	1
2.	(Ph-substituted 2-isoxazoline)	Pd/C, H₂	β-hydroxyketone with Ph (21)[a]	EtOH/HOAc, 25°C, 1 atm. 5 days	1
3.	(Ph-substituted 2-isoxazoline)	Ti^{3+}	β-hydroxyketone with Ph (~100)[a]	MeOH, H₂O, 25°C, 2,5 days	58
4.	(acetyl-substituted 2-isoxazoline)	Ti^{3+}	diketone with OH (82)	MeOH, H₂O, 25°C, 3 days	58
5.	((CH₂)₆COOCH₃ substituted)	Ti^{3+}	CH₃OOC(H₂C)₆-β-hydroxyketone (90)	MeOH, H₂O, 25°C, 3 days	58
6.	(allyl-substituted)	Ti^{3+}	allyl-β-hydroxyketone (ca. 80)	MeOH, H₂O, 25°C, 3 days	58
7.	(AcO-substituted)	Ti^{3+}	β-hydroxyketone with OAc (72)	AcOH, H₂O, 25°C, 7,5 hrs	136
8.	(EtO, OEt, (CH₂)₅CH₃ substituted)	RaNi/H₂	CH₃(CH₂)₅ β-hydroxyketone with OEt, OEt (~100)	EtOH, 25°C, 1 atm. 3 hrs	137
9.	(Ph, acetyl substituted)	RaNi/H₂	Ph-β-hydroxyketone (40)[b]	EtOH, 25°C, 1 atm. 4 hrs	138
10.	(bicyclic R-substituted)	RaNi/H₂	cyclopentane with R and OH (97)[c]	MeOH:H₂O, 5:1, AlCl₃, 4eqn. 1 atm ½ h.	139

2-Isoxazolines, the Masked Aldols. Synthesis of β-Hydroxyketones

Table 5.3. (continued)

2-Isoxazoline	Reducing agent	Product (%) yield	Conditions (solv., temp. etc.)	Ref.
11. [isoxazoline structure]	RaNi/H$_2$	[product structure] (95)[c]	MeOH:H$_2$O, 5:1, AlCl$_3$ or H$_3$BO$_3$ added, 1 atm, 3 h, 23 °C	139, 140
12. [Ph-isoxazoline structure]	RaNi/H$_2$ or Pt/C	[product structure] (70)[c]	MeOH:H$_2$O, 5:1 AlCl$_3$ or H$_3$BO$_3$ added, 1 atm, 3 h, 23 °C	140

[a] The hydroxy compound (as in entry 3) is obtained with RaNi as the catalyst, after 3 h of hydrogenation; yield ca. 90%.
[b] The imino compound was first isolated and then hydrolyzed with AcOH to the ketone.
[c] No isomerizations occurred at the starred positions in the presence of AlCl$_3$ or H$_3$BO$_3$.

involved with the isoxazoline and Claisen routes. The starting materials for the isoxazoline route are styrene and nitroethane or propanal oxime, which regioselectively give the isoxazoline in high yield. The hydroxyketone is obtained in practically quantitative yield by titanous ion reduction at pH 2–4; but when the reduction is carried out at lower pH and over 4 days, minor amounts of 1-phenyl-3-butanone are formed, and this also occurs for catalytic hydrogenation over Pd/C (entry 2). With Raney-Ni as catalyst the isoxazoline is practically quantitatively cleaved to the hydroxyketone. The Claisen reaction starts from benzaldehyde and 2-butanone, eq. (5.48). There are two problems connected with this type of reaction that may be difficult to handle. First, benzaldehyde may attack the methylene group as well; second, the reaction may proceed to the α,β-unsaturated stage. The α-acylation and the α-cyanation procedures of butadiene are more impressive. Formally, compounds **100** and **101** can be considered as products of a presumed aldol reaction between acrolein and a methyl ketone or acetonitrile, respectively, eq. (5.49a, b), but in this case

$$\text{PhCHO} + \text{[methyl ketone]} \xrightarrow{\text{Base}} \text{[Ph-CH(OH)-CH}_2\text{-CO-]} \quad (5.48)$$

$$(5.49)$$

the Claisen route is not applicable. The compounds are obtained in satisfactory yields by the isoxazoline route,[59] eq. (5.50). The isoxazoline and Claisen routes are complementary, and the choice is a matter of synthetic planning.

$$（5.50）$$

Epimerization at the α-carbon (starred positions in entries 10–12 of Table 5.3) has been shown to occur during the catalytic hydrogenation with Raney-Ni. Adding $AlCl_3$ or boric acid to the medium, which enhances the rate of hydrolysis of the imine, prevents epimerization completely.[139,140] $BF_3 \cdot Et_2O$ has also been suggested as an additive for the Raney-Ni–catalyzed reduction of 4,5-dialkyl-substituted isoxazolines.[141]

An isoxazoline-based route to β,β'-dihydroxyketones and aminodiols, which is highly stereoselective in the condensation and reduction steps, is depicted in eq. (5.51).[142] This procedure has been applied to the synthesis of optically active gingerol (**86**), eq. (5.52). Compound **104** ($R^1 = H$, $R^2 = C_5H_{11}$) is first separated into its diastereomers and then alkylated with **105** and reduced[143] (cf. Section 5.5). Reduction of 3-unsubstituted isoxazolines (e.g., **102**) gives β-hydroxyaldehydes, a reaction exploited in the synthesis of 2-deoxyaldoses[60] (Section 5.15).

$$（5.51）$$

5.8 α,β-UNSATURATED CARBONYL COMPOUNDS, β-ACYLATED ACRYLIC ESTERS AND ACROLEIN, 2-ENE-1,4-DIONES, 1,4-DIONES, γ- AND δ-KETOESTERS, α-HYDROXYESTERS, AND KETONES

A variety of functionalized alkanes can be prepared via the isoxazole route, and many of the products are useful as building blocks in organic synthesis. The reduction of isoxazolines gives access to β-hydroxyketones, as shown in Section 5.7, and subsequent elimination of water leads to α,β-unsaturated carbonyl compounds in analogy with the general Claisen condensation scheme. Eliminating the water is best done by acetylation and heating, eq. (5.53).[57-59,137,138,144]

If X = COOR, β-acylated acrylic esters (**108**) are the products. Vinylketones (X = acyl) provide a novel entry into the difficultly accessible ene-1,4-diones.[58] Butadiene (X = vinyl) or substituted conjugated dienes give α,β,γ,δ-unsaturated ketones.[59] Compounds **109–112** are obtained according to this route.

The cycloaddition of silyl nitronates or nitrile oxides to acrolein was not successful, but the acrolein acetal gave the anticipated 5-diethoxymethyl-2-isoxazolines **113** in good yields,[137,145–147] eq. (5.53a). Acid hydrolysis of **113** to the free 5-formyl derivative proceeds in poor yield. The isoxazoline nucleus is cleaved according to eq. (5.54). The dicarbonyl fragment can be trapped by *ortho*-phenylenediamine as 2-methylquinoxaline (**114**).

$$R = H, CH_3, CH(CH_3)_2, C_4H_9, C_5H_{11}, C_6H_{13}, Ph, CH_3CO, (CH_2)_7COOCH_3 \quad (5.53a)$$

$$(5.54)$$

Compounds **113** are useful precursors for several functionalized aldehydes, eq. (5.55), such as α-hydroxyacetals **115**, β-acylated acrolein acetals **116**, β-acylated acroleins **117**, and α-acyloxyaldehydes **118**.[137] Of interest is the facile acid-catalyzed rearrangement of **115** into γ-ketoesters **119**, which seems to be a novel general procedure for this class of compounds. The aldehyde derivatives

$$(5.55)$$

115–118 are fairly unstable and difficult to prepare by other methods. Thus, the cycloaddition and subsequent reduction, eqs. (5.53), (5.55), constitute a general procedure for preparing α-hydroxycarbonyl compounds **107**, X = acyl, COOR, CHO, and CH(OR)$_2$.

In connection with 1,3-carbonyl transposition (Section 5.5) a regioselective reduction of 3,5-disubstituted isoxazoles into α,β-unsaturated ketones was described, eq. (5.37).[115] This work has been extended[97] and is depicted generally in eq. (5.56). It has been used to synthesize dihydrojasmone (**85**),[116] gingerol (**86**),[117,118] eq. (5.38), and prostanoids,[119] (e.g., **121**), eq. (5.57).

$$(5.56)$$

$$(5.57)$$

2-Isoxazolines undergo ring fission on treatment with lithium diisopropylamide (LDA) to give α,β-unsaturated oximes **122**, which on subsequent reduction with Ti^{3+} yield the corresponding α,β-unsaturated ketones **123**, eq. (5.58).[132] The oximes **122** retain the C=N configuration of the starting isoxazolines, and the enones **123** have the E-configuration. Equation (5.58) is thus an alternative to eq. (5.53).

Secondary nitro compounds are O-silylated by chlorotrimethylsilane and triethylamine in acetonitrile, and the silyl nitronates **124** formed cycloadd to acrylonitrile to give the isoxazolidines **125**, eq. (5.59). Subsequent heating with solid potassium fluoride causes fragmentation and formation of β,β-disubstituted α,β-unsaturated aldehydes **126**.[148] The reaction represents a convenient route to senecioaldehyde (**126**) for $R^1 = R^2 = CH_3$. Similarly, nitrocyclopentane yields the unsaturated aldehyde **127**. An isomeric mixture of **128** and **129** is obtained from methyl 4-nitropentanoate in a total yield of ca. 20%.

The homologation of aldehydes by two carbon atoms into α,β-unsaturated aldehydes **132** is accomplished by cycloaddition of its N-methylnitrone **130** to vinyltrimethylsilane and subsequent treatment of the isoxazolidine **131** with 50% aqueous HF, eq. (5.60).[149] The overall yield is ca. 50%.

Substituted isoxazolidines **133** are cleaved to α,β-unsaturated ketones **134** in good yields on heating with trimethyl phosphate in a Hofmann-like elimination reaction, eq. (5.61).[150]

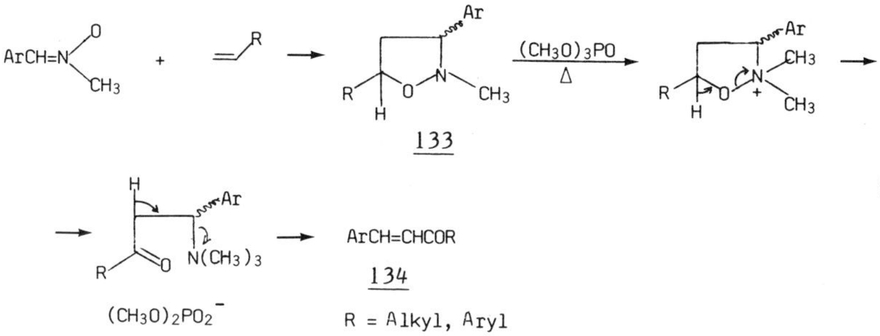

(5.61)

δ-Functionalized α,β-unsaturated aldehydes **135** are produced by adding nitrile oxides to 1-phenylthio-1,3-butadiene and subsequent reduction, eq. (5.62).[151] Most likely 1-acetoxy- and 1-dialkylamino-1,3-butadienes could serve the same purpose and act as precursors for 1,5-diones by reduction and hydrolysis.

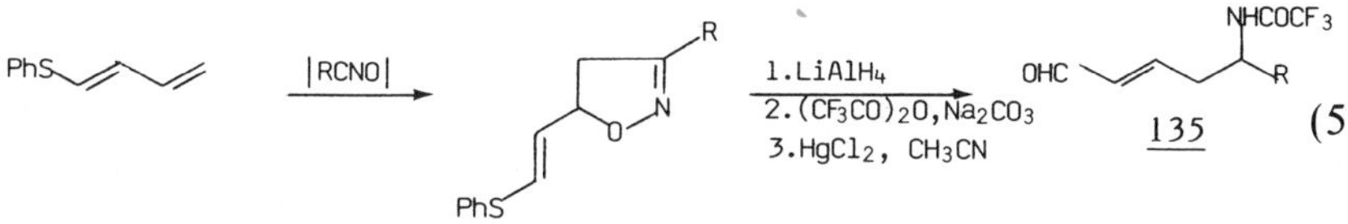

(5.62)

1,4-Diones are important starting materials for cyclopentanoids, and useful syntheses are always in demand.[152] As demonstrated earlier [cf. eqs. (5.52),(5.53),(5.55)] the reduction of 5-acyl-2-isoxazolines **136** by Ti^{3+} or Raney-Ni, H_2, leads to 2-hydroxy-1,4-diones **137**. The reaction shown in eq. (5.63) is a facile, high-yielding procedure. It was observed that zinc powder in the presence of Ti^{3+} ions also caused rupture of the C^5—O bond to give the 1,4-diones **137**.[58,145] This elimination of the 2-hydroxy group was utilized in a short synthesis of dihydrojasmone (**85**) from methyl vinylketone and heptanaloxime or, alternatively, from nitroheptane,[145] eq. (5.63). It was also

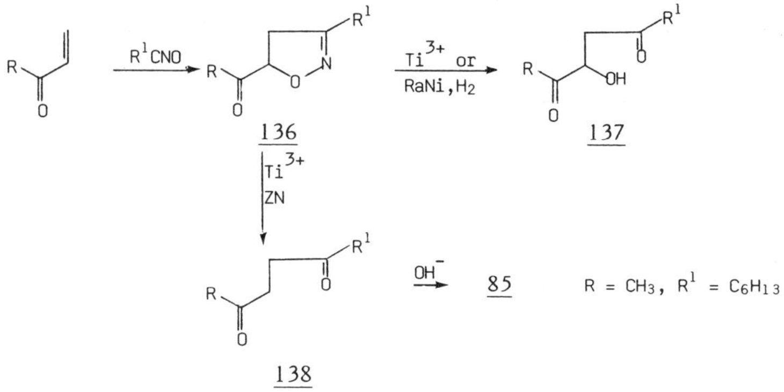

(5.63)

observed that Ti^{3+} reduction of the isoxazoline acetals **113** gave a low yield (ca. 30%) of 4-ketoaldehydes **139**, eq. (5.64),[145] whereas catalytic hydrogenation over Raney-Ni gave the expected 2-hydroxy derivatives **115**.[137,147]

$$\mathbf{113} \xrightarrow[H^+]{Ti^{3+}} \mathbf{139} \text{ (OHC-CH}_2\text{-CO-R)}$$
$$\downarrow \text{Raney Ni, } H_2, \text{ EtOH, } H_2O$$
$$\mathbf{115} \tag{5.64}$$

The 1,4-dione **138** (R = CH$_3$, R^1 = C$_6$H$_{13}$) has been prepared from the isoxazoline **140** and cyclized to **85**,[119] eq. (5.65).

$$\mathbf{140} \xrightarrow[\text{2. Na, NH}_3, t\text{-BuOH}]{\text{1. MnO}_2} \quad \text{(product)} \tag{5.65}$$
$$\downarrow \text{1. [H], 2. H}^+$$
$$\mathbf{85} \xleftarrow{OH^-} \mathbf{138}$$

5-Keto-3-hydroxyesters **141** are available from the reaction shown in eq. (5.66).[137] Oxidation of the isoxazoline nucleus to isoxazole and subsequent catalytic hydrogenation should, presumably, give the corresponding 3,5-diketoester.

$$\text{COOCH}_3\text{-CH}_2\text{-CH=CH}_2 \xrightarrow{|RCNO|} \text{(isoxazoline)} \xrightarrow{Ti^{3+}} \mathbf{141} \tag{5.66}$$

Silylation of primary nitro compounds normally stops at the silyl nitronate stage. However, excess of the reactive trimethylsilyl triflate (**142**) gives further silylation with migration of the trimethylsilyloxy group, leading to trimethyl silylated α-hydroxyaldoximes **143**, eq. (5.67).[153] α-Oxygenation of aldehydes and of some ketones is accomplished by acylation and rearrangement of their nitrones to α-acylimines **144**, which are hydrolyzed to aldehydes, eq. (5.68).[154]

$$RCH_2CH_2NO_2 \xrightarrow[Et_3N]{142} RCH_2\text{-CH=N(O)-OSi(CH}_3)_3 \xrightarrow[Et_3N]{142}$$
$$\rightarrow RCH=N(OSi(CH_3)_3)_2 \rightarrow RCHCH=NOSi(CH_3)_3$$
$$\qquad\qquad\qquad\qquad\qquad\qquad | \text{OSi(CH}_3)_3$$
$$\mathbf{143} \tag{5.67}$$

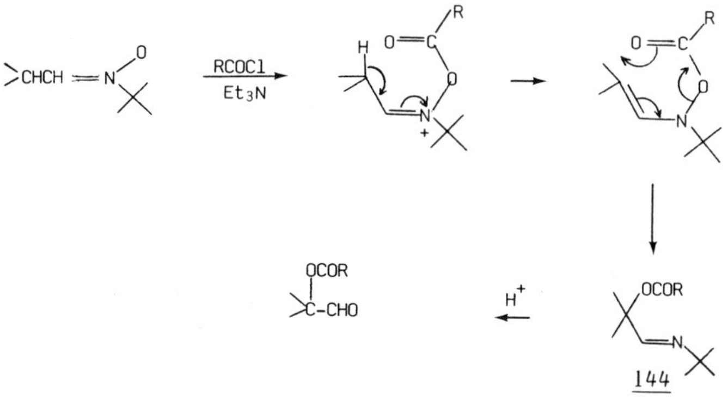

(5.68)

5.9 FURANS, 3-(2H)-FURANONES, AND PYRONES

Nitrile oxides cyloadd to α- and β-hydroxysubstituted acetylenes to give 5-hydroxyalkylisoxazoles **145**, which on catalytic hydrogenation spontaneously cyclize to 3-(2H)-furanones **146** and 2,3-dihydro-γ-pyrones **147**, respectively, eq. (5.69).[120,155,156] The 3-(2H)-furanone nucleus occurs in several naturally occurring compounds, such as the antitumoral jatrophone[157] and related eremantholides,[158] geiparvarin[159] and its dihydro derivatives,[160,161] bullatenone[162] (**146**, R = Ph, $R^1 = R^2 = CH_3$), and several simple alkyl-substituted 3-furanones,[163] giving flavor to berries, soy sauce, bread, onions, and coffee.

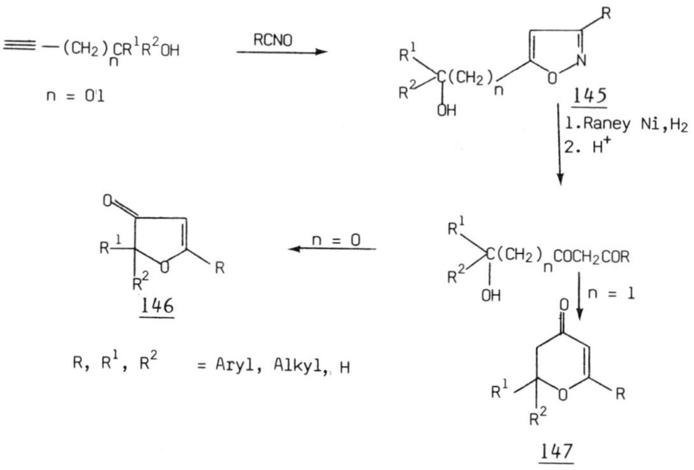

(5.69)

R, R^1, R^2 = Aryl, Alkyl, H

Equation (5.70) represents another simple route to 3-(2H)-furanones **146**. It was discovered that 2-hydroxy-1,4-diones, which are readily prepared from vinylketones and nitrile oxides or silylnitronates, rearrange to 3-(2H)-furanones when heated in acetic acid in the presence of sodium acetate.[58,59,138,145] Two related reactions start from vinyl ethers,[164] eq. (5.71a), and from 2,4-dioxoalkanoates,[165a] eq. (5.71b). In the first reaction the vinyl ether is lithiated to **147** and treated with a carbonyl compound to give the alcohol **148**. Dipolar cycloaddition of **148** with a nitrile oxide followed by reduction and cyclization give the 3-(2H)-furanone **146**. In the procedure shown in eq. (5.71b) the 2,4-dioxoalkanoate is regioselectively treated with hydroxylamine to give the isoxazole **149**, which is alkylated, reduced, and cyclized to **146**.

The 3-(2H)-furanone portion of geiparvarin (**153**) is prepared by cycloaddition of the acetylene **150** to the nitro compound **151**. Coupling of the isoxazole derivative with umbelliferone (**152**), reduction with molybdenum hexacarbonyl, followed by acid-catalyzed cyclization provide the natural product **153**,[165b] eq. (5.72).

When the 2-hydroxy-1,4-dione **154** is heated in toluene with p-TsOH, neither the anticipated furanone nor the enedione are formed. Instead the 2,5-disubstituted furan **155**, formed by migration of the double bond, is isolated in good yield.[58] In connection with derivatization of butadienes a novel furan synthesis was discovered.[59] Butadiene cycloadds to two moles of nitrile oxide. Reduction of the bisadduct **156** gives the 3,4-dihydroxy-1,6-dione **157**, which cyclizes to the furan **158** in an acid-catalyzed reaction, eq. (5.73). The cycloaddition can be performed stepwise so that different R groups are introduced.

(5.73)

Ethyl formylacetate can be oximated and chlorinated to hydroximoyl chloride **159**. Cycloaddition of the derived nitrile oxide to olefins and subsequent reduction of the isoxazoline **160** give the β-ketoester **161**, which cyclizes to ethyl 2-furylacetate (**162**)[166] when treated with acid. The α-pyrone derivative **163** was not observed.

5.10 PYRROLE AND INDOLE DERIVATIVES

Nitrile oxides, generated in situ, add to allyl acetamide, giving 5-(N-aceto-aminomethyl)-2-isoxazolines **164**. Reduction by Ti^{3+} and cyclization of the N-acetylaminoalcohol by heating with sodium acetate in acetic acid give pyrroles, eq. (5.74).[145] The reaction seems to be general, and by using substituted allylic amines, one should be able to prepare 3,4- or 5-substituted pyrroles.

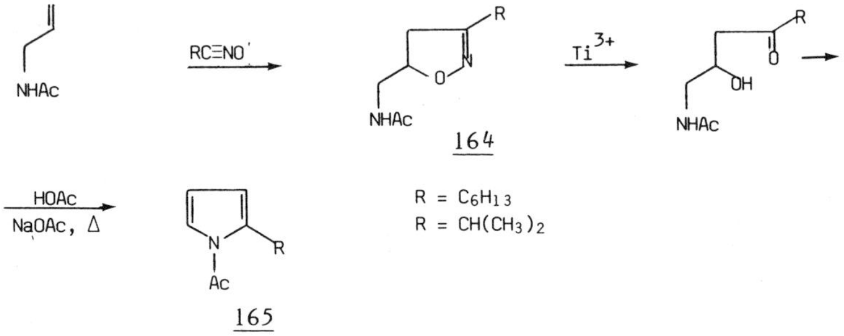

(5.74)

$R = C_6H_{13}$
$R = CH(CH_3)_2$

Reductive ring opening of isoxazoles and subsequent intramolecular reaction of the generated enamino function with suitably located functional groups in the side chain is a general principle for the synthesis of new heterocycles, which have been applied to the preparation of pyrroles, eqs. (5.75),(5.76).

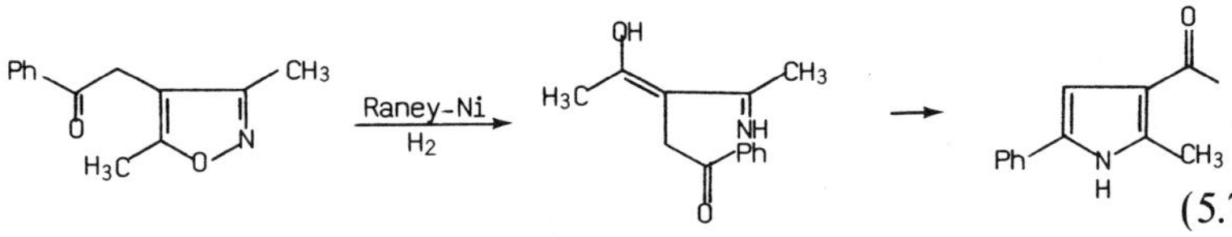

(5.75)[167]

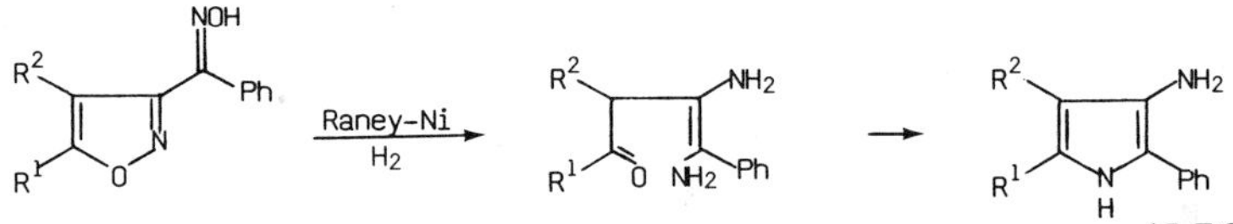

(5.76)[167,168]

The reaction is of special interest in conjunction with synthetic approaches to the corrin skeleton.[169–175] It was visualized that all of the essential features of the A/D-*seco*-corrin molecule **167** could, in principle, be assembled in the triisoxazole **166** by 1,3-dipolar coupling of the four corner stones A–D, eq. (5.77). The novel strategy was first investigated in model studies,[170–173] and then additional reactions were performed with the naturally occurring building blocks.[174,175]

The semicorrin **168** was prepared by cycloaddition of the nitro compound **170** to the acetylene **169**. Reduction of the isoxazole **171** gave the enaminoketone **172**, which spontaneously lactamized. Compound **168** was formed by treatment with ammonia and subsequently by potassium *t*-butoxide, eq. (5.78). Two

(5.77)

(5.78)

molecules of **168** have previously been converted into corrin derivatives.[176] The stepwise synthesis of a *seco*-corrin analogue **173** was carried out according to eq. (5.79), demonstrating the power of the isoxazole approach.[175]

*The formulae in eq. (5.79) are reprinted with permission from Pergamon Books, Ltd.

4-Alkylideneisoxazol-5-ones rearrange on pyrolysis into pyrroles via the azene intermediate **176**, eq. (5.80). At the high temperature the dimethylpyrroles **174** and **175** interconvert. The cyclopentylidene derivative **177** is pyrolyzed to **178**.[177]

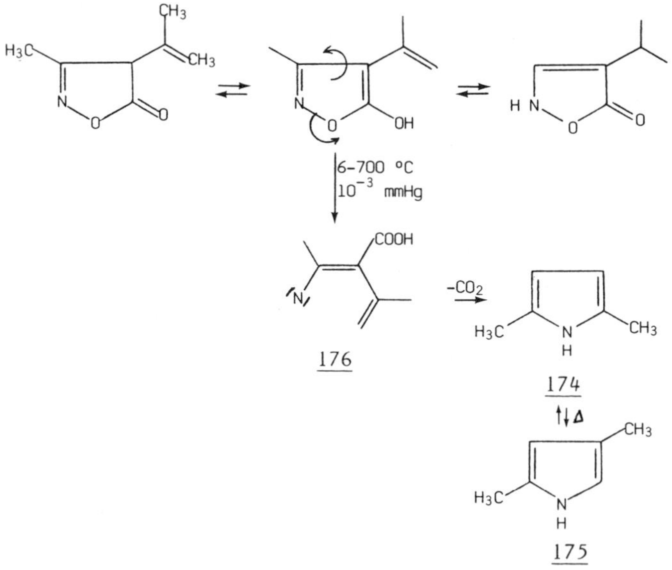

(5.80)

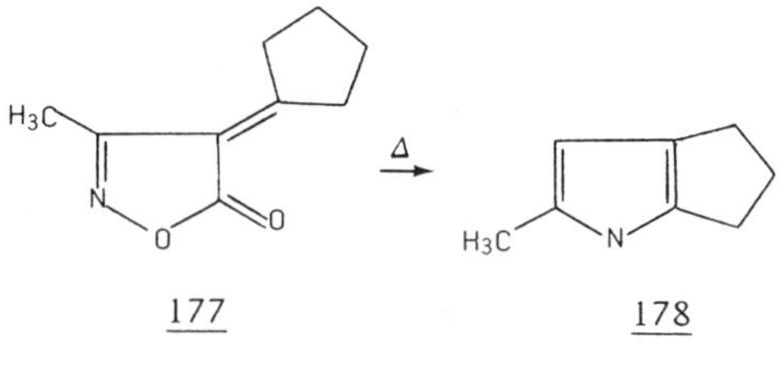

(5.81)

3-Acyl-2,4-pyrrolidones **180** (3-acyltetramic acids) are obtained by cleavage of isoxazolium salts **179** with weak bases, eq. (5.82).[178]

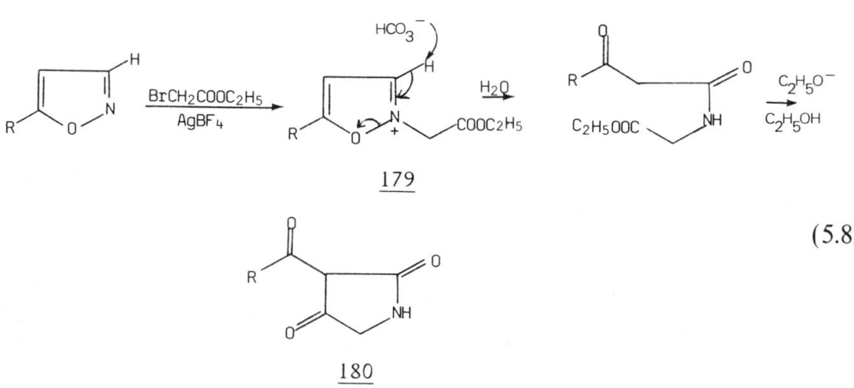

(5.82)

The addition of silyl nitronates **181** to methyl acrylate gives the *N*-silyloxy-isoxazolidine **182**, which on catalytic reduction and cyclization is converted to the pyrrolidone **183**, eq. (5.83).[57,148] The addition of acrylonitrile to **181** takes a different course, eq. (5.84).[57] The fragmentation of **184** into the 5-silyloxyxazoline **185** is the major reaction (5.84 route *a*), but a considerable amount of **186** is also formed according to 5.84 route *b*.

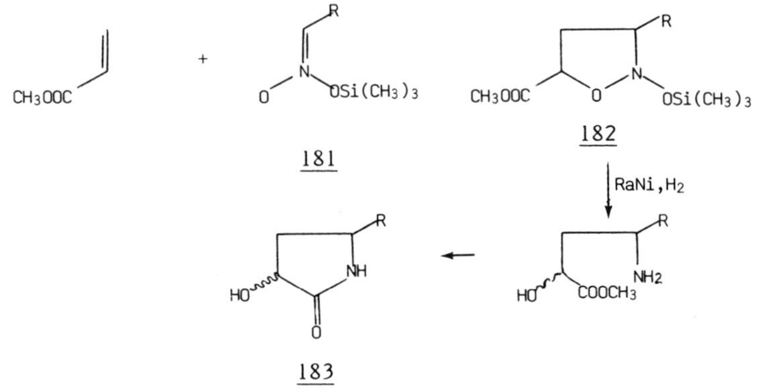

(5.83)

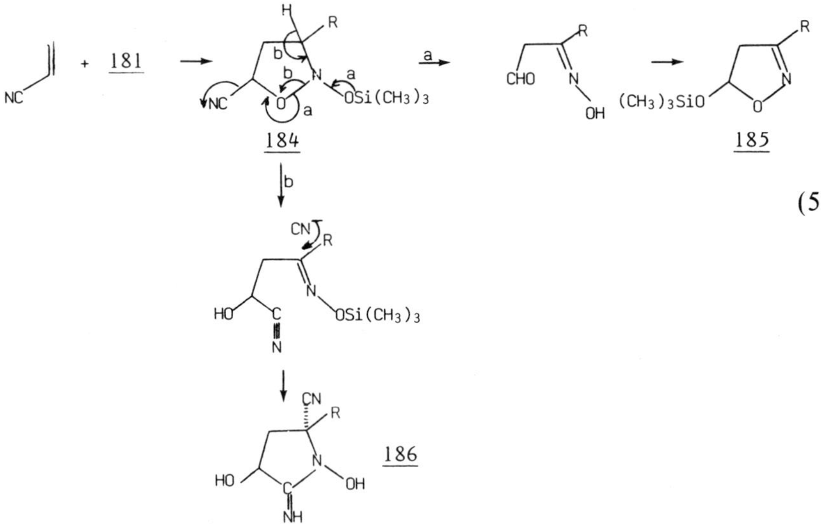

(5.84)

4-Isoxazolines, which are obtained by cycloaddition of nitrones to alkynes are characteristically unstable and undergo thermal rearrangement.[179-188] The nitrone **187** gives, e.g., with methyl propiolate the 4-isoxazoline **188**, which on heating is converted to a mixture of pyrroles **189**, **190**.[181,182] The former is derived via a hetero-Cope rearrangement (5.85 route *b*), and the latter pyrrole via the 2-acyl-aziridine route (5.85 route *a*). In a similar fashion the 4-isoxazolines in reactions shown in eqs. (5.86),(5.87) are converted to pyrrol derivatives **191**[184] and **192**.[187] The reaction has been extended to a number of phenyl-substituted pyrroles by cycloaddition of nitrones to phenylacetylene.[188]

The cycloaddition of nitrones with allenes leads to 3-pyrrolidinones,[189-191] eq. (5.88). For allene ($R^3 = R^4 = H$) and nitrone ($R^1 = R^2 = Ph$) two regiochemical modes of addition are observed leading to **193** or to a mixture of **194**, formed as shown in eq. (5.88), and **195**, formed by ring closure involving the N-phenyl group.[191]

N-Aryl-substituted nitrones are acylated by ketenes or keteneimines and rearrange in a hetero-Cope-type reaction to indole derivatives[192-200] **196**. The reaction with dichloroketene ($R^1 = R^2 = Cl$) gives, after hydrolysis, isatins (**196**: $R^1, R^2 = O$) in good yields,[195] eq. (5.88a).

Certain 4-phenyl-substituted isoxazoles rearrange into indoles when irradiated or heated,[201] eq. (5.89). Indoles are effectively synthesized from α- or β-allylpyrroles and nitrile oxides. Hydrogenation of the isoxazoline and cyclization of the intermediate β-hydroxyketone with zinc or magnesium triflate give 4- or 7-substituted indoles, eq. (5.90).[202] The reaction has been used in the synthesis of Lyngbyatoxin A. A number of condensed systems, such as benzothiophenes, benzofurans, and naphthalines, etc. could conceivably be synthesized by this procedure.

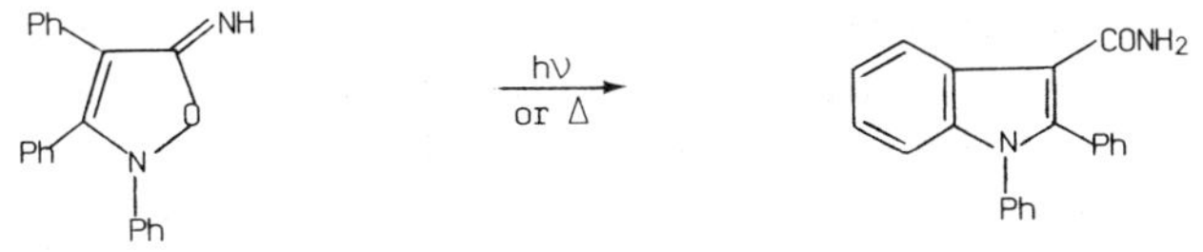

(5.89)

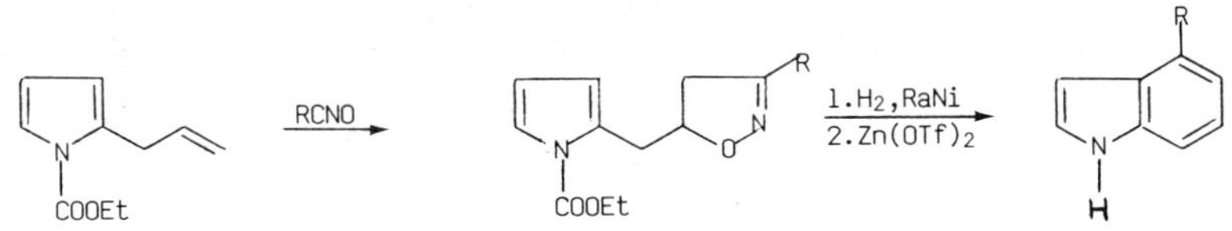

(5.90)

5.11 AZIRIDINES AND 1-AZIRINES

Reduction of 2-isoxazolines with lithium aluminum hydride is a convenient route to 1,3-aminoalcohols, but, depending on substituents and reaction conditions, aziridines can be produced and, occasionally, constitute the major product. Lithium aluminum hydride reduction of 3,5-diphenyl-2-isoxazoline (**197**) in ether gives 1,3-diphenyl-3-aminopropanol (**198**) in 62% yield,[203] but when tetrahydrofuran is used as the solvent, **198** (26%) is accompanied by cis-2-phenyl-3-benzylaziridine (**199**) in 31% yield,[204] eq. (5.91). Similarly **200** gives a mixture of cis-aziridine **201** and aminoalcohol **202**, eq. (5.92). Since both

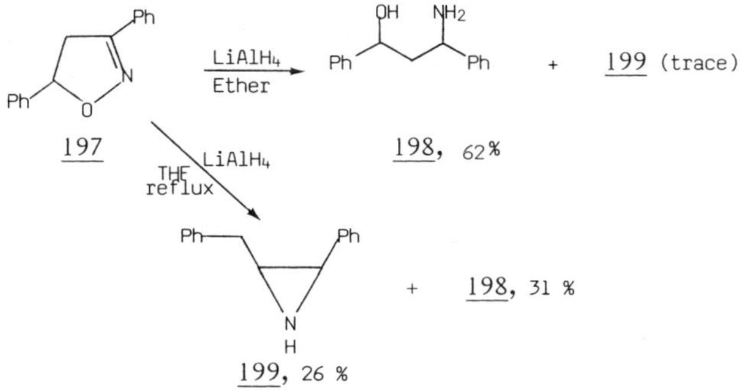

(5.91)

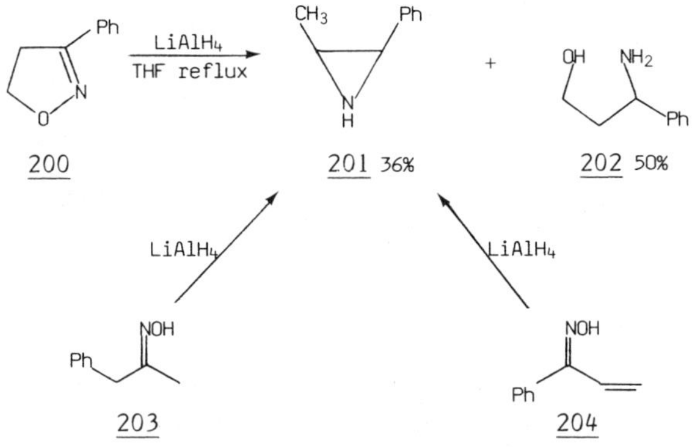

(5.92)

methylbenzylketoxime (**203**) and phenylvinylketoxime (**204**) give *cis*-aziridine **201**, it is likely that the aziridine formation from 2-isoxazoles starts with a base-catalyzed hydrogen abstraction at C^4 followed by ring opening. Hydride addition and recyclization complete the reaction shown in eq. (5.93). This mechanism is further substantiated by the observation that anion stabilizing substituents at C^4 (e.g., R^2 = phenyl) favor the formation of arizidines. When R^2 = methyl, only minute amounts of the aziridine are formed. The formation of *cis*-aziridine via the azirine intermediate **205** is guided by steric factors.

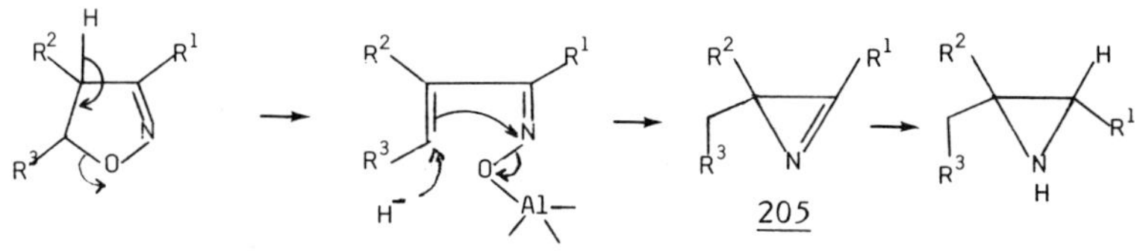

(5.93)

Isoxazolin-5-ones rearrange to aziridines on reduction with lithium aluminum hydride,[205] eq. (5.94).

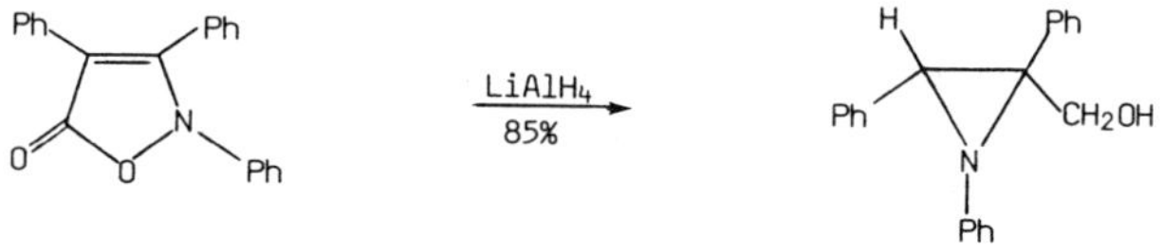

(5.94)

2-Acylaziridines are suggested as intermediates in the rearrangement of 4-isoxazolines to pyrroles (Section 5.10). They have been isolated in good yields in several cases,[180,206-212] eqs. (5.95),(5.96).

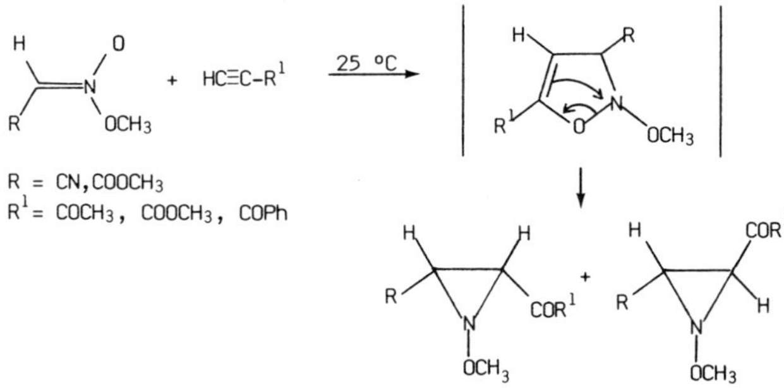

(5.95)[207]

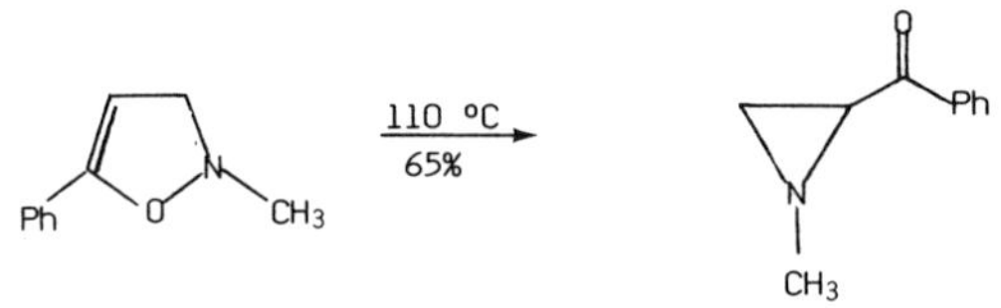

(5.96)[208]

1-Azirines **206** having various substituents at C^1 and C^3 are obtained by thermolysis or photolysis of isoxazolines in moderate to high yield,[211,213-228]

$$\text{(5.97)}$$

X = OR, SR, NHR, Ar
R = Aryl, Alkyl

eq. (5.97). They can be reduced to the corresponding *cis*-aziridines with sodium borohydride.[220] The reactions of lithio compounds generated in situ from 2-[(trimethylsilyl)methyl]pyridine (**207**), *N,N*-dimethyl(trimethylsilyl)-acetamide (**208**), ethyl(trimethylsilyl)acetate (**209**), and lithium diethylamide in tetrahydrofuran with *N*-alkylnitrones give aziridines **210**, eq. (5.98). The *N*-arylnitrones give *E*-alkenes in a Peterson-type elimination reaction, eq. (5.99).[229]

$$\text{(5.98)}$$

$$\text{(5.99)}$$

5.12 PYRAZOLES, IMIDAZOLES, OXAZOLES, AND ISOTHIAZOLES

The reaction of hydrazines with 1,3-difunctional compounds is a standard method for the synthesis of pyrazole derivatives. Thus, reductive cleavage of isoxazoles followed by treatment with hydrazines constitutes a general pyrazole synthesis from olefinic or acetylenic compounds and nitrile oxide precursors,[230-233] eq. (5.100). There are several reports on the conversion of isoxazoles into pyrazoles[226,234-249] by direct nucleophilic attack of hydrazines at the C^3 and C^5 carbon atoms of the ring, eqs. (5.101), (5.102), or by a preceding conversion of the C^4 acyl group to the corresponding hydrazone, eq. (5.104).

Ring cleavage to cyanoketones followed by cyclization with phenylhydrazine gives 3-aminopyrazoles, eq. (5.103). 5-Trimethylsilyloxy-2-isoxazolines **211**, prepared from acrylonitrile and silyl nitronates, react in an acid-catalyzed reaction with hydrazines to form pyrazoles **212**, **213** in high yields,[250] eq. (5.105). It is to be expected that addition of nitrile oxides to silyl- or alkylvinyl ethers and subsequent treatment of the 5-trimethylsilyloxy- or 5-alkoxyisoxazolines with hydrazines, should give general access to a number of pyrazoles.

The isoxazolium salt **214** gives with phenyl hydrazine the pyrazoles **215**–**217**, depending on the site of nucleophilic attack and reaction conditions, eq. (5.106).[251,252] Imidoyl-substituted oxosulfonium ylides react with nitrile oxides to give pyrazole 2-oxides.[253]

$$(5.106)$$

A few examples of the synthesis of imidazoles from nitrones are given in Section 3.2.E. Thermolysis of 5-amino-3,4-dialkylisoxazoles **218** affords 4,5-dialkylimidazolones **219**, eq. (5.107).[254] 5-Imino-2,3-diphenyl-4-aryl (or alkyl)-isoxazolines **220** similarly rearrange photolytically to the imidazolinones **221**, eq. (5.108).[255,256]

$$(5.107)$$

$$(5.108)$$

A general synthesis of 4(5)-acylimidazoles **224** has been accomplished from 4-acylaminoisoxazoles **222**, which are available by nitration, reduction, and acylation of isoxazoles, eq. (5.109).[257] Treatment of the enamine **223** with an excess of a primary amine gives selectively the N^1-substituted compound **225**. N^3-substitution can be accomplished by N-alkylation of the intermediate 4-aminoisoxazole before acylation.

Isoxazoles are transformed into oxazoles by thermolysis or by irradiation with UV light in the 3000-Å region.[213,215,216,219,220,223,224,227,258–268] It has been demonstrated that the reaction proceeds via azirines, which often have

$$(5.109)$$

been isolated; cf. Section 5.11. The photolytic reaction, initiated by homolytic N-O-fragmentation, is quite general and appears to be reversible, eq. (5.110). It can also be formulated as an intramolecular (photo) Beckmann rearrangement; the true nature of the reaction is not completely known. The reaction can be used to prepare condensed isoxazoles,[269-276] such as benzisoxazolin-3-one **226**, which gives benzoxazolin-2-one **227**, eq. (5.111).[269]

$$R^1, R^2, R^3 = \text{Alkyl, Aryl, NH}_2, \text{OH}$$

$$(5.110)$$

$$(5.111)$$

226   **227**

The 1,3-dipolar addition of nitrones with ketenes leads occasionally to oxazolidones, depending on the regioselectivity of addition. N-Alkylnitrones chiefly give isoxazolidones **228** and N-arylnitrones give oxazolidones **230** via the dioxazoles **229**, eq. (5.112).[193,277-281]

$$(5.112)$$

The cycloadduct **231** obtained from *t*-butylnitrone and dimethyl acetylenedicarboxylate rearranges at 80°C to 4-oxazoline **233**,[180] presumably via the 2-acylaziridine **232**, eq. (5.113).

$$\text{231} \longrightarrow [\text{232}] \longrightarrow \text{233} \tag{5.113}$$

Acyl-*N*-benzylnitrones undergo base-catalyzed intramolecular cyclization to oxazoles.[282] As masked enaminones, isoxazoles serve as starting material for the synthesis of isothiazoles,[283,284] eq. (5.114).

$$\tag{5.114}$$

$R^1 = R^3 = $ Alkyl, Aryl
$R^2 = $ H, Acyl, CN

5.13 SYNTHESIS OF HETEROCYCLES BY RING TRANSFORMATION INVOLVING THE SIDE CHAIN OF ISOXAZOLES

There are many examples of heterocyclic ring transformations involving the side chain denoted in eq. (5.115)[265,266,285,286]; however, certain requirements have to be met: (a) the rearrangements are limited to heterocycles with D=O and S; (b) Z must be a good nucleophile, such as O, S, N, or C; $x=y$ can be N=C, C=N, C=C, N=N. The reaction is reversible only when Z = O. The known ring conversions of isoxazoles, i.e., A=B is equal to C=C are systematized in eq. (5.116), demonstrating the usefulness of isoxazoles for the generation of other heterocycles. The atoms over the arrows represent x, y, and z of the side chain. Thus, NCN depicts an amidine side chain, C^3—N=C—NR, which cyclizes to the 1,2,4-triazole; CNO could depict an oxime, C^3—C=NOH, which cyclizes to the 1,2,5-oxadiazole, etc.

It has been reported that heating the oximes of 3-acylisoxazoles **234** with aqueous or alcoholic potassium hydroxide[287-289] or with copper powder[290]

$$\tag{5.115}$$

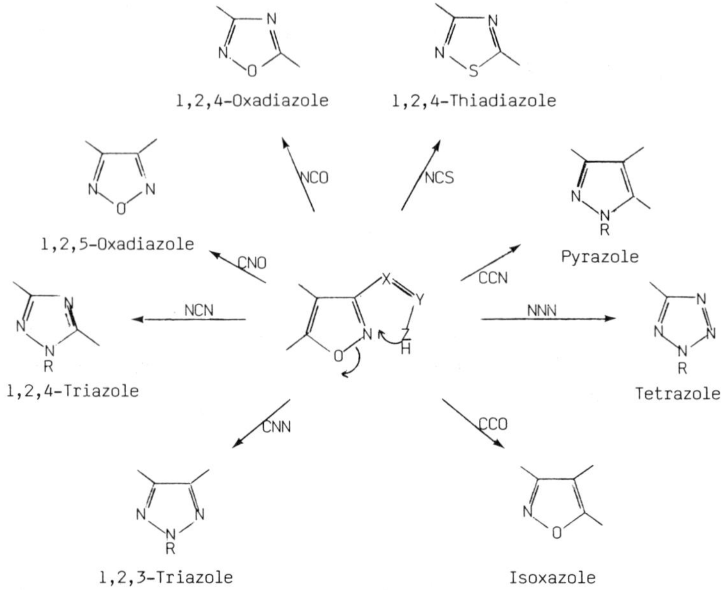

(5.116)

gives 1,2,5-oxadiazole derivatives **235**, eq. (5.117). Related ring transformations also afford the 1,2,5-oxadiazole nucleus.[291]

The benzisoxazoles **236** and **238** are reported to rearrange into the 1,2,4-oxadiazole derivatives[292,293] **237**, eq. (5.118), and pyrazole derivatives[294] **239**, respectively, eq. (5.119).

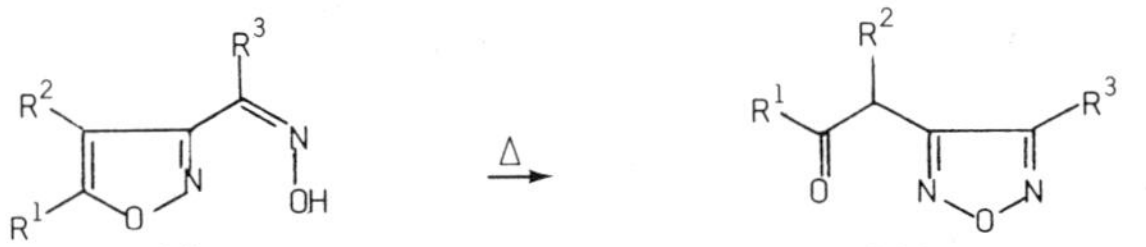

234 235

(5.117)

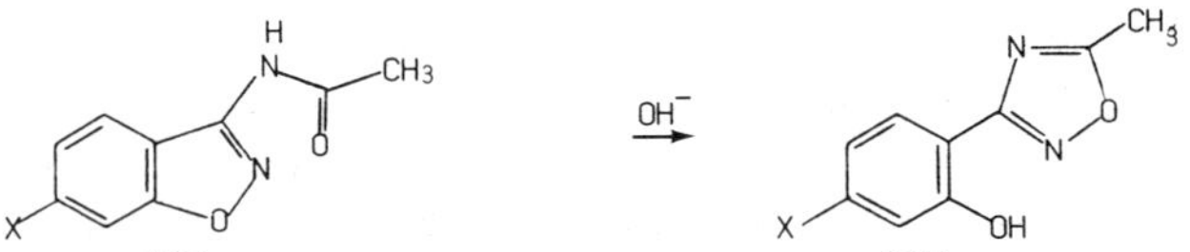

236 237

X = H, CH3CONH, Cl, NO2

(5.118)

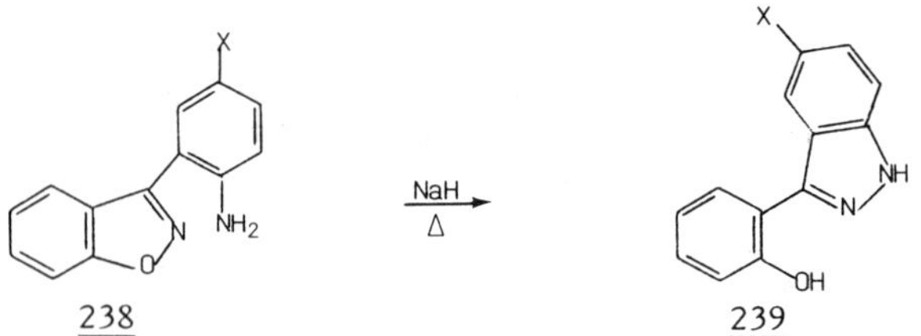

238 239

X = H, Cl

(5.119)

3-Diazoamino-5-methylisoxazole rearranges readily in the presence of base to the tetrazole **240**, eq. (5.120),[295] and 3-*N*-(arylformamidino)aminoisoxazole rearranges in refluxing ethanol with sodium hydroxide to the 1,2,4-triazole **241**,[296] eq. (5.121). The tetrazoles **242** and **243** have been obtained by treatment of 4-ethoxycarbonyl-2-methyl-3-isoxazolin-5-one[297] and 2-methyl-5-phenylisoxazolium methylsulfate,[298] respectively, with aqueous sodium azide.

A number of 1,2,3-tirazoles **245** are obtained by rearrangement of the hydrazones of 3-acylisoxazoles **244** with or without copper powder,[291,299-303] eq. (5.122). Diazotization of 4-amino-5-*t*-butylisoxazole (**246**) gives 1-hydroxy-1,2,3-triazole **247**,[304] eq. (5.123).

3-Amino-5-methylisoxazole (**248**) reacts with phenylisothiocyanate to give the 3-thioureido derivative **249**, which cyclizes on heating to the 1,2,4-thiadiazole **250**,[305] eq. (5.124).

5.14 PYRIDINES, PYRIMIDINES, PYRIDAZINES, OXAZINES, QUINOLINES, AND QUINOXALINES

Catalytic reduction of 4-(3-oxoalkyl)isoxazoles and subsequent controlled cyclization of the β-enaminone afford 3-acylpyridines, eq. (5.125).[306,307] 3-Acyl-γ-pyridones are prepared by a similar pathway,[308,309] eq. (5.126). α- And γ-pyridones are obtained from the base-catalyzed reaction of isoxazolium

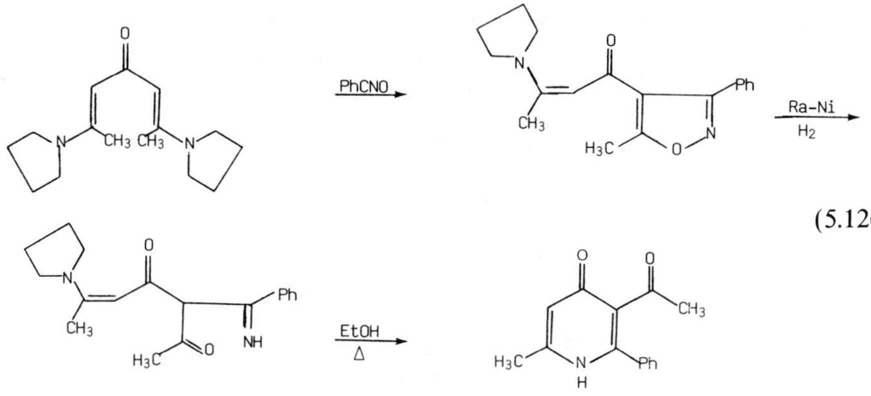

(5.126)

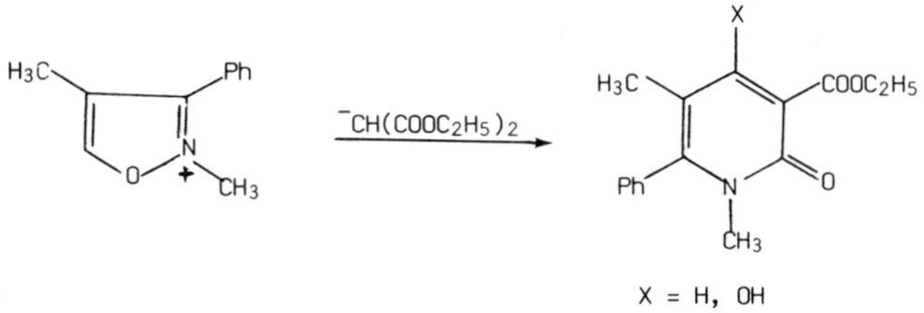

(5.127)

salts with alkyl malonate, eq. (5.127),[310] and from the reaction of diphenylcyclopropenone with isoxazoles.[311] It has been reported that hexacarbonylmolybdenum induces formation of pyridines from isoxazoles and dimethyl acetylenedicarboxylate, which adds across the C^4—C^5 double bond,[312] eq. (5.128).

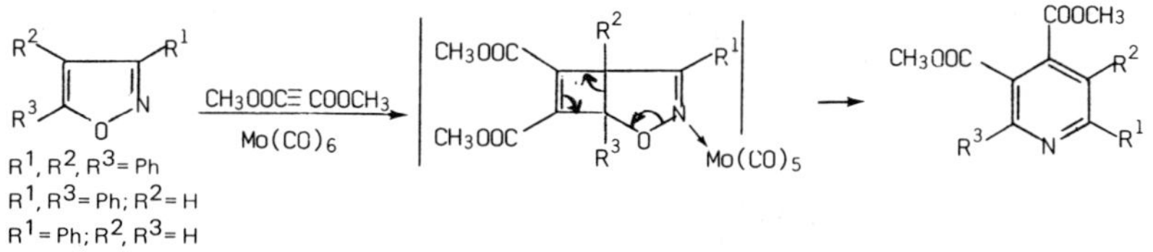

(5.128)

Unlike simple isoxazoles, 2,1-benzisoxazole behaves as a diene, giving rise to quinoline derivatives,[313–315] eq. (5.129 route a). The reaction with active methylene compounds gives quinoline-N-oxides,[316] eq. (5.129 route b). Reduc-

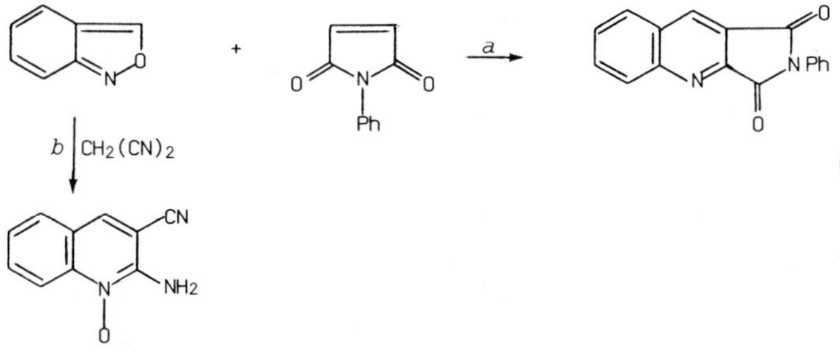

(5.129)[313]

tive cleavage of the isoxazole **253** leads to the quinoline **254**,[120] eq. (5.130). Several other examples of using isoxazoles for preparing certain pyridines and quinolines as well as various heterocycles are reviewed in Ref. 243.

$$\text{253} \xrightarrow{\text{Ra-Ni}, H_2} \text{254} \quad (5.130)$$

A few pyrimidines have been prepared by the catalytic reduction of 5-acylaminoisoxazoles followed by base-catalyzed cyclization, eq. (5.131).[317,318] The β-eneaminones obtained by reductive cleavage of isoxazoles react with N-substituted ureas to give a mixture of 2-pyrimidinones, eq. (5.132).[319]

$$\quad (5.131)$$

$$R^1, R^2, R^3 = \text{Alkyl, Aryl} \quad (5.132)$$

Catalytic hydrogenation of 5-methylisoxazole-3-carboxhydrazide (**255**) and the condensed isoxazole **257** afford directly the corresponding pyridazones **256** and tetrahydrocinnolone **258**, eqs. (5.133),(5.134).[320-322]

$$\text{255} \xrightarrow{\text{Ra-Ni}, H_2} \rightarrow \text{256} \quad (5.133)$$

$$\text{257} \xrightarrow{\text{Ra-Ni}, H_2} \text{258} \quad (5.134)$$

Pyridazines **260** are prepared by the general route depicted in eq. (5.135).[58,137] The intermediate 5-acyl-2-isoxazoline **259** can also be directly transformed into **260** in a low-yielding process in eq. (5.135), route a.

5-Acyl-2-isoxazolines **259** are fragmented into α-diketones and nitriles in

$$(5.135)$$

a base- or acid-catalyzed process.[145,323,324] In the presence of o-phenylenediamine, quinoxalines are produced, eq. (5.136).

$$(5.136)$$

The 1,3-oxazine ring system is obtained by reacting isoxazolium salts with a base, eq. (5.137).[325] Ring cleavage has been used to synthesize curcumin.[112] Related are the thermal rearrangements of 3-benzoyl-2,1-benzisoxazole, eq. (5.138),[326,327] the reaction of 2-ethyl-1,2-benzisoxazolium tetrafluoroborate with cyanate, thiocyanate, and thiourea,[328] and the transformation of the isoxazolidine **261** with t-C_4H_9OK to the 1,3-oxazine **262**,[329] eq. (5.139).

$$(5.137)$$

$$(5.138)$$

$$(5.139)$$

5.15 AZETIDINES, β-LACTAMS, AND β-LACTONES

Reduction of 3,5-dimethylisoxazole with sodium in 1-pentanol followed by tosylation gives a stereoisomeric mixture of the aminoalcohol derivatives **263**, from which the *threo*-isomer separates. It gives the *cis*-azetidine **264** on treatment with sodium ethoxide, eq. (5.140).[330] The unstable azetinone **266** is formed by base-catalyzed ring cleavage of benzisoxazolium salt **265**,[331] eq. (5.141).

$$\text{(5.140)}$$

$$\text{(5.141)}$$

Routes to β-lactams are of outmost interest since this functionality is essential to a number of potent antibiotics. Copper acetylide reacts with nitrones to give a mixture of *cis*- and *trans*-β-lactams, eq. (5.142).[332,333]

$$\text{(5.142)}$$

Nitrones add to *trans*-1-cyano-2-nitroethylene to give a mixture of the regioisomers **267** and **268**. The major isomer **267** is thermally or photolytically contracted to the *cis*- and *trans*-β-lactams **269** and **270**, respectively, eq. (5.143).[334]

Several research groups have explored nitrone and nitrile oxide-based routes to preparing thienamycin (**271**) and analogues. In model reactions the problems concerning the relative stereochemistry at C^5, C^6, and C^8 and the synthesis of the gross structure of the bicyclic system have been solved. The isoxazolidine **272** is obtained as the major isomer by adding methyl crotonate to 1-pyrroline-

1-oxide, eq. (5.144).[335] Catalytic reduction and cyclization give the β-lactam skeleton **273**. The functionalized β-lactams **278** and **279** are obtained along similar lines, eq. (5.145).[336] The isoxazolidines **276** and **277** are formed in a ratio of 5:1.

Catalytic reduction and cyclization with dicyclohexylcarbodiimide (DCC) afford a mixture of β-lactams, from which the desired isomer **278** was isolated by chromatography. By using nitrone **280** researchers have prepared the appropriate precursor **281** for carbapenem antibiotics, eq. (5.146).[337]

$$ (5.146) $$

Formal syntheses of thienamycine (**271**) and analogues have been developed by employing 1,3-dipolar addition of the nitrile oxide from 3-nitropropanal dimethyl acetal (**282**) to methyl crotonate, eq. (5.147).[338-340] Chiral β-lactams are obtained by using menthyl crotonate or chiral nitrones.[341,342] The conversion of four-membered cyclic nitrones into β-lactams has been reported.[343]

$$ (5.147) $$

3,4-Diphenylisoxazole fragments into benzonitrile and lithium phenylethynolate (**283**) on treatment with butyl lithium, eq. (5.148). The ethynolate **283** can be trapped with chlorotrimethylsilane to give the ketene **284**. It also reacts with aldehydes and ketones to give β-lactones **285**.[344]

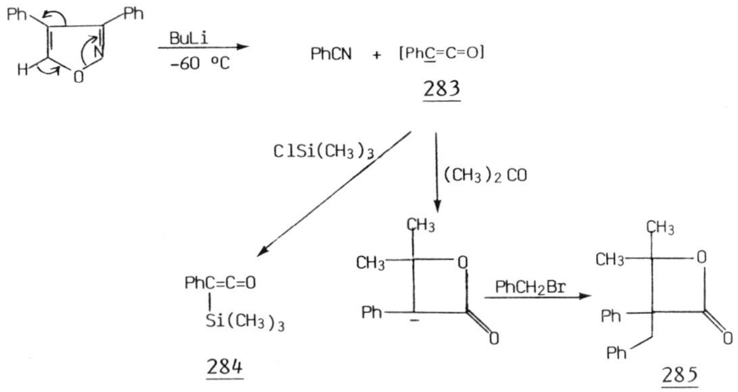

[Scheme for eq. (5.148) showing reaction of diphenyl isoxazole with BuLi at −60 °C giving PhCN + [PhC=C=O] (283), then with ClSi(CH₃)₃ to give PhC=C=O–Si(CH₃)₃ (284), or with (CH₃)₂CO to give a β-lactone, followed by PhCH₂Br to give 285.]

(5.148)

5.16 MISCELLANEOUS HETEROCYCLES. CONDENSED SYSTEMS. CYCLOADDITIONS TO HETERO DIPOLAROPHILES

Sections 5.9–5.14 demonstrate the power of the reductive ring cleavage of isoxazoles for constructing various basic heterocyclic ring systems. This principle has been extended to isoxazoles replaced with heteroaromatic substituents to allow formation of condensed systems by intramolecular reaction of the β-enaminone and the substituent. A selection of ring systems that have been synthesized is shown in eqs. (5.149)–(5.152).

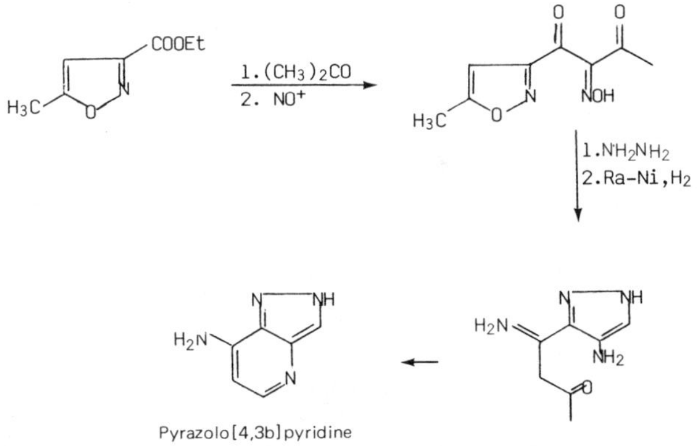

[Scheme for eq. (5.149): 3-carbethoxy-5-methylisoxazole treated with 1. (CH₃)₂CO, 2. NO⁺ gives the diketone oxime intermediate, then 1. NH₂NH₂, 2. Ra-Ni, H₂ yields the aminopyrazolyl pyridine precursor and finally pyrazolo[4,3-b]pyridine.]

Pyrazolo[4,3-b]pyridine

(5.149)[345]

A different approach to heterocycles involves adding nitrones or nitrile oxides to hetero dipolarophiles. Their reactivities decrease in the order $C=S > C=C$, $C=N > C=O$, and the orientation of the cycloaddition agrees with that predicted by frontier orbital calculations.[349] Highly polarized hetero atomic multiple bonds, such as sulfur or phosphorus ylides, probably undergo stepwise reactions.

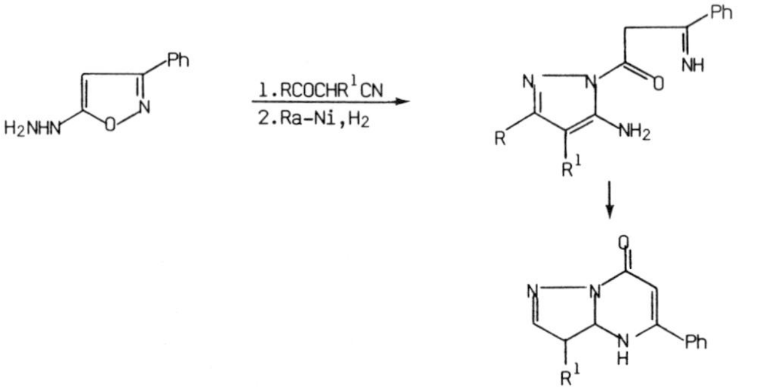

(5.150)[346]

(5.151)[347]

7-Oxopyrazolo[1,5a]pyrimidine

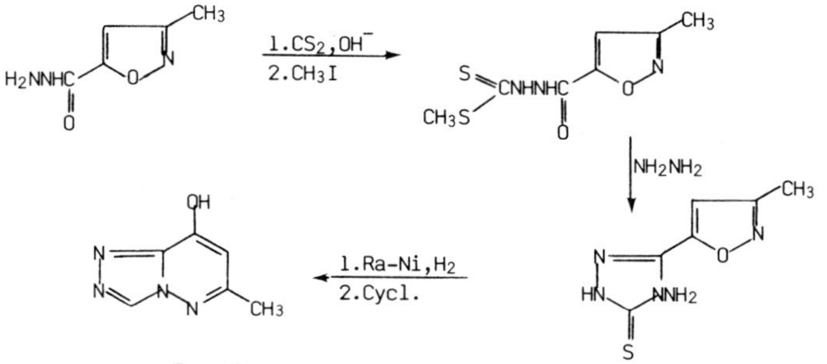

S-Triazolo[4,3b]pyridazine

(5.152)[348]

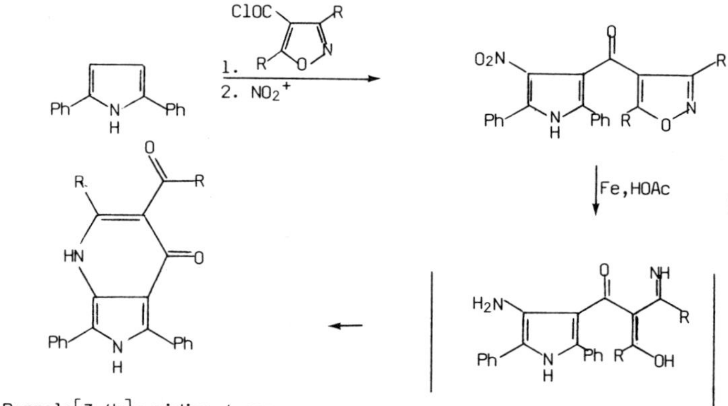

Pyrrolo[3,4b]pyridine-4-one

Nitrones react very sluggishly or not at all with nitriles. It has been reported that 1-methylbenzimidazole-N-oxide (**286**) reacts with benzonitrile to give the amide **288**, supposedly via the cycloadduct **287**, eq. (5.153).[350] Activated nitriles occurring in cyanic esters ROCN or tetracyanoethylene give Δ^4-1,2,4-oxadiazolines, as shown in eq. (5.154).[351] Nitriles cycloadd easier to nitrile oxides than to nitrones,[125,352-383] eq. (5.155). Nitriles are generally considerably less reactive than structurally related olefins and acetylenes, but more reactive than carbonyl compounds.[380,381] The reactivity of the cyano groups

$$\text{(5.153)}$$

$$\underset{286}{} \quad \underset{287}{} \quad \underset{288}{}$$

$$PhCH=N\overset{O}{\underset{Ph}{}} + (CN)_2C=C(CN)_2 \longrightarrow (CN)_2C=\overset{CN}{\underset{O-NPh}{C-C\overset{Ph}{\underset{}{}}}} \quad \text{(5.154)}$$

$$\underset{289}{}$$

$$RCNO + R^1CN \longrightarrow \underset{R,R^1 = Alkyl, Aryl}{} \quad \text{(5.155)}$$

increases if they are attached to an electron withdrawing group or complexed with BF_3 etherate.[358,360] The competition between the C=C and C≡N bonds has been studied in β-aminocinnamonitriles. Cycloaddition of the nitrile oxides to the C≡N bond predominates in non- or weak-hydrogen bond acceptor solvents, such as benzene, whereas in methanol addition to the C=C bond is what predominates.[356,357]

The reactivity of the C=N bond in isocyanates, isothiocyanates, carbodiimides, oximes, amidines, and imino esters is, on the whole, comparable to that of the corresponding olefinic bond towards nitrile oxides and nitrones. Borontrifluoride etherate catalysis is often needed to achieve satisfactory yields, which indicates a two-step polar process. See Refs. 125 and 383–385 for a review of this area. Imines (Schiff's base) yield Δ^2-1,2,4-oxadiazolines[353,355,386-402] with nitrile oxides, eq. (5.156). There are no reports on the cycloaddition of nitrones to simple imines.

$$RCNO + R^1N=C< \longrightarrow \quad \text{(5.156)}$$

N^4-Hydroxy-1,2,4-oxadiazolines **290** or related 1,2,4-oxadiazoles **291** are formed from oximes and nitrile oxides, eq. (5.157).[353,355,403-409] The C=N bond in 2-isoxazolines reacts analogously forming condensed systems, eq. (5.158).[410-417]

$$RR^1C=NOH + R^2CNO \longrightarrow \underset{290}{} \xrightarrow{R=H} \underset{291}{} \quad \text{(5.157)}$$

RR^1 = H, Alkyl, Aryl
R^2 = Alkyl, Aryl

$$\xrightarrow{RCNO} \quad \text{(5.158)}$$

Imino esters **292** and amidines **293** form 3,5-substituted 1,2,4-oxadiazoles **294** with nitrile oxides, eq. (5.159);[125,358,389,418-421] amide oximes give rise to 1,2,4-oxadiazol-N^4-oxides.[422]

$$R^1C=NH \quad \xrightarrow{RCNO} \quad \cdots \quad \longrightarrow \quad \mathbf{294} \qquad (5.159)$$
$$\quad | \quad$$
$$\quad X$$

292, X = OR, SR
293, X = NR^2

The cumulated C=N bonds in isocyanates and carbodiimides react sluggishly with nitrile oxides, giving rise to 1,2,4-oxadiazolin-5-ones and 1,2,4-oxadiazolin-5-imines, eq. (5.160).[353,387,395,423-425] The 5-imino compound rearranges in an acid-catalyzed reaction into the 1,2,4-triazolin-5-one **295**.[424]

$$RN=C=X \quad + \quad R^1CNO \quad \longrightarrow \quad \cdots \quad \longrightarrow \quad \mathbf{295} \quad \text{(for X= NR)} \qquad (5.160)$$
$$X = O, NR$$

Nitrones react quite easily with isocyanates and produce rather stable 1,2,4-oxadiazolidin-5-ones,[426-447] which, during thermolysis, eliminate carbon dioxide and yield amidines **296**, eq. (5.161).

$$\cdots + R^4NCO \longrightarrow \cdots \xrightarrow{-CO_2, R^2=H} \mathbf{296} \qquad (5.161)$$

$$\cdots + R^4NCS \longrightarrow \cdots \xrightarrow{-OCS, R^2=H}$$

Isothiocyanates are less prone to react with nitrones. Unstable 1,2,4-oxadiazolidine-5-thiones are formed, which undergo further fragmentations, eq. (5.161).[428,431,439,448-450] The addition often takes place at the C=S bond, giving rise to thioamides. The cycloaddition of nitrones to carbodiimides leads to 1,2,4-oxadiazolidin-5-imines, which subsequently rearrange to triazolidin-5-ones **297** or amidines, eq. (5.162).[451,452] The cycloaddition of nitrile oxides to imidazoles,[421] 2-pyrazolines,[453] and 1,2-diazepines[454,455] gives rise to new condensed heterocyclic systems. C-Aryl nitrones react in a BF_3-catalyzed reaction with nitrile oxides to give the 2,3-dehydro-1,4,2,5-dioxadiazine ring system **298**.[456]

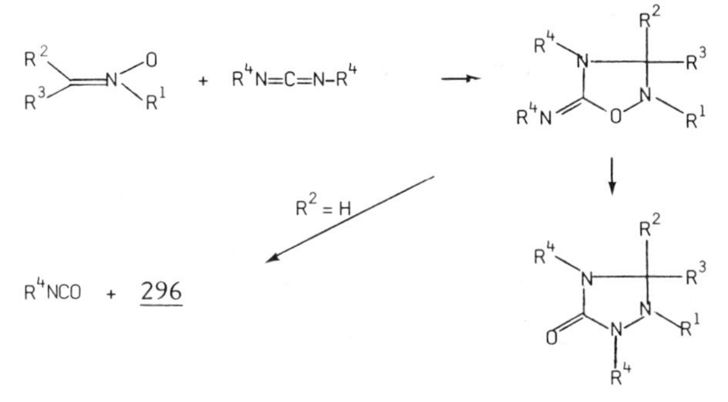

(5.162)

Nitrile oxides cycloadd to aliphatic aldehydes and ketones activated by electron withdrawing groups, such as α-keto esters, chloral, aromatic aldehydes, α-diketones, and quinones to form 1,3,4-dioxazolidines, eq. (5.163).[355,360,380,390,396,397,457-469] Simple aliphatic aldehydes do not react but can be brought into reaction as BF_3 complexes: however, the ester function is inert.

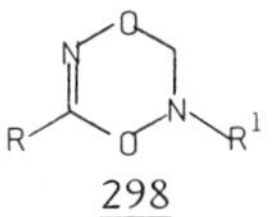

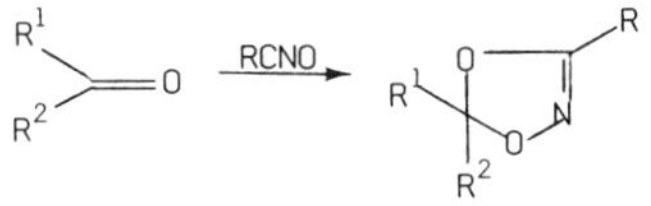

(5.163)

Only one report has been published on the cycloaddition of a nitrone to a carbonyl compound. It has been shown that 2,5-di-*t*-butyl-4-phenyl-1,3,4-dioxazolidine (**299**) undergoes cycloreversion, eq. (5.164).[470]

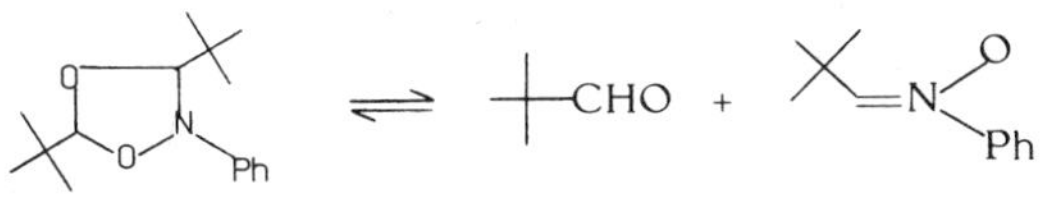

(5.164)

The high reactivity of the C=S bond in thioaldehydes, thioketones,[471-481] thiono esters,[471,472,482-484] thioamides,[485,486] thioketenes,[487-490] isothiocyanates,[428,439,491,492] and carbon disulfide[428,434,438,439,493-495] with nitrile oxides and nitrones is well documented, eqs. (5.165),(5.166).

The 1,4,2-oxathiazolines **300** and 1,4,2-oxathiazolidines **301** are unstable and fragment spontaneously or when heated. The fragmentation of **300**,

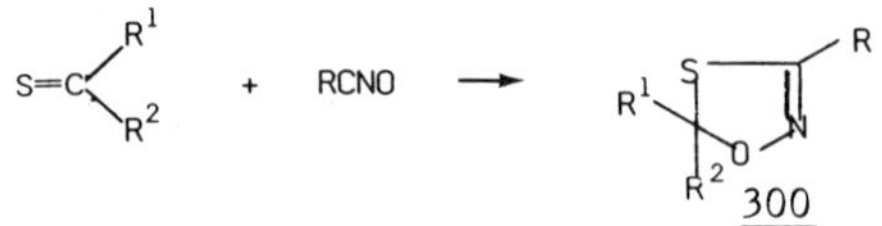

$$S=C\begin{matrix}R^1\\R^2\end{matrix} + RCNO \longrightarrow \underset{\underset{\mathbf{300}}{R^2}}{\overset{S}{\underset{O}{\bigsqcup}}}\overset{R}{\underset{N}{}}\quad (5.165)$$

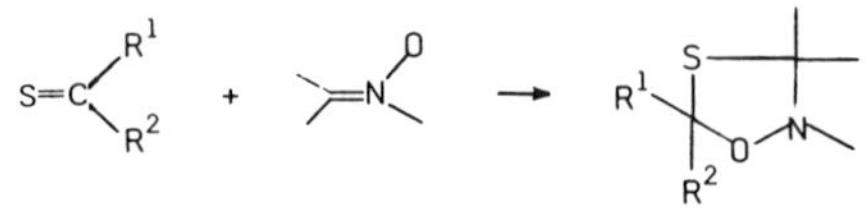

$$S=C\begin{matrix}R^1\\R^2\end{matrix} + \overset{O}{\underset{N}{\bigtriangledown}} \longrightarrow \underset{\underset{\mathbf{301}}{R^2}}{\overset{S}{\underset{O}{\bigsqcup}}}\overset{}{\underset{N}{}}\quad (5.166)$$

$R^1 = R^2 = H$, Alkyl, Aryl

$R^1 = $ Alkyl, Aryl; $R^2 = $ OR, OS, $N{<}\begin{matrix}R\\R\end{matrix}$

$S=C{<}\begin{matrix}R^1\\R^2\end{matrix}$ = thioketene, thioisocyanate, carbondisulfide

eq. (5.167), is used to convert thiones into carbonyls.[496] Compound **301** gives imines or thioamides, depending on the nature of the R group, eq. (5.168).

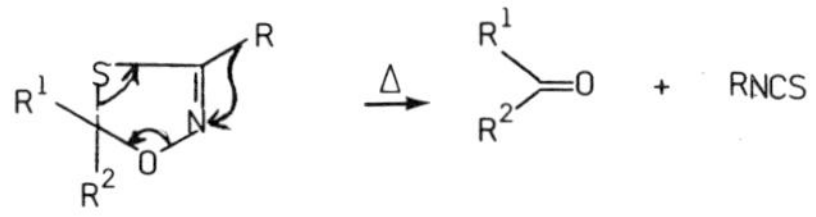

$$\underset{\mathbf{300}}{\text{[structure]}} \xrightarrow{\Delta} \underset{R^2}{\overset{R^1}{}}{=}O + RNCS \quad (5.167)$$

$$\underset{\mathbf{301}}{\text{[structure]}} \longrightarrow \begin{matrix} COS + S + \overset{R}{\underset{R}{N}} & R= Aryl \\ \\ COS + \underset{HN{-}R}{\overset{S}{\underset{}{\bigsqcup}}}\overset{R}{} & R= Alkyl \end{matrix} \quad (5.168)$$

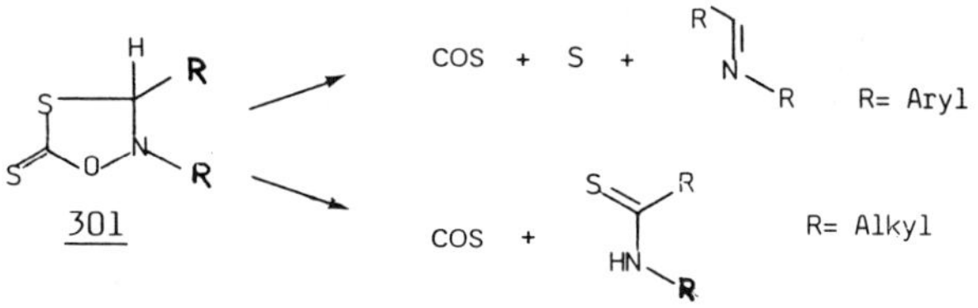

Cycloadditions of nitrile oxides or nitrones to $C=P$,[497-506] $C=Se$,[507] $N=P$,[423] $N=N$,[508-510] $N=B$,[511] $N=S$,[512-521] and $S=O$[522-525] bonds have been carried out to give several unusual, more or less stable, heterocycles, such as 1,2,5-oxazaphosphole **302**, 1,2,5,3-oxadiazaphosphole **303**, 1,2,3,5-oxatriazole **304**, 1,3,5,2-oxadiazaborole **305**, 1,2,3,5-oxathiadiazole **306** and 1,3,2,5-dioxathiazole **307**. They frequently undergo spontaneous ring fission and form various products.[125,383-385,526]

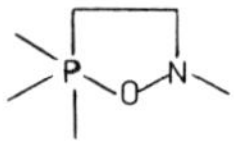

302 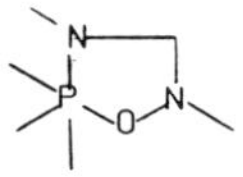 **303** 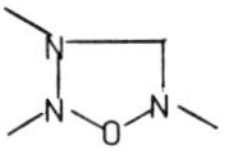 **304**

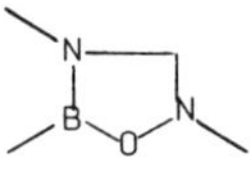

305 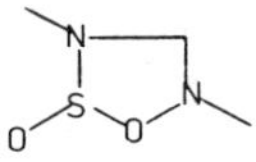 **306** 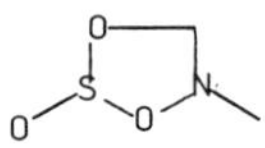 **307**

5.17 REACTIONS WITH α-CHLORONITRONES AND α,β-EPOXYNITRONES

The α-chloro- and α,β-epoxynitrones have recently been introduced as useful reagents in organic synthesis.[527-536] α-Chloronitrones are prepared either by chlorination of alkylnitrones with N-chlorosuccinimide (NCS) or by reaction of α-chloroaldehydes with N-alkylhydroxylamines.[527] The α,β-epoxynitrones are prepared from α,β-epoxyaldehydes with N-alkylhydroxylamines, eqs. (5.169), (5.170).[536] Cyclohexylhydroxylamine is commonly chosen as reaction component because it gives reasonably stable nitrones.

$$\text{(5.169)}$$

R = H, Alkyl
R^1 = t-Bu, c-Hexyl
308

$$\text{(5.170)}$$

R^1 = c-Hexyl
309

Reaction of nitrone **308** with silver ions produces the N-alkyl, N-vinylnitrosonium ion **310**, which participates smoothly in a cis-1,4-cycloaddition with olefins to give the 1,2-oxazinium ion **311**. Treating **311** with potassium cyanide gives the stable nitrile **312**.[527] The diastereoisomeric mixture **312** undergoes base-promoted ring contraction to the γ-lactone **313**.[528] If R = CH_2Cl, α-methylidene-γ-lactone **313** (R = CH_2=) is obtained.[530] Deprotonation of **311**, cycloreversion of the intermediate **314**, and hydrolysis lead to cleavage of the olefinic bond and formation of the functionalized α,β-unsaturated aldehyde **315**, eq. (5.171).[529] When the addition of **310** to olefins is carried out in a polar solvent, substitution takes place to give stereospecifically the β,γ-unsaturated nitrone **316**, eq. (5.172).[531] The cycloaddition products **317** from the nitrosonium ion **310** and acetylene fragment along a different pathway. The basic reagent adds to the oxazinium ion **317**, and the acetylene is eventually transformed into an α,β-unsaturated ketone **318**, eq. (5.173).[532]

α,β-Epoxynitrones **309** can be transformed into N-alkyl, N-vinylnitrosonium ions **319** with trimethylsilyl triflate, eq. (5.174).[536] They undergo 1,4-cycloadditions with olefins and rearrange in analogy to the reaction shown in eq. (5.171).

The interest in the synthetically useful 1,2-oxazines is manifested in the search for other routes to these compounds, i.e., alkylation of 1,2-oxazines[536,537] or Diels-Alder reactions with nitrosoalkenes.[538,539]

Applications in Synthesis

5.18 CYCLOADDITION OF OXIMES WITH OLEFINS

There are some reports in the literature on the cycloaddition of oximes to olefins or acetylenes.[540-550] Acrylonitrile and methyl acrylate react with formaldoxime to give a mixture of the isoxazolidines **320** and **321**, eq. (5.175).[540] The formation of **320** is explained by the supposition that the oxime reacts in its tautomeric nitrone form, which may be present in minute amounts. The order of events in the reaction leading to the 2:1 adduct **321**, eq. (5.175), is not fully known, but route b seems to be the most favored pathway.[547] Since formaldoxime is prone to polymerize and this process passes a nitrone stage **323**, it is quite conceivable that **320** is formed as depicted in eq. (5.176). This mechanism is supported by the isolation of compound **324**.[545] A number of 2,3,5-substituted isoxazolidines are prepared from various aldoximes or ketoximes and olefins substituted with electron withdrawing groups. It is suggested that the acid catalyzed intramolecular cycloaddition eq. (5.177) proceeds via a cationic intermediate, since the thermal cycloaddition fails. The corresponding phenyl derivative does not give any cycloadduct.[550]

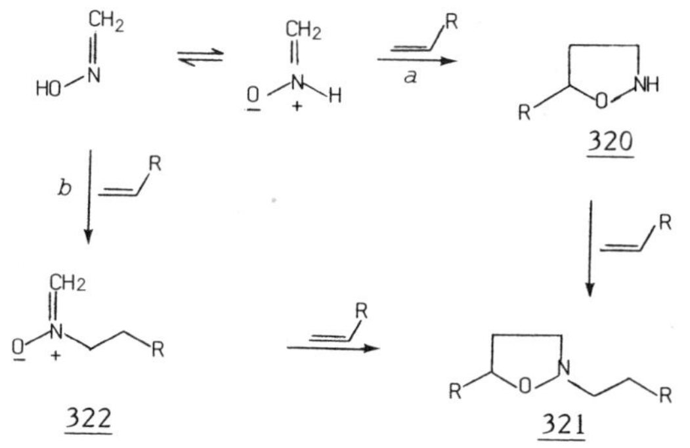

R = CN, COOCH$_3$, COCH$_3$

(5.175)

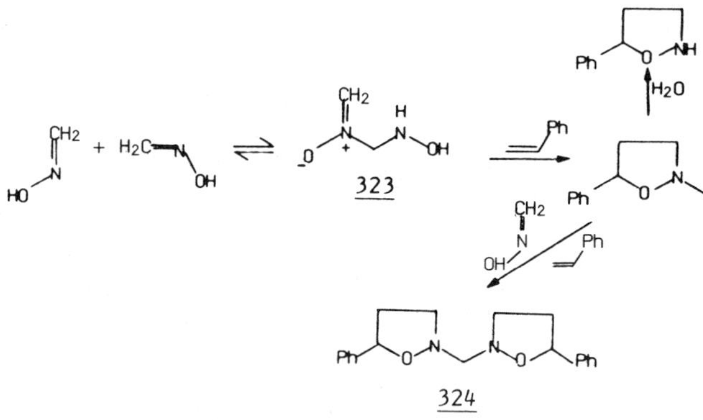

(5.176)

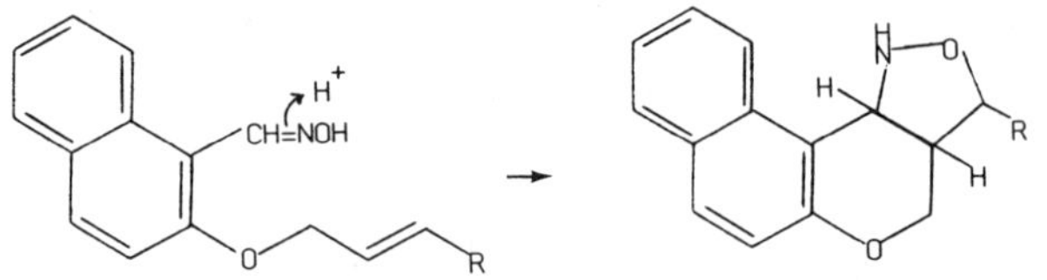

(5.177)

5.19 CYCLOPENTANE DERIVATIVES, PROSTAGLANDIN PRECURSORS, AND TERPENOIDS. RING ANNULATION, INTRAMOLECULAR CYCLIZATION, BRIDGED CYCLOALKANES

Numerous routes are available for preparing cyclopentane derivatives, but simple procedures are always in demand because cyclopentanes are structural units in many important natural products, such as rethrolones (insecticides), prostaglandins (hormones), and cyclopentanoid terpene derivatives.[551] One method for synthesizing 2-cyclopentenones is based upon the base-catalyzed cyclization of 1,4-diones, which now are available via cycloaddition of silyl nitronates or nitrile oxides to vinyl ketones and subsequent reduction of the 3-alkyl-5-acyl-2-isoxazolines produced.[58,136,137,145,552] Allethrolone (**325**) is synthesized from 5-nitro-1-pentene and methyl vinyl ketone according to eq. (5.178).[58] In an analogous way calythrone (**326**),[58] eq. (5.179), dihydrojasmone (**327**),[145] tetrahydrorethrolone (**328**),[145] and dihydrocinerolone (**329**)[552]

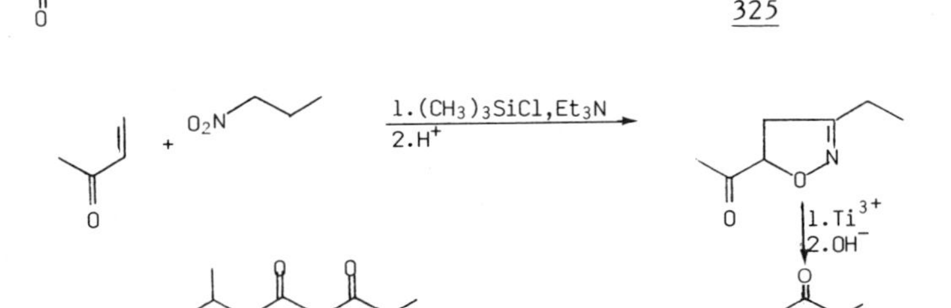

(5.178)

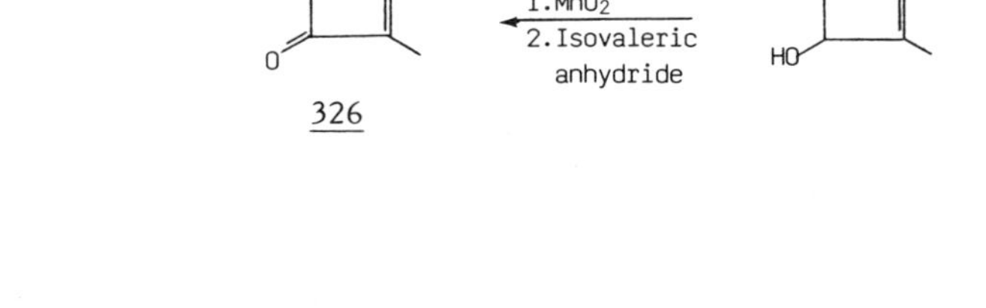

(5.179)

are prepared, eq. (5.180). This methodology opened a novel route to prostanoids, and this was borne out in practice by the synthesis of the prostaglandin intermediates **330**,[136,145] and **331**,[137,552] eqs. (5.181),(5.182).

(5.180)

(5.181)

(5.182)

The lower β-chain of prostanoids has been elaborated via isoxazole derivatives,[119] eq. (5.183). Isoxazoles have also been applied as intermediates in the synthesis of 8-aza-11-deoxyprostaglandins.[553]

Intramolecular cycloaddition of C-alkenyl nitrones[554–558] and alkenyl nitrile oxides[555,559–561] is an efficient and general route to 5-, 6- and higher-membered carbocyclic and heterocyclic rings. This methodology has been used to synthesize prostanoids, terpenoids, various bridged and condensed cycloalkanes, alkaloids (Section 5.20), carbohydrates (Section 5.21), and miscellaneous natural products (Section 5.23).

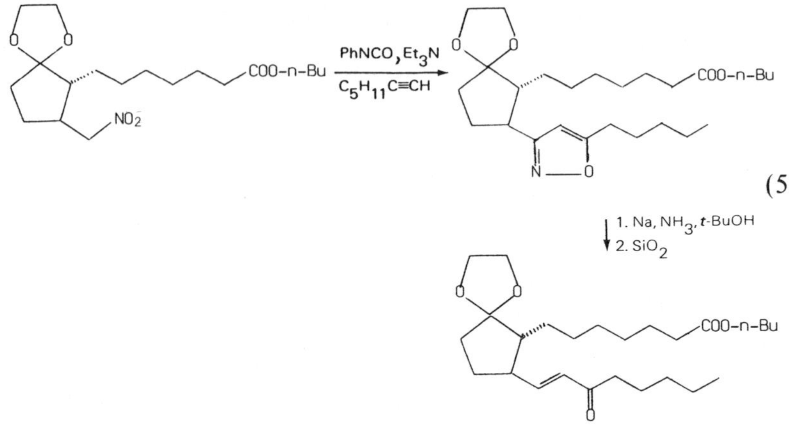

(5.183)

Chiral cyclopentanoles have been prepared by reductive cleavage of 6-bromo-6-deoxy- and 6-iodo-6-deoxyglucosides with zinc followed by ring closure with N-methylhydroxylamine and catalytic reduction, eq. (5.184).[562-565] The tosylated isoxazoline **332** is directly hydrogenated over Raney-Ni to the aziridine **333**, which in subsequent steps can be transformed to the chiral epoxy lactone **334**, a prostaglandin intermediate.[563-565] The cyclization is characterized by being highly diastereoselective. The large groups at $C^{6,8}$ are located on the less congested convex side.[566]

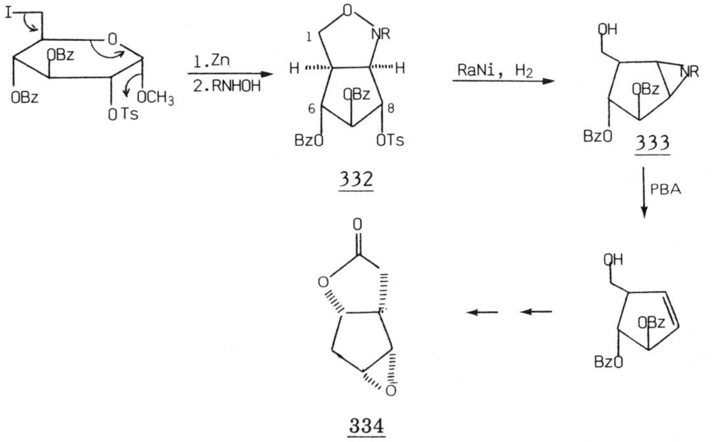

(5.184)

The cyclopentane ring of sarcomycin **336**[567] and of the prostaglandin $F_{2\alpha}$ intermediate (**337**)[568] are assembled according to the same principles, eqs. (5.185),(5.186).

The configuration at $C^{5,6}$ in **335** is controlled by steric interactions in the transition state, as demonstrated in eq. (5.187).[569]

Intramolecular cycloaddition of C-alkenyl nitrones **338** with a propylene chain gives the cis-fused bicyclo [3.3.0] octanes (**339**, route a), which are less strained than the trans-fused isomers (**340**, route b). When both R^1 and R^2 are alkyl groups, steric interaction in the eclipsed transition state favors formation of the bridged products (**341**, route c), eq. (5.188).[554] The configuration of the olefin is retained in the product **339** with R^1 and R^3 in the trans-position. The homologous nitrone **342** gives a cis-trans mixture of [4.3.0] nonanes **343**, **344** and minor quantities of the bridged [4.2.1] nonane derivative **345**, eq. (5.189). This result is explained by the higher flexibility of the six-membered ring in the transition state.[570-575] It is difficult to predict the steric outcome of the reaction because of uncertainty about the stability of the syn-anti forms of the nitrone, but if this barrier of rotation is higher than the activation energies for the transition states represented by **342**, the ratio of **343**:**344**:**345** is determined kinetically by the conformational energies of **342**. At higher temperatures (200–300°C) considerable retro condensation occurs, which is evident from the observation that the composition of the products **343–345** slowly changes to the thermally most stable mixture.

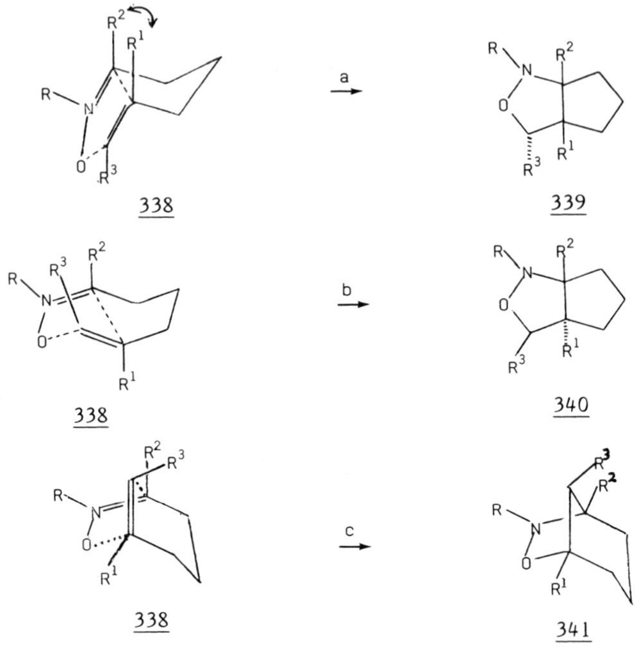

(5.188)

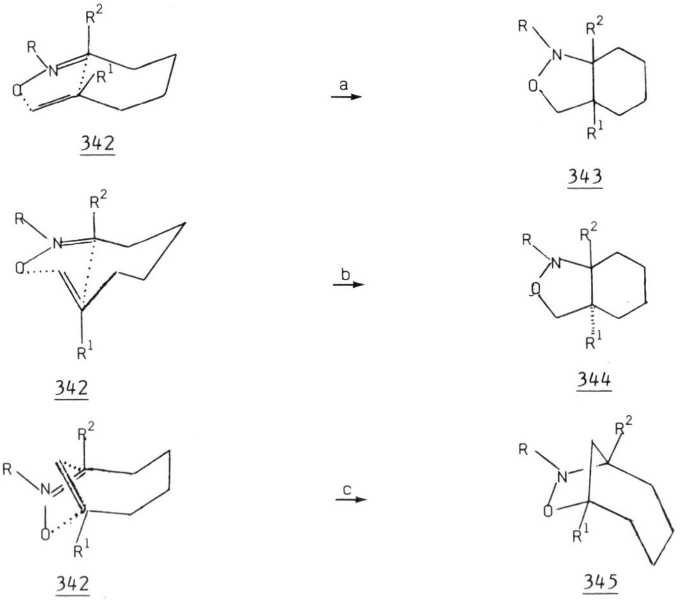

338 →a→ 339
338 →b→ 340
338 →c→ 341

(5.189)

342 →a→ 343
342 →b→ 344
342 →c→ 345

The usefulness of the intramolecular nitrone cyclization is demonstrated by the syntheses of a variety of polycyclic structures,[555] eqs. (5.190)–(5.221). Selective reduction of the N—O bond produces a skeleton functionalized with hydroxy and amino groups. If the alkenyl group is attached to the nitrogen atom, it is built into the ring. This case is treated in more detail in connection with the synthesis of alkaloids (Section 5.20).

$(5.190)^{573}$

$(5.191)^{576}$

$(5.192)^{575,576}$

$(5.193)^{577}$

$(5.194)^{571}$

$(5.195)^{571}$

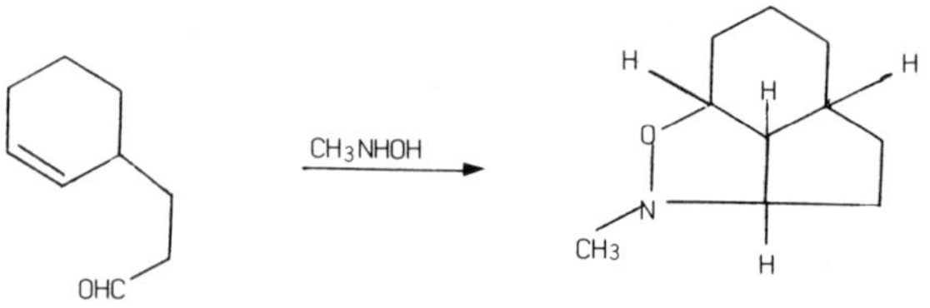

Applications in Synthesis

(5.196)[571]

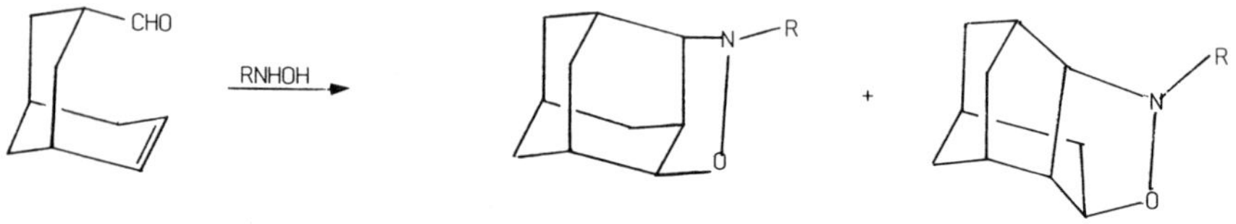

(5.197)[578]

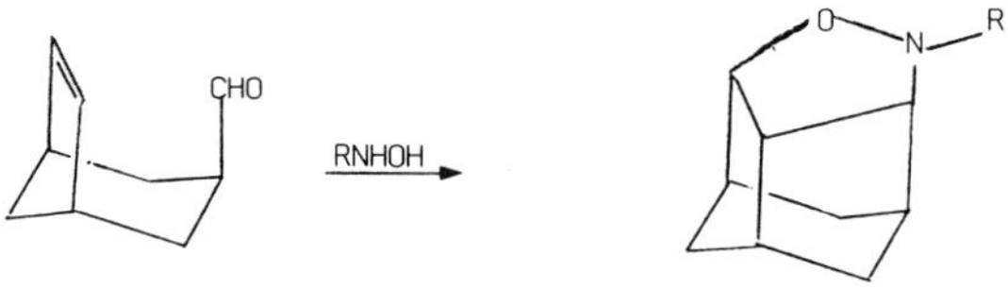

(5.198)[579]

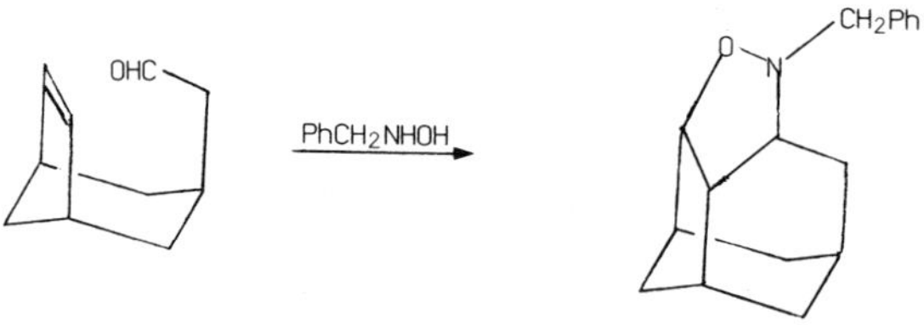

(5.199)[579]

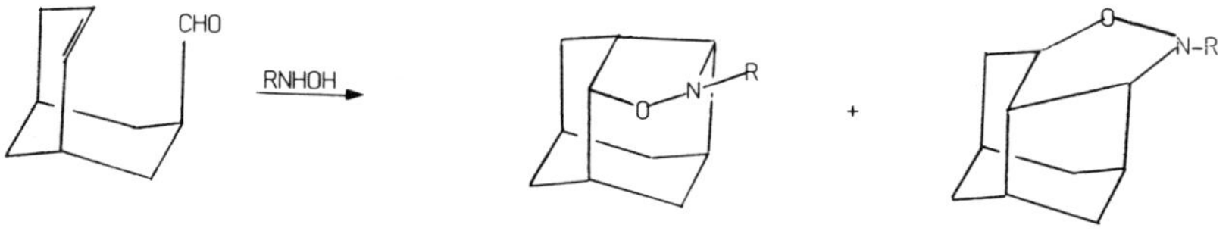

(5.200)[579]

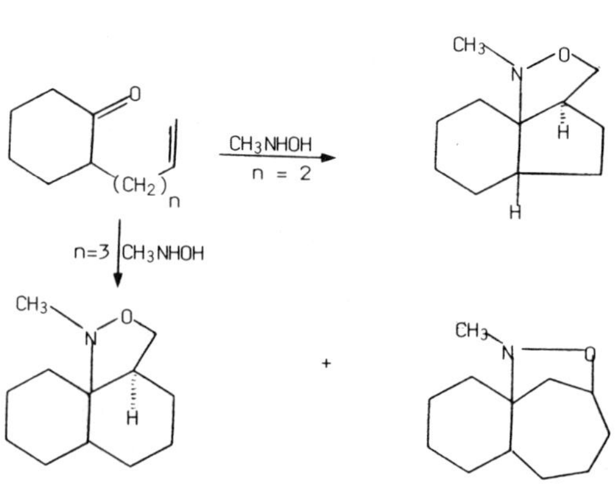

(5.201)[580]

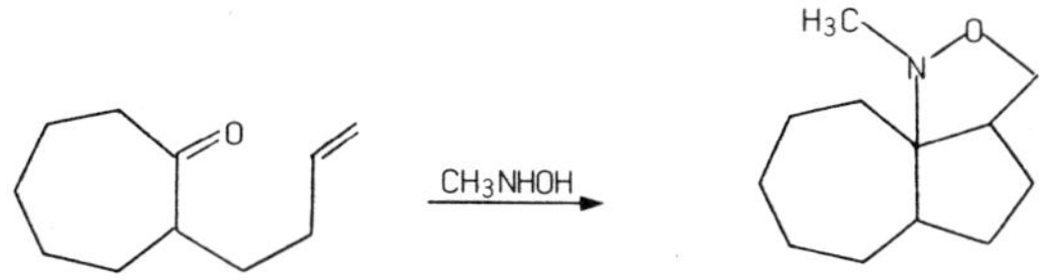

(5.202)[581]

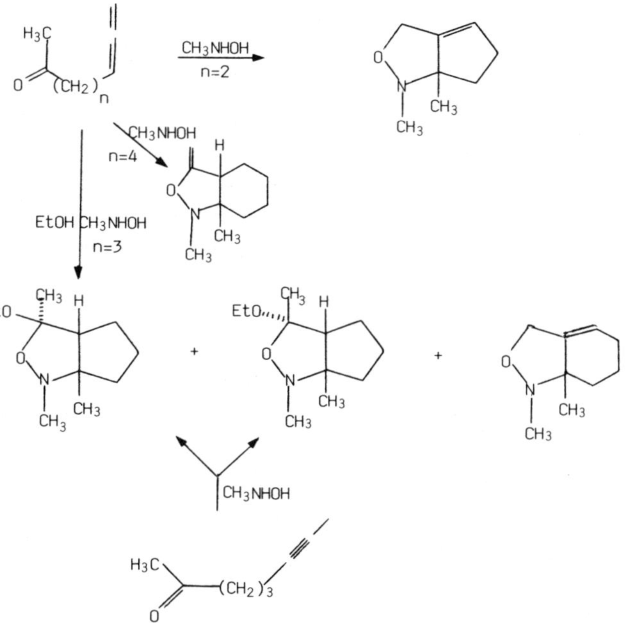

(5.203)[582]

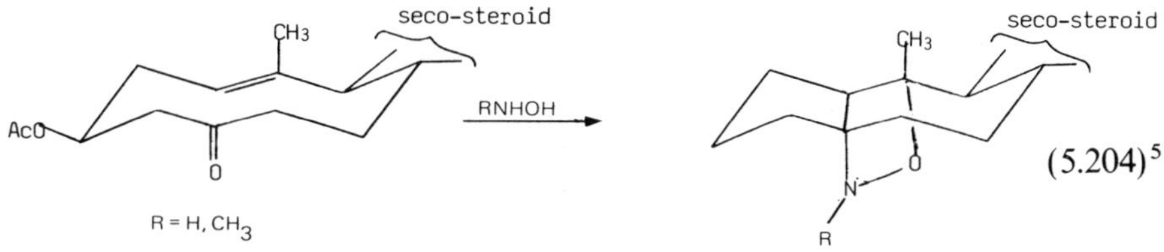

(5.204)[583]

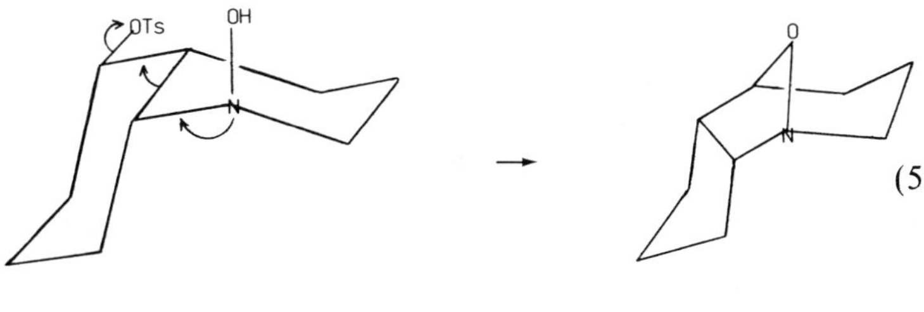

R = H, CH$_3$

(5.205)[584]

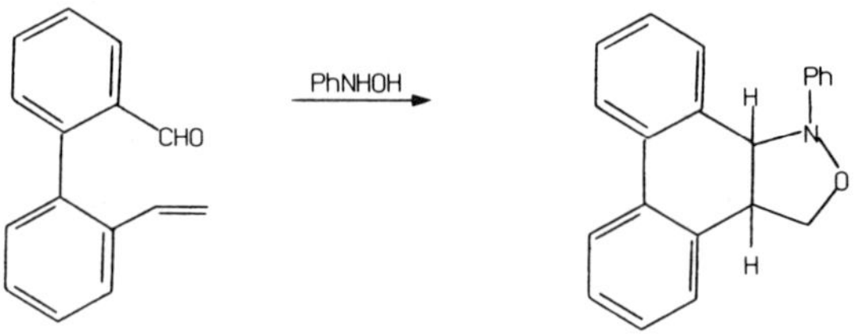

(5.206)[585]

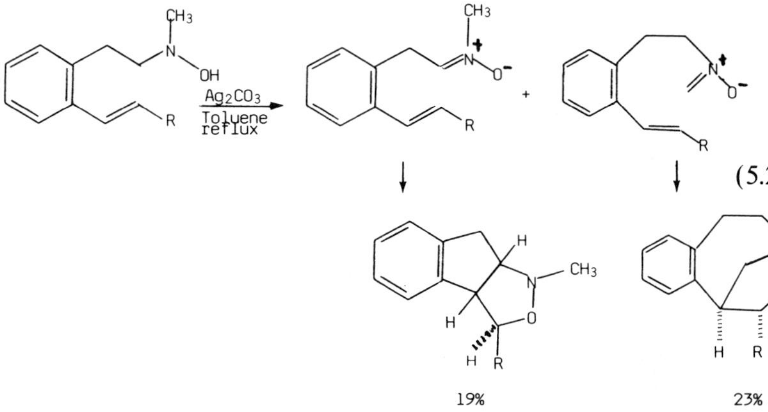

(5.207)[586]

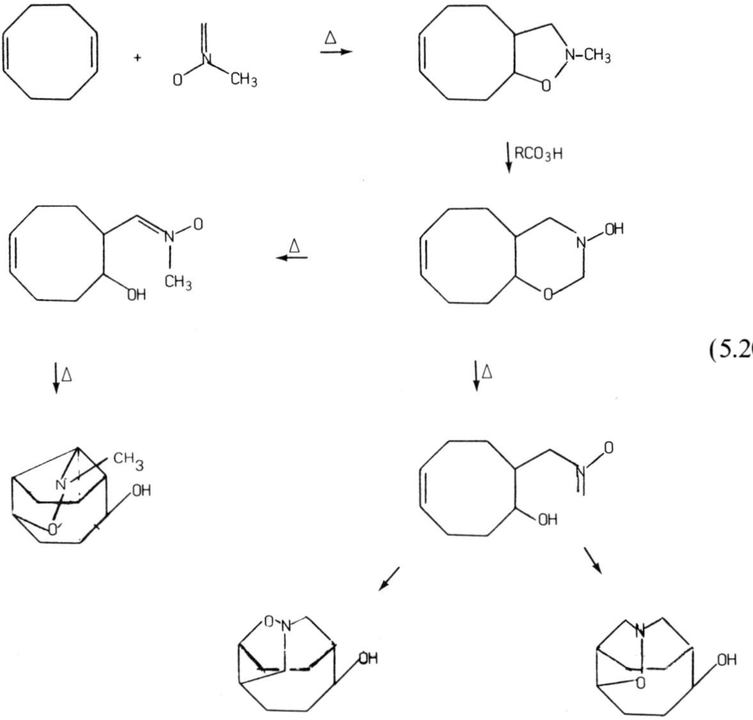

19%   23%

(5.208)[587]

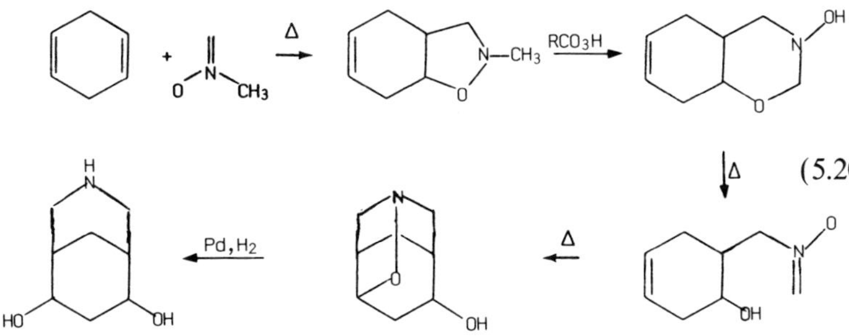

Cyclopentane Derivatives, Prostaglandin Precursors, and Terpenoids

(5.209)[587]

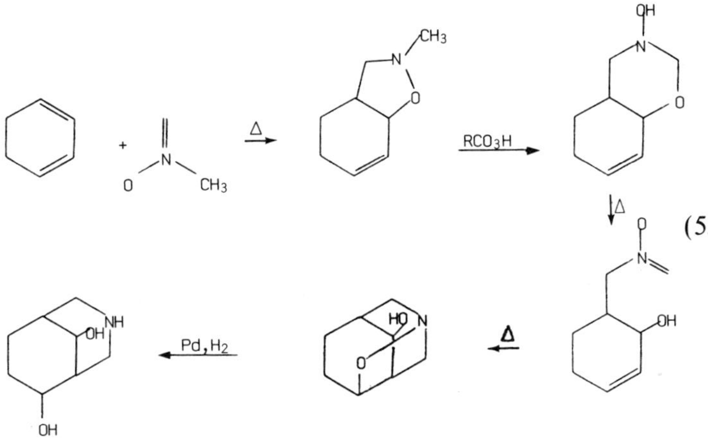

(5.210)[587]

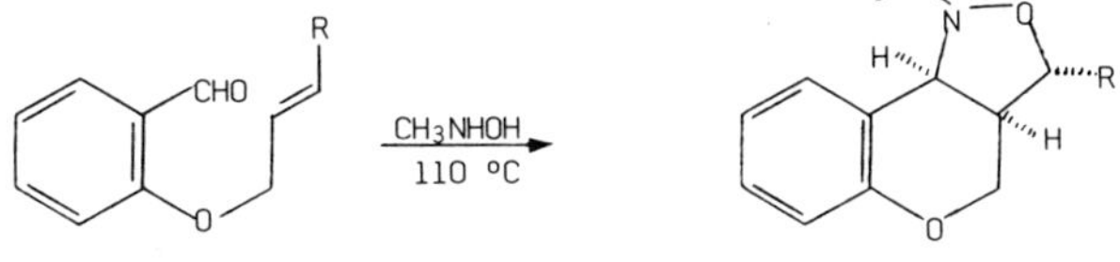

R = H, CH₃, Ph, COOCH₃, CH₂Cl

(5.211)[549,588]

X = O, NCHO

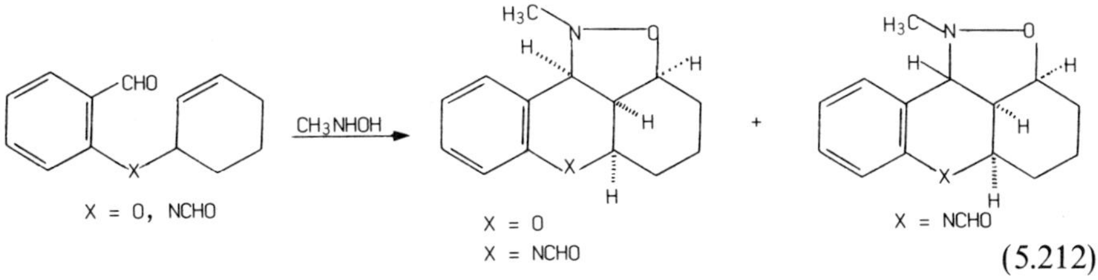

(5.212)[549]

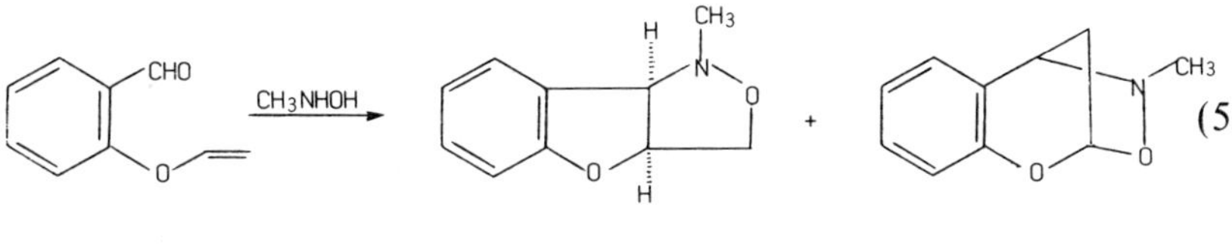

(5.213)[549]

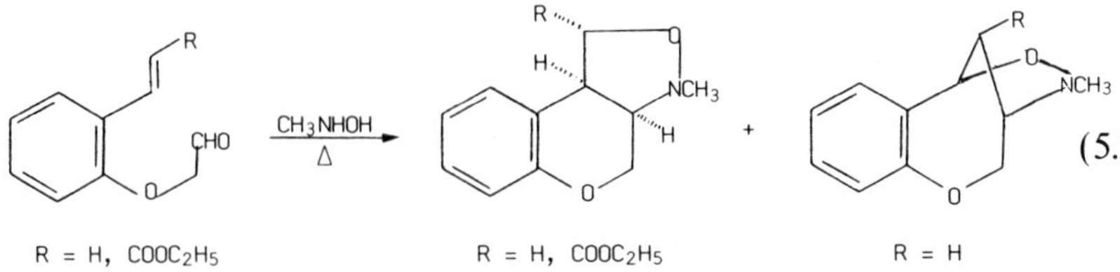

Applications in Synthesis

R = H, COOC$_2$H$_5$ → CH$_3$NHOH, Δ → R = H, COOC$_2$H$_5$ + R = H (5.214)^{589}

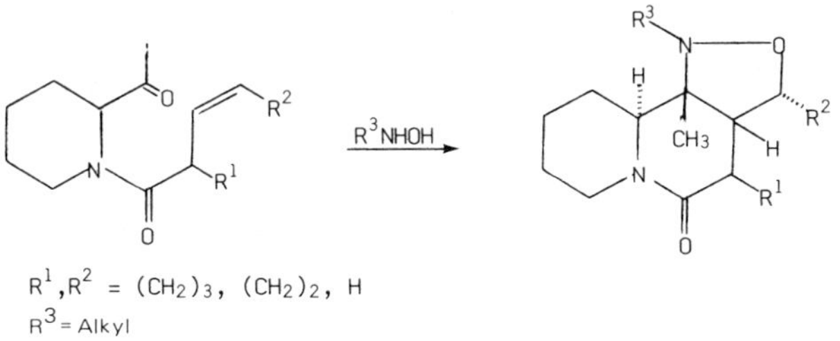

R^1,R^2 = (CH$_2$)$_3$, (CH$_2$)$_2$, H
R^3 = Alkyl

R^{3}NHOH → (5.215)^{590}

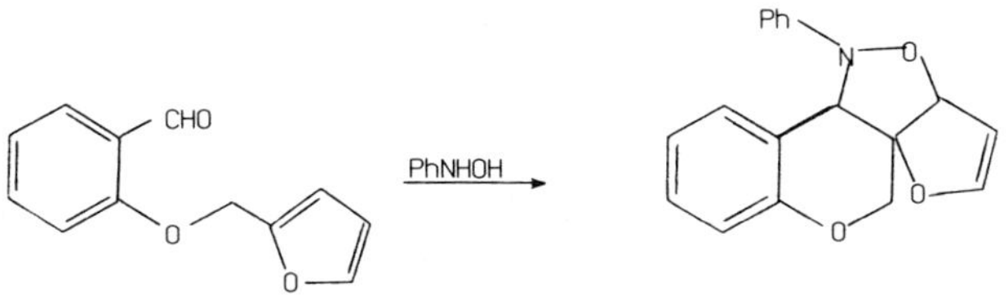

PhNHOH → (5.216)^{591}

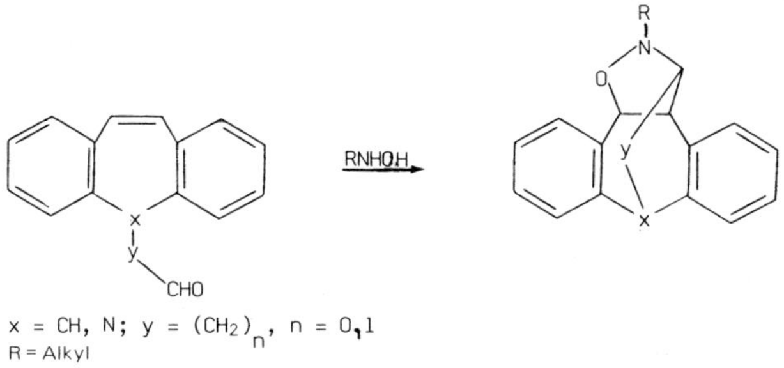

x = CH, N; y = (CH$_2$)$_n$, n = 0,1
R = Alkyl

RNHOH → (5.217)^{592}

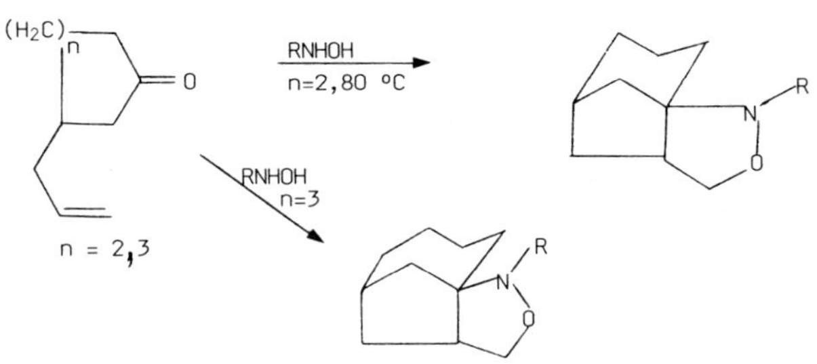

n = 2,3

RNHOH, n=2, 80 °C →
RNHOH, n=3 →
(5.218)^{593}

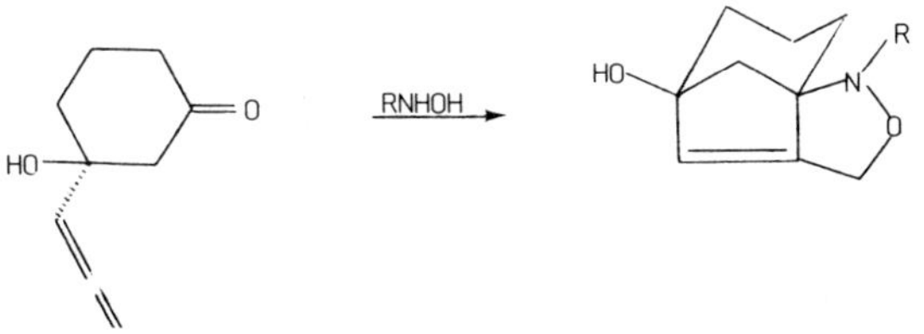

RNHOH → (5.219)^{593}

The nitrone cycloaddition has been used as key step for constructing the linearly fused tricyclopentanoids hirsutene (**346**) and coriolin (**347**) and the bridged sesquiterpene 7,12-secoishwaran-12-ol (**348**),[593,594] eqs. (5.222),(5.223), and cf. eq. (5.218).

The diastereomeric racemic (6S, 7S-6R,7R)-α-bisabolol (**351**) and (6S,7R-6R,7S)-α-bisabolol are stereospecifically synthesized from (6Z)- and (6E)-farnesol (**349**), respectively, using intramolecular cycloaddition of the corresponding N-methylnitrones as the key step controlling ring formation and stereochemistry, eq. (5.224).[595,596] The isoxazolidine ring is cleaved by quaternization of the nitrogen atom with methyl iodide. Treating the salt with a base affords the 1,3-oxazine **350**. Subsequent quaternization with methyl iodide and reductive deamination with lithium in ammonia gives α-bisabolol (**351**). In initial exploratory work citral was transformed into α-terpineol (**352**),[595] eq. (5.225). In an analogous way diterpenes α- (**353**) and β-eudesmol (**354**) and their diastereomers have been synthesized from methyl farnesate, eq. (5.226).[597]

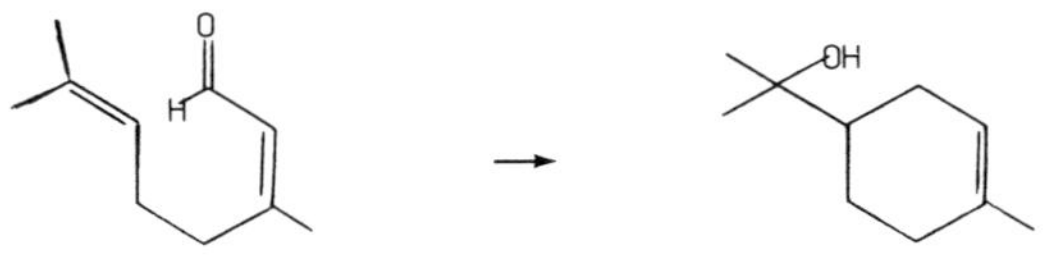

(5.225)

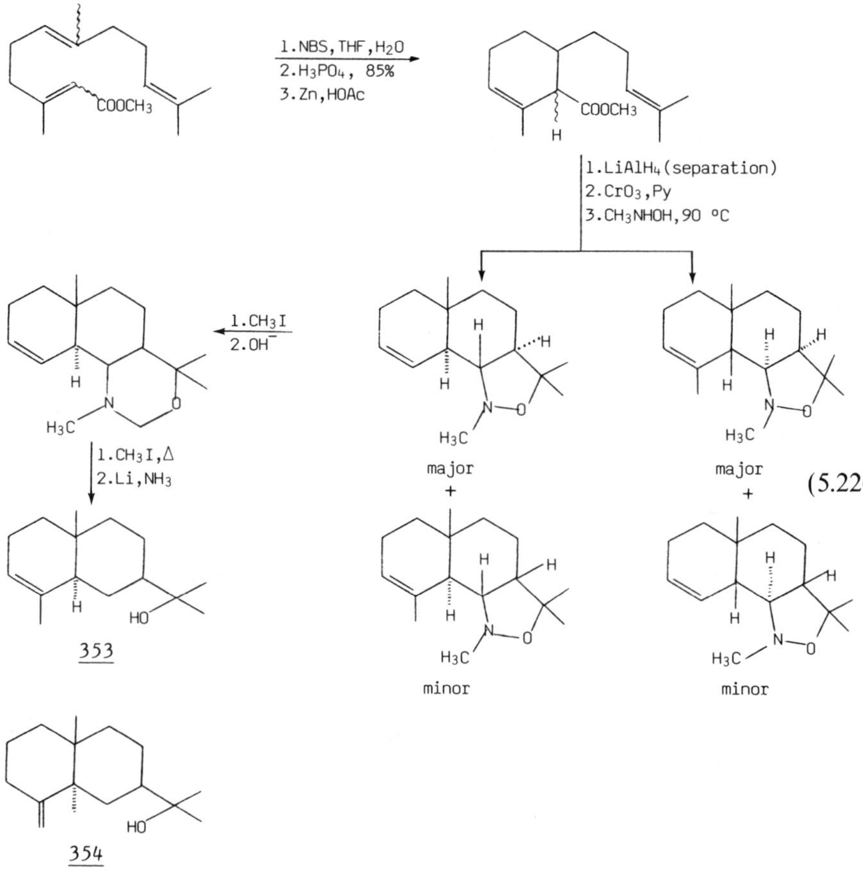

(5.226)

The remote chiral center at C^{25} in 1α,25S,26-trihydroxycholecalciferol (**359**) is introduced regio- and, to a certain extent, stereoselectively by cycloaddition of the C^{23}-nitrone **355** with methyl methacrylate. The diastereoisomeric (23S, 25S)-isoxazolidine **356** is obtained in 36% yield, but after separating the desired **356** the remaining diastereomeric mixture can be thermally equilibrated to give a total yield of 71% of **356**. Compound **356** is transformed into **358** via **357** by conventional procedures,[598] eq. (5.227), and finally into **359**.

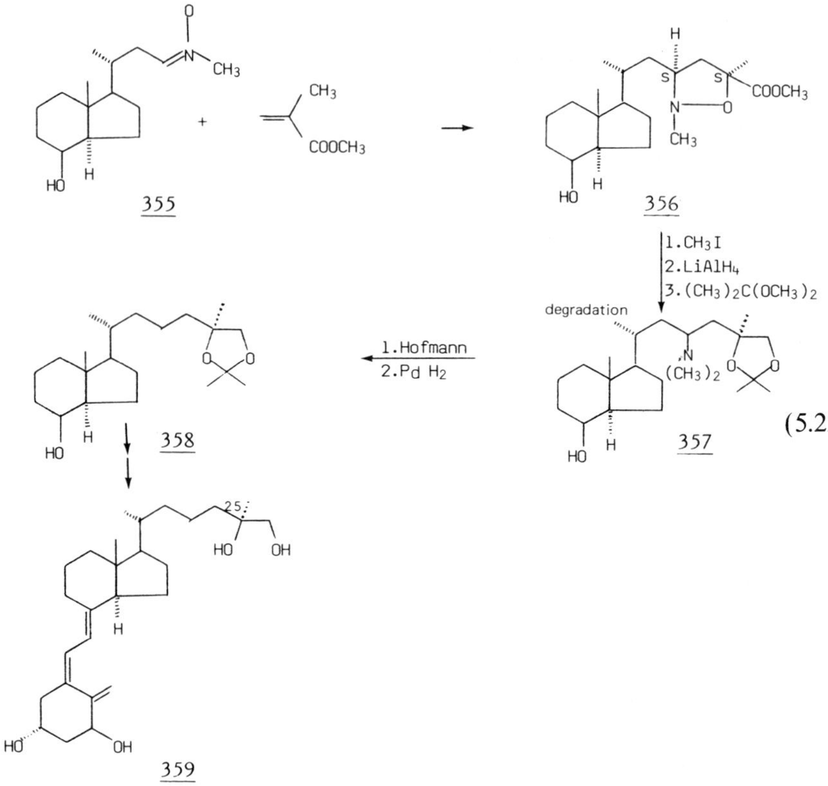

(5.227)

359

The attempt to construct the carbocyclic skeleton of the antitumor sesquiterpene quadron (**360**) by intramolecular nitrone cyclization,[599] eq. (5.228), met with failure, possibly as a result of congestion around the double bond.

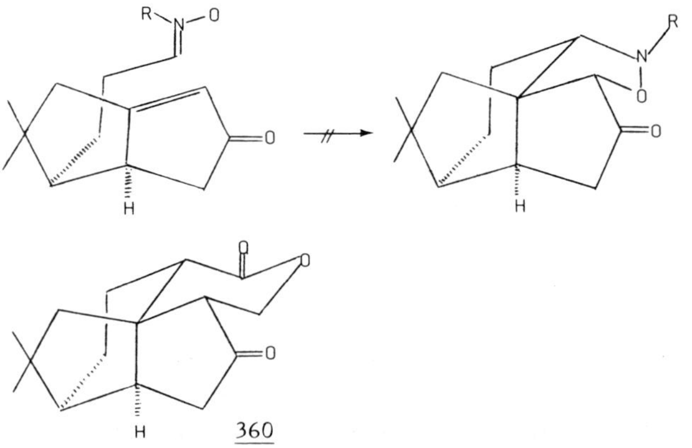

(5.228)

Ring annulation by intramolecular nitrile oxide cycloaddition is used to construct the basic iridodial ring skeleton **361** with steric control,[560,600]

Cyclopentane Derivatives, Prostaglandin Precursors, and Terpenoids 207

$$(5.229)$$

eq. (5.229). The hydroazulenenone ring system **362** is prepared analogously,[601] eq. (5.230). Via tandem Diels-Alder dipolar cycloadditions, multiply fused carbocyclic systems can be constructed with regio- and stereochemical control.[602] The initial Diels-Alder reaction generates a double bond suitably located for the subsequent dipolar addition, eqs. (5.231),(5.232). The inter-

$$(5.230)$$

$$(5.231)$$

$$(5.232)$$

mediate **363** offers two reactive sites for capturing the nitrile oxide. The preference for forming the more-strained eight-membered ring product **364** reflects the higher reactivity of the α,β-unsaturated carbonyl system. When this bond is reduced, the anticipated six-membered ring system **365** is formed. Then reactions shown in eqs. (5.233)–(5.235) provide further examples of construction of five- and six-membered carboxylic rings via intramolecular nitrile oxide additions serving as key steps in the synthesis of α-methylene-γ-lactones **366**,[603] δ-lactones **367**,[68] and the hexahydronaphthalene moiety of compactin (**368**).[68,735]

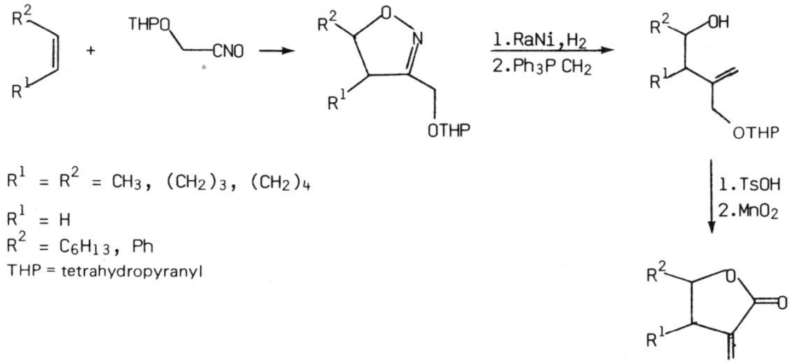

(5.233)

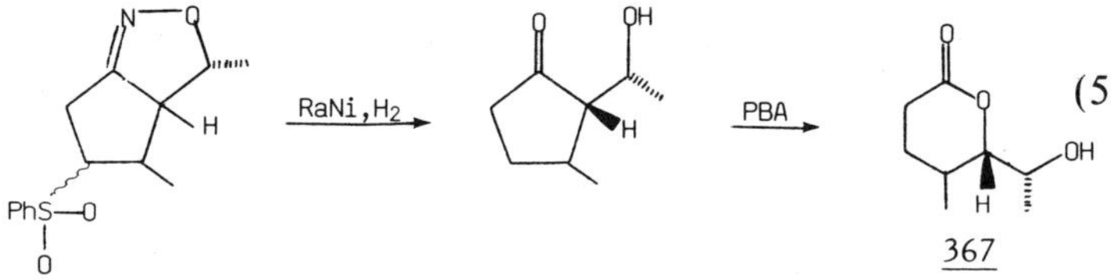

(5.234)

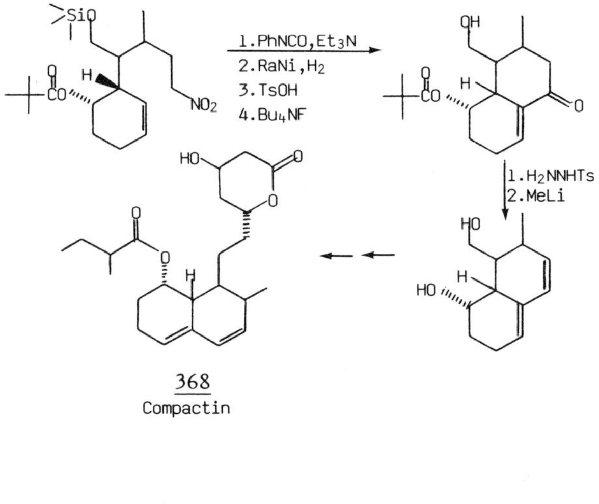

(5.235)

368
Compactin

The oxidative homolytic fragmentation of β-stannyl oximes by the action of lead tetraacetate at −60°C leads to stereospecific formation of unsaturated nitrile oxides. It offers a new approach to isoxazolines involving ring contraction,[604] eq. (5.236). A variety of oxygen-containg ring systems have been prepared,[559,605] eq. (5.237); cf. eq. (2.28).

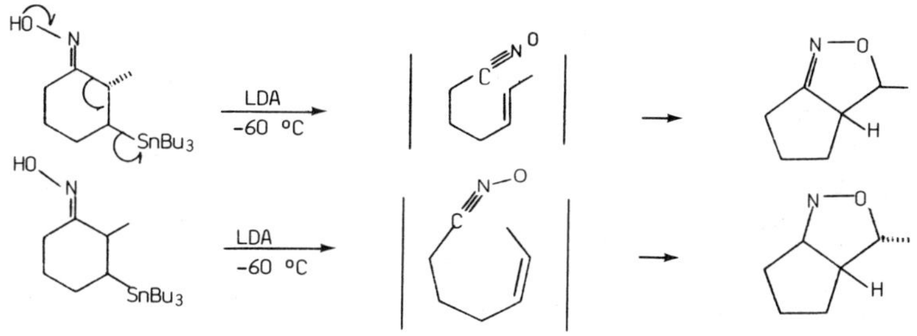

(5.236)

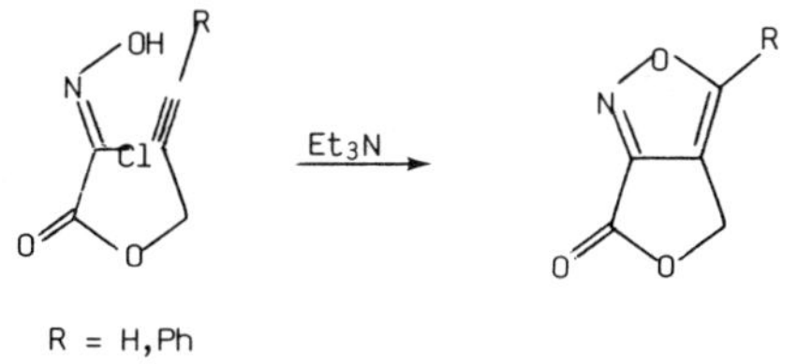

(5.237)

R = H, Ph

5.20 1,3-AMINOALCOHOLS. ALKALOIDS. INTRAMOLECULAR CYCLIZATION

The 1,3-dipolar cycloaddition of nitrones, nitronates, and nitrile oxides to olefins is especially suited for the synthesis of alkaloids because the introduction of the nitrogen atom occurs simultaneously with the assembly of the carbon skeleton. When intramolecular cyclization of C-alkenyl-substituted nitrones occurs, the nitrogen atom becomes a substituent of the carbocyclic ring (cf. Section 5.19), whereas N-alkenyl nitrones give rise directly to the N-heterocycle. The oxygen atom of the dipole is incorporated as a highly desirable hydroxy function, which either can be oxidized to a carbonyl group or eliminated to give unsaturation. Finally, both the regio- and stereoselective characteristics of the cycloaddition are to a great extent known, which enables us to make important predictions about the steric outcome of the reaction.

The isoxazolidines and isoxazolines are masked 1,3-aminoalcohols, and therefore valuable intermediates for a general synthesis of this class of compounds and the potentialities of the procedure were rapidly realized. The reductions of isoxazolidines and 2-isoxazolines proceed smoothly with lithium aluminum hydride to 1,3-aminoalcohols, whereas isoxazoles are hardly affected by this reagent. Catalytic hydrogenation and reduction with titanous ions cleave selectively the N—O bond. The reductions of the isoxazole and their hydrogenated derivatives are discussed and referenced in Sections 1.4–1.6. Amino acids and carbohydrates, which contain the 1,3-aminoalcohol functionality, are discussed separately in Sections 5.21 and 5.22.

The steric structure of the 1,3-aminoalcohol is already embedded in the isoxazolidine to be reductively cleaved and is determined by the structure of the interacting dipole and dipolarophile and, to a minor extent, by the cyclization conditions. Furthermore, it should be kept in mind that the cycloaddition is strictly a cis-addition; thus (Z)- and (E)-olefins give entirely different products. The reduction of 2-isoxazolines creates a new asymmetric center at C^3, and the diastereomeric ratio is controlled by the reducing agent and the substituents at the C^4-, C^5-, and C^3-C^α-carbons. Since 2-isoxazolines **369** are readily available, it is important to find experimental conditions allowing steric control in the reductive step, eq. (5.238). It is shown that $BH_3:S(CH_3)_2$, $NaBH_4$, Zn/HOAc, Na,Hg/H_2O, Na/C_2H_5OH, Na/t-C_4H_9OH, Pd/H_2, and $NaAlH_2(OCH_2CH_2OCH_3)_2$ give poor stereoselectivity,[608–612] whereas $LiAlH_4$ or diisobutyl aluminum hydride (DIBAL) in ethyl ether turn out to be suitable reducing agents both with regard to yield (ca. 80–90%) and to steric discrimination,[203,606,607,613] (ee ca. 70–90%),[606,607] giving the *erythro*-isomer **370** as the major product.

$$\text{(5.238)}$$

			370:371
$R^1, R^3 = H$	$R^2 = CH_3$	$C^\alpha = Ph$	69:31
$R^1, R^3 = H$	$R^2 = i\text{-}C_3H_7$	$C^\alpha = Ph$	85:15
$R^1, R^2 = CH_3$	$R^3 = H$	$C^\alpha = Ph$	90:10
$R^1 = CH_3$	$R^2, R^3 = H$	$C^\alpha = CH_3$	85:15
$R^1 = C^\alpha = Ph$	$R^2, R^3 = H$		95:5

It is observed that the asymmetric induction of the C^5-substituent R^1 is stronger than that of the C^4-substituent $R^{2,3}$. The related acyclic β-hydroxy-O-benzyloximes are stereoselectively reduced by $LiAlH_4$ in THF, eq. (5.239).[615] The *syn*-oxime gives throughout higher enantiomeric ratios: for $R^1 = R^2 = Bu$, **372:371** (from syn) = 95:5, (from anti) = 79:21; for $R^1 = Ph$, $R^2 = CH_3$, **372:371** (from syn) = 88:12, (from anti) = 85:15.

$$\text{(5.239)}$$

Replacement of ether by THF as solvent for the LiAlH$_4$ reduction of 2-isoxazolines strongly influences the product composition in that aziridines become the major product.[204]

Phytosphingosin (**374**) is prepared as depicted in eq. (5.240).[616] Hydroxylation at C^4 and reduction of the C^3=N bond occur predominantly trans to the C^5-alkyl chain. Stereospecific reduction and cleavage of the isoxazoline **375** have been used as one of the key steps in the synthesis of the antibiotic (−)-vermiculine,[614] eq. (5.241).

$$(5.240)$$

$$(5.241)$$

A route to homoallylamines is shown in eq. (5.242).[617] Allyltrimethylsilane adds regioselectively to nitrones and nitrile oxides to give 5-substituted isoxazolidines and isoxazolines. The 1,3-aminoalcohols obtained by reductive cleavage are readily converted into homoallylamines by Peterson-type elimination.

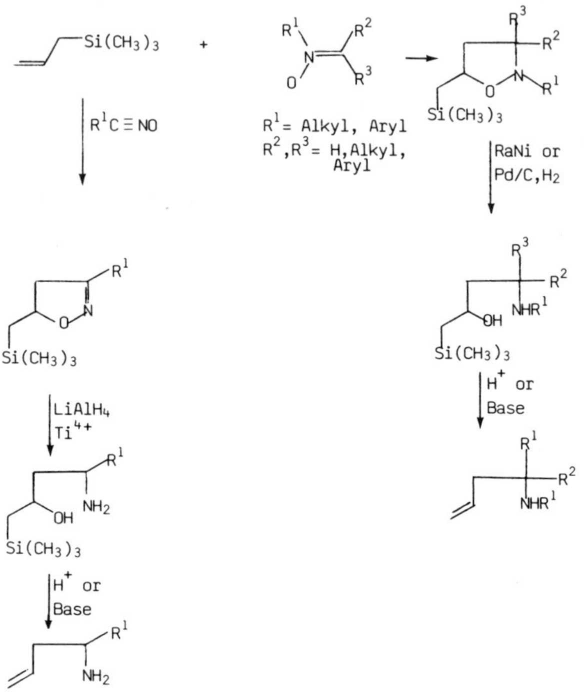

(5.242)

The hepatotoxic pyrrolizidine or *Senecio* alkaloids were early target molecules for the nitrone-based methodology. Methyl 3-hydroxycrotonate cyclo adds regio- and stereospecifically to 1-pyrroline-1-oxide to give the adduct **376**, eq. (5.243). Reductive cleavage of the mesylate and concomitant cyclization leads to the hydroxypyrrolizidine **377**, which on dehydration and reduction with the lithium aluminum hydride–aluminum chloride couple gives (±)-supinidine (**378**).[618] Minor products from the final reduction step are presumably (±)-trachelantamidine (**379**) and (±)-isoretronecanol (**388**).

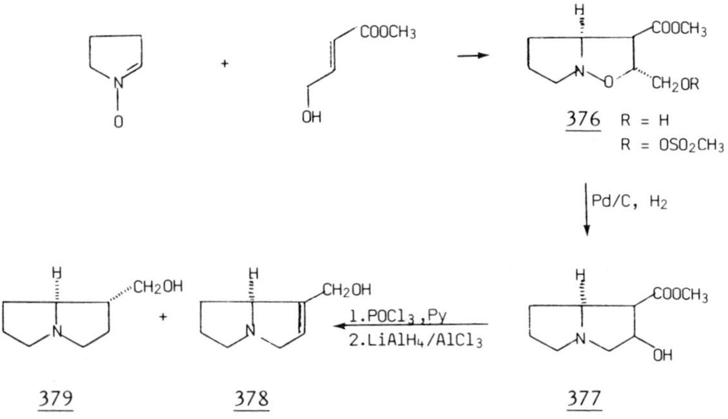

(5.243)

This work was extended to the syntheses of (±)-retronecine (**383**) and (±)-croalbinecine (**384**)[384,619] by using the functionalized ketal nitrone **380** as starting material, eq. (5.244). The pyrrolizidine **381** is readily dehydrated, hydrolyzed, and reduced to **383**. The synthesis of (±)-croalbinecine (**384**) involves an ester epimerization, reduction, and acetylation to **382** before final hydrolysis and reduction to **384**.

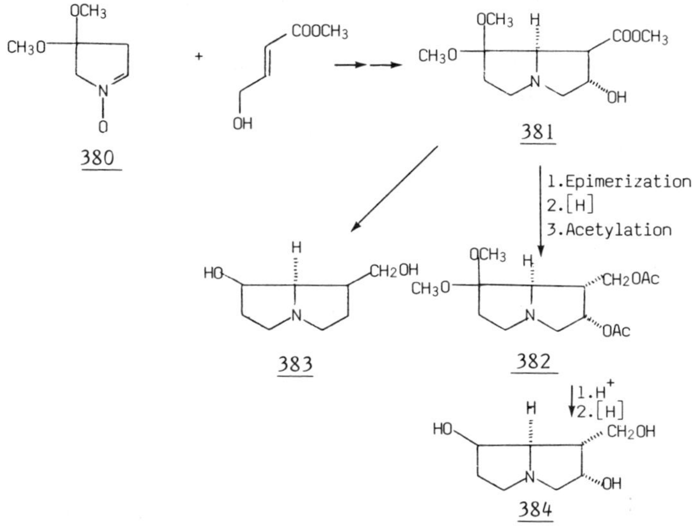

(5.244)

(±)-Isoretronecanol (**388**) is stereoselectively synthesized by the 1,3-dipolar addition of 1-pyrroline-1-oxide to dihydrofuran, eq. (5.245). The exo-transition state **385** is highly favored over the endo-transition state **386** and gives the adduct **387**. Reduction with lithium aluminum hydride in THF followed by silylation, selective iodination, and fluoride ion–induced cyclization give **388**.[620]

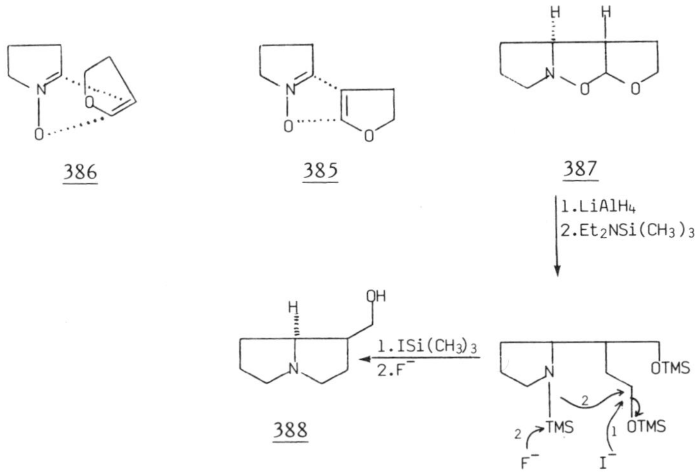

(5.245)

Nitrone methodology has successfuly been used to synthesize indolizidine alkaloids occurring in *Elaeocarpus* species of the rain forests of New Guinea. The readily available 1-pyrroline-1-oxide, which turns out to be useful in a number of alkaloid syntheses, is added to the styrene **389**. The isoxazolidine **390** is reductively cleaved. Oxidation of the hydroxy group gives the amino-ketone **391**, which on treatment with acrolein and base-catalyzed cyclization is converted to **392**. Demethylation and subsequent spontaneous Michael ring closure form a separable mixture of (±)-elaeocarpine (**393**) and (±)-isoelaeocarpine (**394**),[621] eq. (5.246). (±)-Elaeokanine-A (**396**) and (±)-elaeokanine-C (**397**) are synthesized analogously from 1-pentene and 1-pyrroline-1-oxide,[622] eq. (5.247). The stereostructural relationship at C^3 and C^5 in **390** and

395 demonstrates that the reactants are predominantly exo oriented in the transition state.[623] (±)-Eleaokanine-A (**396**) and -C (**397**) have also been synthesized by a slightly different route,[624] eq. (5.248).

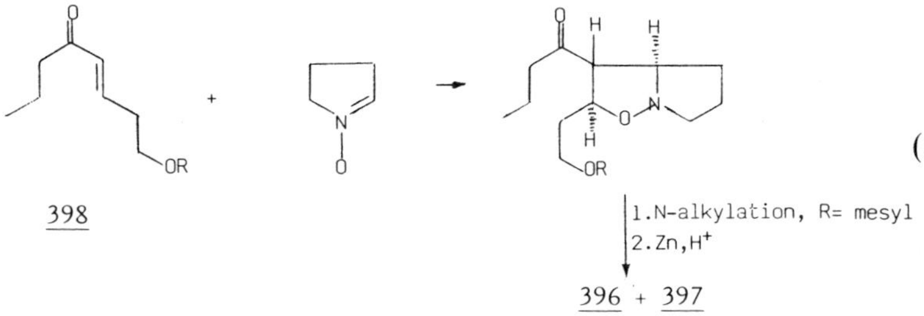

(5.248)

1-Pyrroline-1-oxide is added to the α,β-unsaturated ketone **398** to give the isoxazolidine **399**, which is converted into **396** and **397** by intramolecular N-alkylation by the mesylate and reductive ring cleavage with zinc powder in acetic acid.

(±)-Septicine (**403**), isolated from *Ficus septica*, is prepared by cycloaddition of 1-pyrroline-1-oxide to 2,3-bis(3,4-dimethoxyphenyl)butadiene (**400**) to give a 1:2 mixture of the isoxazolidines **401** and **402**, eq. (5.249). The latter compound is processed further to **403**,[625] which changes into the antitumoreactive tylophorine (**404**) by UV irradiation. An alternative synthesis of **403** and of the photocyclization product **404** is shown in eq. (5.250). Cycloaddition of

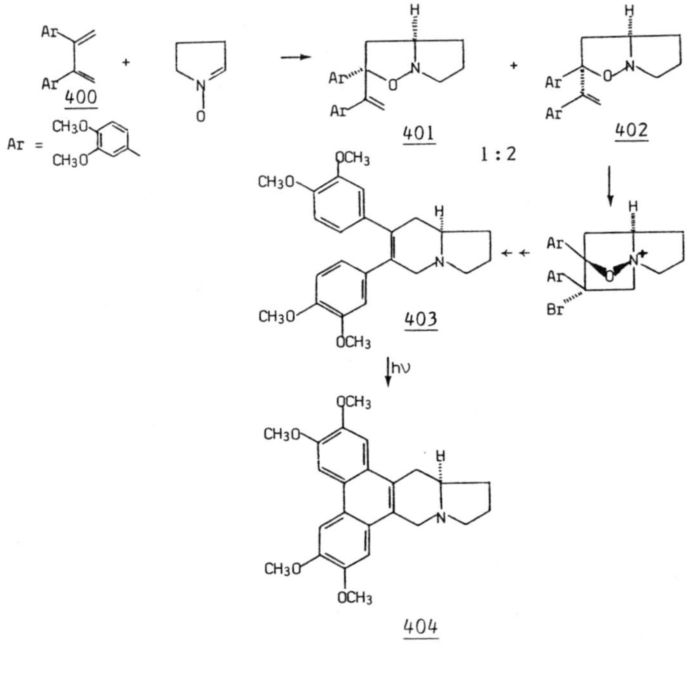

(5.249)

1-pyrroline-1-oxide to 3,4-dimethoxystyrene and sequential reductive ring cleavage and acylation affords **405**. Chromic acid oxidation, aldol condensation, photocyclization, and LiAlH$_4$ reduction complete the synthesis of **404**.[626]

($\pm$)-Ipalbidine (**406**), the aglycone of ipalbine and ipomine, isolated from seeds of *Ipomoea* species, have been synthesized analogously to eq. (5.250) from *p*-allylanisole and 1-pyrroline-1-oxide.[627,628]

The tropane skeleton **407** is available by intramolecular cycloaddition of 5-allyl-1-pyrroline-1-oxides eq. (5.251).[629,630] The addictive alkaloid cocaine, found in *Erythroxylum coca*, which is indigenous to South America, has been synthesized with a high degree of stereochemical control by initial cycloaddition of 1-pyrroline-1-oxide to methyl 3-butenoate in refluxing toluene and subsequent peracid cleavage[570] to a second nitrone **408**,[631-633] eq. (5.252). Since elimination of water cannot be achieved in the presence of the nitrone moiety in **408**, this function is protected with methyl acrylate as the isoxazolidine **409**. Thermal cycloreversion and intramolecular cycloaddition give **410**, which is converted to ($\pm$)-cocaine (**411**).

The bloom-forming cyanophyte *Anabaena flos aquae* produces a highly toxic alkaloid, anatoxin-a (**412**), causing losses of cattle. This alkaloid is structurally related to the tropane alkaloids, and the synthesis is carried out analogously to eq. (5.252) from 1-pyrroline-1-oxide and the dienol **413**,[634] eq. (5.253).

1,3-Aminoalcohols. Alkaloids. Intramolecular Cyclization

(5.252)

A major constituent of the poison gland secretion of the *Solenopsis* thief ants is the *trans*-dialkylsubstituted pyrrolidine **414**. The synthesis of the *trans*-isomer is achieved by double nitrone cycloaddition, eq. (5.254).[384]

A number of alkaloids have been prepared by utilizing Δ^1-piperidine-1-oxide (**415**) as a building block, which gives an entrance to the six-membered ring alkaloids of the piperidine and quinolizidine groups. The methodology is the same as developed for the pyrrolidine analogues. The structurally simple *Sedum* alkaloids ($\pm$)-sedridine (**416**) and ($\pm$)-allosedamine (**417**) are constructed stereoselectively according to eq. (5.255).[623]

If Δ¹-piperidine-1-oxide (**415**) is substituted for 1-pyrroline-1-oxide in the reaction shown in eq. (5.250), the closely related alkaloids (±)-julandine (**418**) and (±)-cryptoleurine (**419**) are obtained,[635] eq. (5.256).

(±)-Lupinine (**420**), a constituent of several *Lupinus* species, is prepared analogously to **388** from **415** and dihydropyran,[620] eqs. (5.245) and (5.257).

Another route to **420** and (±)-epilupinine (**423**) starts from **415** and methyl-(E)-5-mesyloxy-2-pentenoate to give the salt **421**, which is reductively cleaved and dehydrated to the ester **422**, which is reduced to **420**,[636] eq. (5.258).

A route to 4-substituted quinolizidines **424**, **425** has been devised,[384,637] eq. (5.259). Under equilibrating conditions **424** converts into the more stable **425**

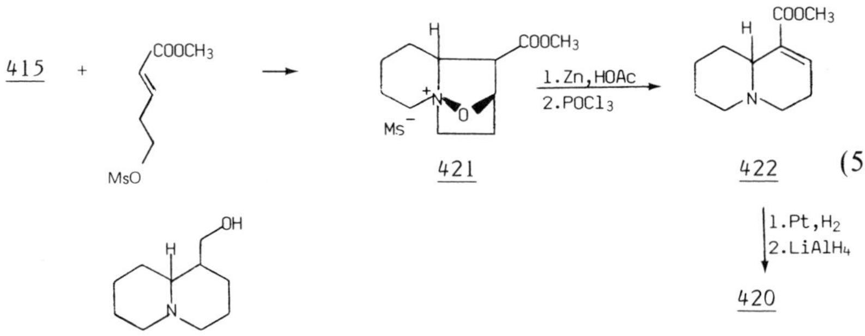

1,3-Aminoalcohols. Alkaloids. Intramolecular Cyclization

(5.258)

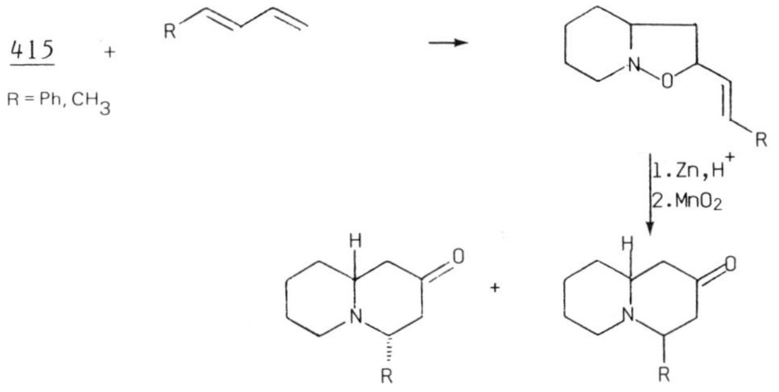

(5.259)

(R = Ph). Similar nitrone methodology has been applied in the synthesis of 4-aryl-2-hydroxyquinolizidines (**426, 427**),[638] (±)-lasubine I,II (**428, 429**),[639] (±)-decaline (**430**), and (±)-vertaline (**431**),[640] having a unique macrolide structure.

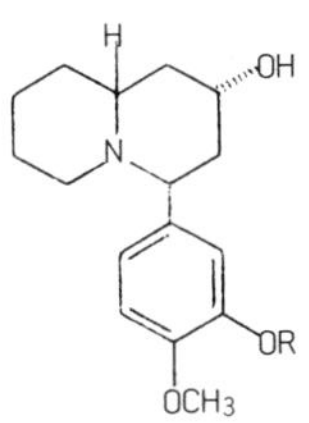

R = H, **426**
R = CH$_3$, **428**

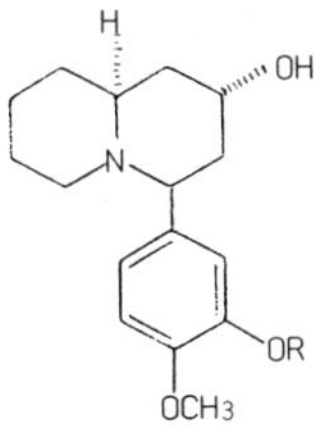

R = H, **427**
R = CH$_3$, **429**

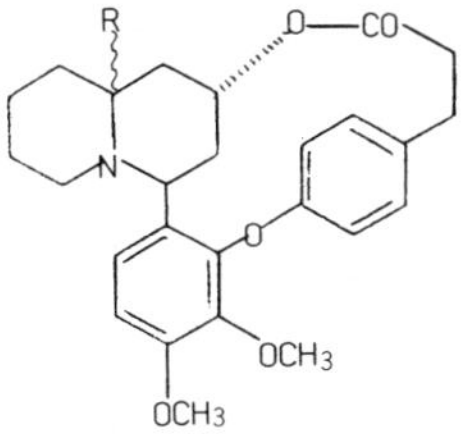

R = α-H, **430**
R = β-H, **431**

The synthesis of **426** and **427** is shown in eq. (5.260). On refluxing the homoallyllic alcohol **432** with **415** in toluene, one obtains the isoxazolidine **433** as a diastereomeric mixture, which is mesylated, cyclized, and reduced to the two quinolizidines **434** and **435**. They could be separated as the acetates. Compound **435** is hydrogenolyzed to the naturally occurring alkaloid **426**. Compound **434** is converted into **427** by inversion of its C^2 center by using the Mitsunobu reaction, (Ph$_3$P, diethyl azodicarboxylate, and benzoic acid) and subsequent hydrogenation to remove the protecting benzyl group in the aromate.

A stereo controlled synthesis of (±)-porantheridine (**436**) and the related alkaloid (±)-porantherilidine (**437**), isolated from *Poranthera corymbosa* (*Euphorobiaceae*), has been reported,[641] eq. (5.261). The nitrone **438**, prepared from **415**, is condensed with 1-pentene to give the *trans*-alkylated adduct **439** on subsequent catalytic reduction over Raney-Ni. Inversion of the C^2 center by using the Mitsunobu reaction, elimination of the benzyl group by catalytic hydrogenation, chromic acid oxidation, and enamination give **440**, which finally is cyclized to **436**.

($\pm$)-Adaline (**441**), extracted from the ladybug *Adalia bipunctata*, has been synthesized according to eq. (5.262).[642]

$$\qquad\qquad\qquad\qquad\qquad\qquad\qquad\qquad\qquad\qquad\qquad\qquad\qquad (5.262)$$

A stereoselective synthesis of α-isospartein (**442**), found in several species of the *Leguminosae* family, has been achieved by adding 2 moles of nitrone **415** to 4*H*-pyran,[643] eq. (5.263). Catalytic reduction of the adduct **443** gives the natural product directly.

$$\qquad\qquad\qquad\qquad\qquad\qquad\qquad\qquad\qquad\qquad\qquad\qquad\qquad (5.263)$$

The electrophysiologically active histrionicotoxins, isolated from the skin of poisonous Columbian frogs,[644] have an unusual spiropiperidine skeleton with acetylenic or allenic functions in the side chain **444a,b**. The alkaloid skeleton has

been assembled from the nitrone **445**, which chiefly gives the adduct **446** and minor amounts of the desired isomer **447**. When **446** is refluxed in toluene, it rearranges slowly into **447**. Catalytic reduction over Raney-Ni gives the model spiro compound **448**,[645] eq. (5.264). Unfortunately the nitrone **449**, which is suitably substituted to be elaborated to perhydrohistrionicotoxin itself, gives the wrong cycloadduct **450**, which refuses to equilibrate to the desired isomer **451**,[645,646] eq. (5.265).

Attempts to assemble the skeleton of certain *Amaryllidaceae* alkaloids by the nitrone route have also been met with resistance. It was assumed that the antitumor agent pretazettine (**452**) could be prepared from the nitrone **453** by retrosynthesis, as shown in eq. (5.266).[647] Model reactions show that 5- and 6-alkenylnitrones give predominantly the desired fused isoxazolidines **454, 455, 456**, but when 6-alkenylnitrones carrying an aryl group at C^6 are tested only bridged isoxazolidines **457** are obtained. Thus, these *Amaryllidaceae* alkaloids are not directly available by this particular method of cycloaddition, eqs. (5.267),(5.268).

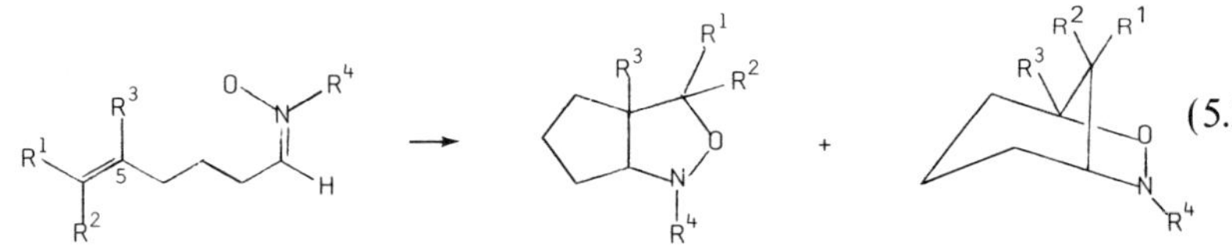

$R^1 = CO_2C_2H_5$, $R^2 = H$, $R^3 = Ph$, $R^4 = CH_3$

454 : **455** = 99 : 1

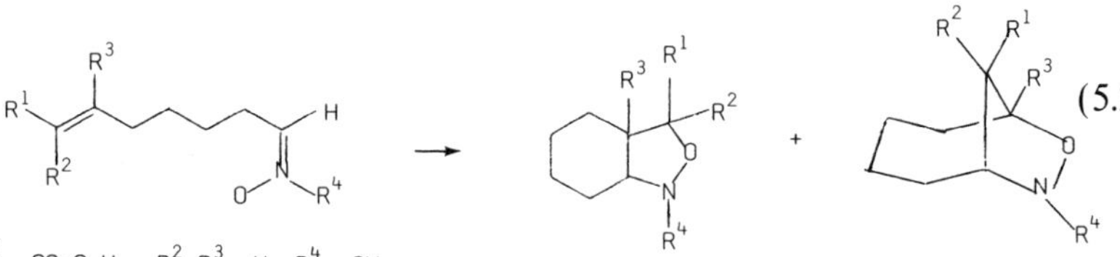

(5.267)

$R^1 = CO_2C_2H_5$, $R^2, R^3 = H$, $R^4 = CH_3$

456 : **457** = 99 : 1

$R^1 = CO_2C_2H_5$, $R^2 = H$, $R^3 = Ph$, $R^4 = CH_3$

456 : **457** = 1 : 99

(5.268)

(±)-Pumiliotoxin-C (**458**), another venomous metabolite in the skin of Columbian frogs, has been synthesized according to eq. (5.269).[557]

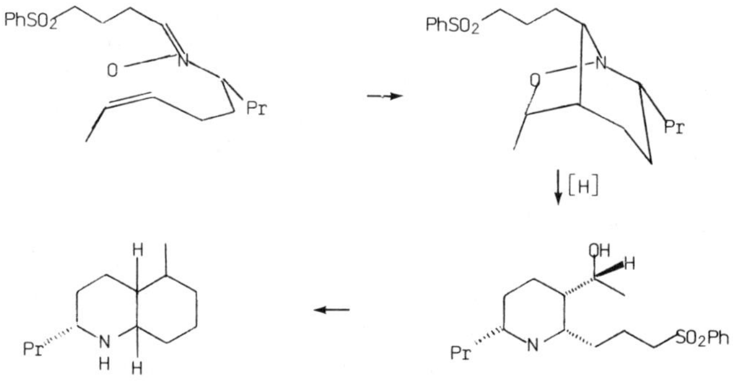

(5.269)

C-Alkenylnitrones have been used in most of the intramolecular cycloadditions described in the preceding sections. The research on intramolecular cycloadditions of N-alkenylnitrones [cf. eqs. (5.251) and (5.269)] is of somewhat later date,[629,630,648] but the characteristic features of these reactions are now known to a great extent.[649] In connection with the synthesis of the *Lycopodium* alkaloid luciduline (**459**), it was found that the nitrone C$^\alpha$-carbon attacked exclusively the nearest olefinic carbon atom,[650] eq. (5.270). The regioisomer **461** was not observed. Steric factors and electronic factors as controlled by frontier orbital overlap are delicately balanced in these cycloadditions. Table 5.4 shows intramolecular cycloadditions of 1-alkenylnitrones.

The observed regioselectivity grossly agrees with the assumptions that in the transition state the carbon-carbon bond is slightly more developed than the carbon-oxygen bond, and that the ring strain is higher in five- and seven-membered transition states, leading to structures **461, 463, 465, 472** ($n = 1$), **475** ($n = 1$), **479**, and **482**, than in six-membered transition states, leading to structures **460, 462, 466, 471, 474, 480**, and **483**. However, in nitrone **464** with an end double bond ($R^1 = H$), ring strain effect is counterbalanced by the more favorable frontier orbital overlap in the transition state, leading to **465**. The outcome of the reaction agrees with the general observation that monosubstituted olefins cycloadd to nitrones and nitrile oxides to give 5-substituted isoxazolidines and isoxazolines. In the nitrone **461** both effects cooperate.

The isoxazolidine **472** ($n = 2$) has been reduced to the alkaloid ($\pm$)-nitramine (**487**),[651] which is structurally related to histrionicotoxin, eq. (5.271).

Isonitramine (**488a**) and sibirine (**488b**) have also been obtained in modest yields by the nitrile oxide route,[653] eq. (5.272).

Table 5.4. Intramolecular Cycloadditions of *N*-Alkenylnitrones

Nitrone	Cycloaddition products Yield (%)		Refs.
461	**462** $R = Ar, R^1 = H$ (72) $R = Ar, R^1 = CH_3$ (85) $R = H, R^1 = CH_3$ (76)	**463** (−) (−) (−)	648 649 649
464	**465** $R, R^1 = H, R^2 = CH_3$ (77) $R, R^1, R^2 = H$ (23) $R = Ar, R^1 = H, R^2 = CH_3$ (93) $R, R^2 = H, R^1 = CH_3$ (−) $R, R^1 = Ar, R^2 = H$ (−)	**466** (10) (48) (−) (95) (95)	649
467	**468** $R = H, R^1 = CH_3$ (23)	**469** (75)	649
470	**471** $n = 1, R = H$ (74) $n = 1, R = Ar$ (97) $n = 2, R = H$ (20)	**472** (−) (−) (50)	649 649 651
473	**474** $n = 1, R = Ar$ (64) $n = 2, R = Ar$ (48)	**475** (−) (32)	649

Table 5.4. (continued)

Nitrone	Cycloaddition products Yield (%)		Refs.
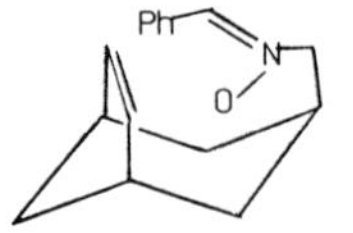 **476**	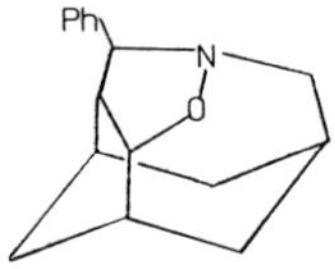 **477**		652
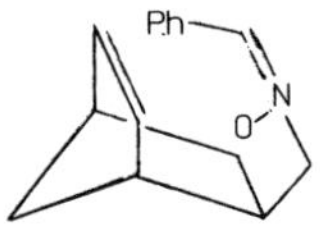 **478**	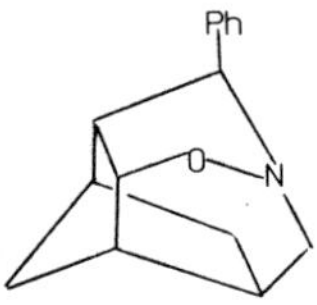 **479** (−)	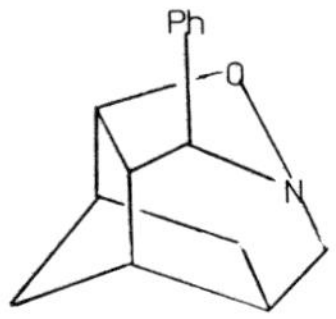 **480** (67)	652
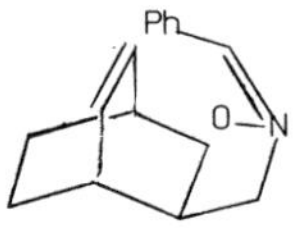 **481**	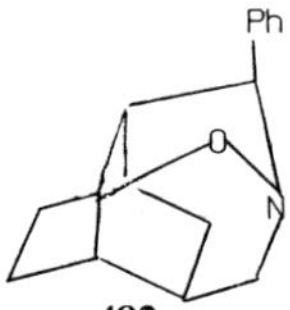 **482** (−)	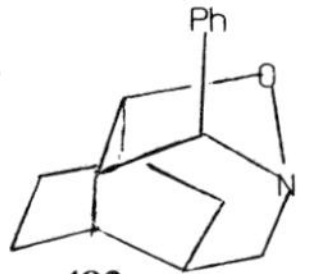 **483** (59)	652
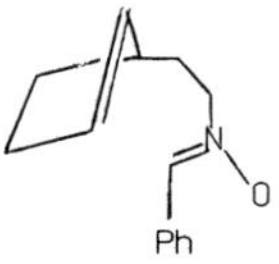 **484**	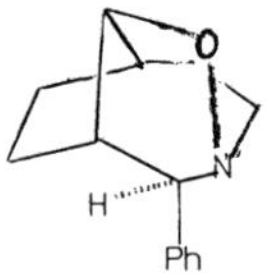 **485** (27)	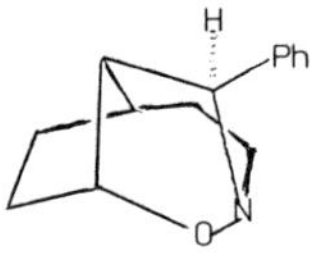 **486** (28)	649

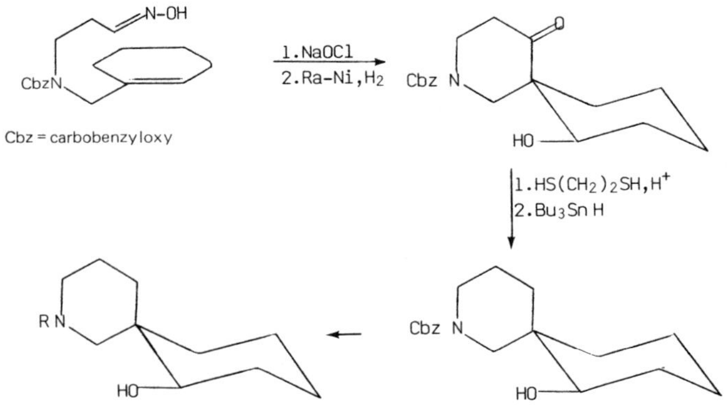

(5.272)

488a R=H
488b R=CH₃

Cbz = carbobenzyloxy

1,3-Aminoalcohols. Alkaloids. Intramolecular Cyclization 227

The 3-pyridylnitrone **489** adds to methyl acrylate to give the isoxazolidine **490**, which has been transformed into the hydroxycotinine (**491**), a metabolite of nicotine, isolated from urine,[654] eq. (5.273). Acetylnornicotyrine (**492**) is prepared from 3-formylpyridine and allylacetamide according to eq. (5.274).[166]

(5.273)

(5.274)

In a search for a new analgesic several analogues of β-prodine (**493**) have been synthesized by utilizing nitrone cycloaddition methodology.[655] Methylenenitrone is added to esters of the general structure **494**, and the isoxazolidines produced are then reductively cleaved and cyclized to tricyclic prodine analogues **495**, eq. (5.275).

The (−)-enantiomer of the antimicrobial and cytotoxic metabolite ptilocaulin (**496**), isolated from the orange Caribbean sponge *Ptilocaulis aff. P. spiculifer*, has been synthesized by employing intramolecular cycloaddition as the key step, eq. (5.276). Guanidation of **497** with 1-guanyl-3,5-dimethylpyrazole nitrate at ca. 150°C gives **496** as the major product.[656]

The pharmacologically highly active ergot alkaloids first isolated from *Claviceps*-type fungi have a structure ideally suited to be assembled by nitrile oxide or nitrone cyclization as the key step. Chanoclavine I (**498**), isochanoclavine I (**499**), 6,7-secoagroclavine (**500**), paliclavine (**501**), chanoclavine II (**502**), and costaclavine (**503**) have been synthesized by nitrone methodology.[657] In reaction eq. (5.277) the synthesis of chanoclavine I (**498**) is depicted. Indole-4-carboxaldehyde (**504**) is subjected to a Wittig reaction, giving the unsaturated ester **505**. The conventional Manich reaction, cyanide displacement and selective reduction of the nitrile with Raney-Ni and sodium hypophosphite give the aldehyde **506**. Treatment of **506** with *N*-methylhydroxylamine hydro-

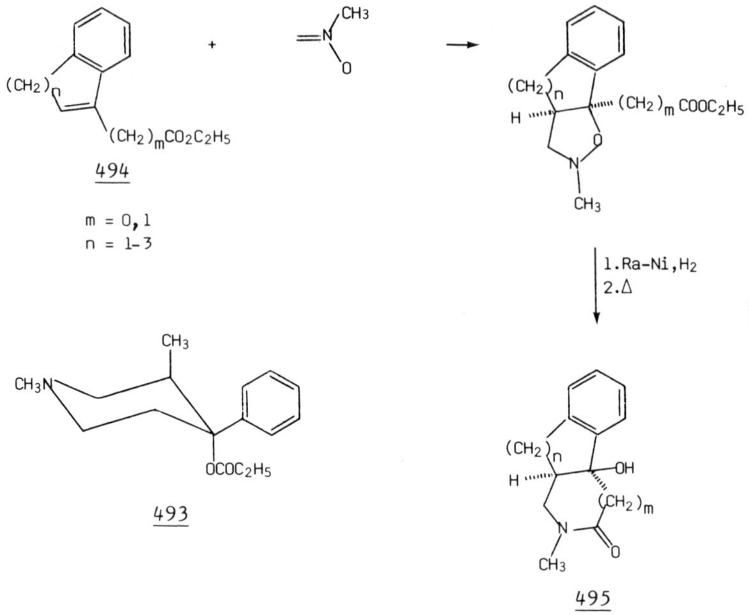

228   Applications in Synthesis

(5.275)

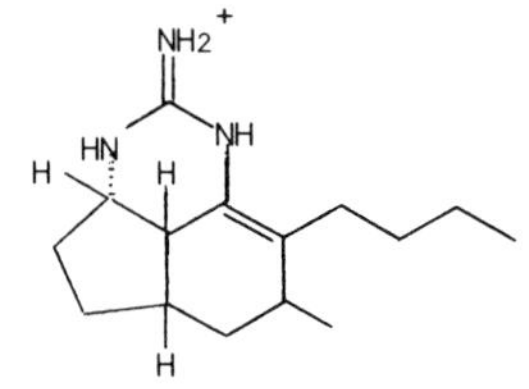

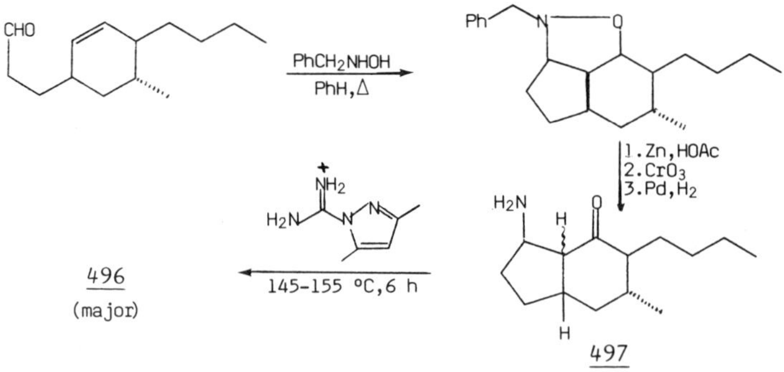

(5.276)

chloride leads to an isomeric mixture of isoxazolidines **507** and **508** in a ratio of 4:1. Each isoxazolidine cycloreverts into the same equilibrium mixture in dichlorobenzene at reflux. Reduction of the ester group in **507** with lithium aluminum hydride, cleavage of the N—O bond by catalytic reduction over

Raney-Ni, introduction of a protecting group, and periodate cleavage of the diol give the *cis*-aldehyde **509**, which slowly epimerizes to the more stable trans-form. Wittig's reaction, removal of the carbo-*t*-butoxy group with trifluoroacetic acid and reduction with DIBAL afford (±)-chanoclavine I (**498**). The intermediate **509** has also been converted to alkaloids **499–503**.

Agroclavine I (**510**) is analogously assembled from the indole derivative **511** by intramolecular cycloaddition with methylhydroxylamine to the tetracyclic isoxazolidine **512**, which on exposure to the ketene silyl acetal of methyl propionate and zinc triflate as Lewis acid gives a stereoisomeric mixture of isoxazolidines **513**,[658] eq. (5.278). Catalytic cleavage of the N—O bond over Raney-Ni and cyclization of the γ-amino ester leads to the lactam **514**. Subsequent elimination of water and lithium aluminum hydride reduction of the lactam in refluxing THF give the desired C,D-cis-fused ergot alkaloid agroclavine I (**510**).

(5.278)

Nitrile oxide cyclization has been used in a slightly different approach to the preparation of chanoclavine I (**498**)[659] and paliclavine (**501**),[660,661] eq. (5.279). The synthesis of **501** starts from the indole **504**, which is treated with the chiral phosphorane **515** to give **516**. Protection of the hydroxyl group as tetrahydropyranyl ether and addition of nitroethylene give an intermediate suitably functionalized for a Mukaiyama cyclization to the isoxazolines **518a,b**. It was hoped that the chiral substituent would induce an asymmetric cyclization and

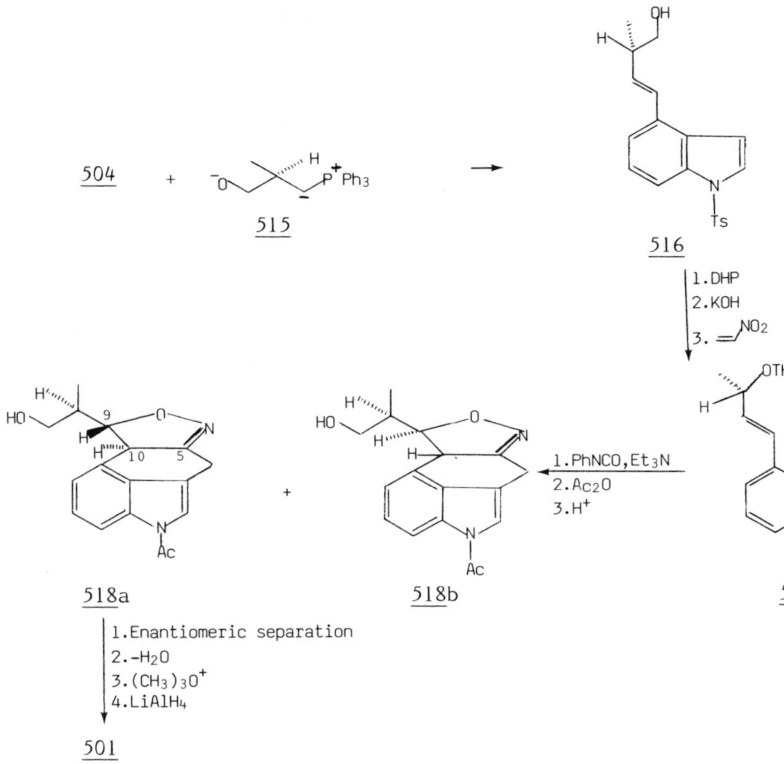

(5.279)

favor the desired enantiomer **518a**. Unfortunately the diastereomeric ratio was close to one, but **518a** and **518b** could be separated into antipodes as mesylates. The reduction of the isoxazoline with sodium borohydride occurs preferentially from the convex face of the molecules to give $C^{5,10}$-cis-fused 5-*epi*-paliclavine, but with lithium aluminum hydride the C^9 substituent exerts sufficient steric hindrance to give some of the desired $C^{5,10}$-trans-fused product which, leads to natural (+) paliclavine (**501**).

An approach is described to the 7-azaindene skeleton of strepazolin (**519**), a

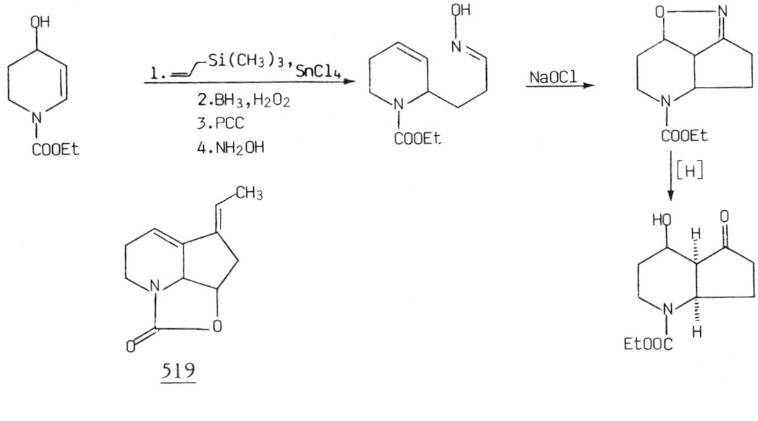

(5.280)

metabolite of *Streptomyces viridochromogenes* with antimicrobial activity,[68,662] eq. (5.280). The *cis*-decahydroquinoline ring system has been prepared in a similar way as a model study toward the synthesis of gephyrotoxin (**520**),[663] eq. (5.281).

(5.281)

5.21 CARBOHYDRATES

For synthetic work in the carbohydrate field the reactions used must be sterically controlled. As discussed earlier, alkylation at C^4 and reduction of the imino bond of 2-isoxazolines show considerable stereoselection, primarily as a result of the nature and number of the substituents at the $C^{4,5}$ positions. The cycloaddition exhibits a high degree of predictability with regard to substituent effects on the stereo structure of the isoxazolines and isoxazolidines formed. Finally, secondary functionalizations, such as hydroxylation of C^5 alkenylisoxazolines are also submitted to significant stereoselection. In several cases these reactions have now been successfully applied to stereospecific syntheses of carbohydrates.

The antitumor, active, anthracyclinone antibiotics adriamycin, daunomycin, and carminomycin contain glycosidically bound L-daunosamine (**521**), which is a 2,3,6-trideoxy-3-aminohexose. Compound **521** and its C^4 epimer, L-acosamine (**522**), have been synthesized by employing the intramolecular cyclization of the chiral nitrone **524** as the key step, eq. (5.282).[664] The nitrone **524** is prepared by heating β-dimethylaminoacrylate (**523**) with chiral (S)-(−)-N-α-phenylethylhydroxylamine. An 82:18 mixture of diastereomeric lactonic isoxazolidines **525a,b** is formed in refluxing xylene. Studies of the steric interactions in the two conformations **524a,b** by Dreiding models indicate that **524b** is slightly favored, which gives rise to the major isomer **525a**. Compound **525a** is reduced with DIBAL, converted to the methyl acetal and hydrogenolyzed over Pd/C, which results in the cleavage of both the N—O and N-benzyl bonds. The anomeric mixture **526** is finally hydrolyzed to acosamine (**522**). The C^4 hydroxyl group is inverted by heating the C^4 mesylate of **522** with aqueous dimethylformamide to give daunosamine (**521**). The preference of the nitrone oxygen to add to the oxygen-bearing carbon atom is reversed. Ring strain in the transition state leading to the bridged [3.2.1]-bicyclooctane system seems to overwhelm the usual regiochemistry. Racemic daunosamine (**521**) has been prepared in a stereospecific fashion by reacting the nitrone **527** with ethyl vinyl ether to give a single isoxazolidine **528**, which is catalytically cleaved in acidified methanol and acetylated to give the methyl glycoside of **521** quantitatively,[665] eq. (5.283). The 2,3,6-trideoxysugar ($\pm$)-rhodinose (**531**) is also prepared from nitrone **527** and vinyltrimethylsilane in a similar fashion. The isoxazolidine **529** is cleaved by hydrofluoric acid, and the α,β-unsaturated aldehyde **530** is reduced and hydrolyzed to **531**,[666] eq. (5.283).

The antibiotic nojirimycin (**536**), isolated from *Streptomyces* and *Bacillus* species, is synthesized by reacting the oxime of 2,3:5,6-diisopropylidene-α-D-mannose **532** and *t*-butyl glyoxalate in an excess of furan. It turned out that the isoxazolidine **533** with desired chirality is formed as the major product,[667] eq. (5.284). Osmylation, protection of the vicinal diol as acetone ketal, catalytic reduction, disposal of the auxiliary mannose derivative, *N*-benzyloxycarbonylation, and lactonization with trifluoroacetic acid give **534**. Reduction of lactone **534** with sodium borohydride and reductive cleavage of the benzyloxycarbonyl group over palladium give the acetonyl derivative **535** of nojirimycin (**536**).

The direction of approach of dipolarophiles and nucleophiles to nitrones has been studied in a few cases.[668-670] The dialkylphosphite anion and methyl methacrylate (COOCH$_3$ in endo position) attack both the configurationally defined spironitrone **537** and the conformationally undefined nitrone **538** predominantly from the front, whereas methyl acrylate attacks **537** and **538** predominantly from the back (COOCH$_3$ in endo position). The formation of **533** conforms with a frontal attack of furan in the endo position. The reason for the observed induction is not well understood.

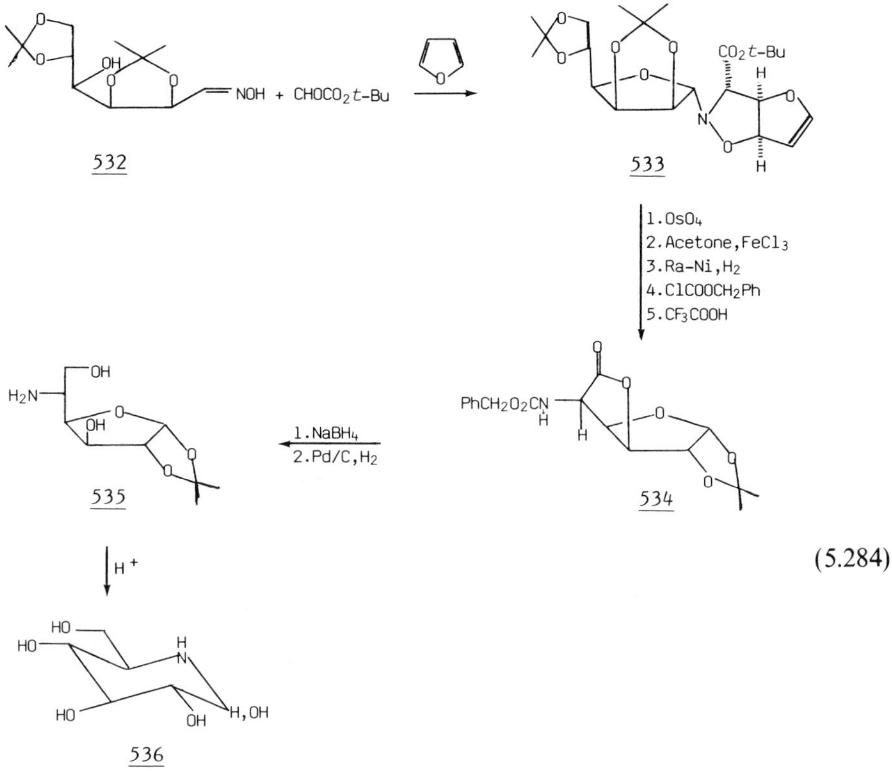

NOH + CHOCO2t-Bu
532
CO2t-Bu
533
1.OsO4
2.Acetone,FeCl3
3.Ra-Ni,H2
4.ClCOOCH2Ph
5.CF3COOH
PhCH2O2CN
534
1.NaBH4
2.Pd/C,H2
H2N
OH
535
H+
HO
N
H,OH
536

(5.284)

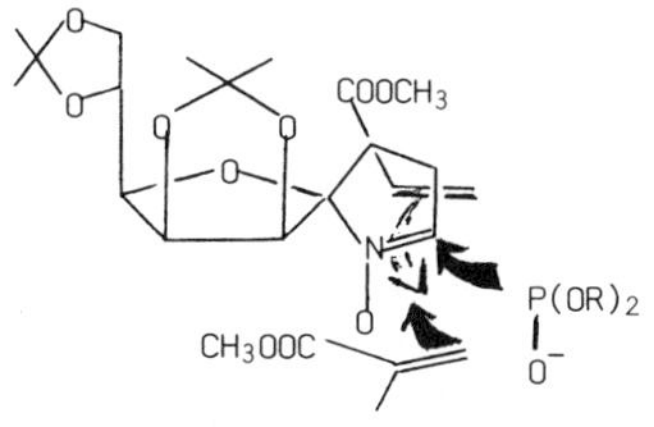

COOCH3
P(OR)2
CH3OOC

537

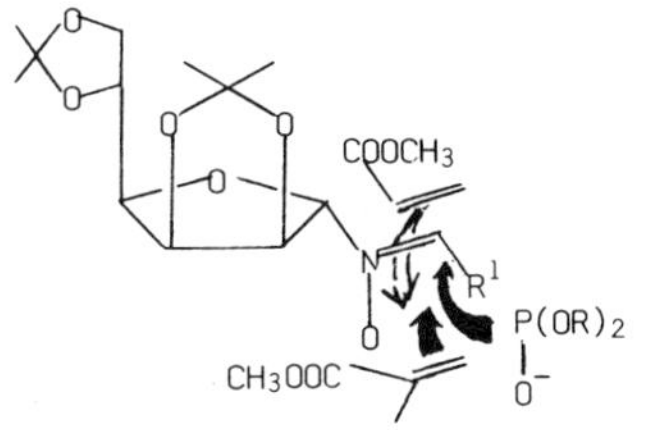

COOCH3
P(OR)2
CH3OOC

538

The condensation of furans with nitrile oxides, generated in situ[414,591,671] has been used for the synthesis of amino sugars.[672–675] Furan reacts sluggishly with nitrile oxides, but the yields of monoadducts **539** are improved by using a dilution technique, large excess of furan, and catalytic amounts of borontrifluoride etherate, eq. (5.285). *m*-Chloroperbenzoic acid oxidation of **539** (R = H, $R^1 = CH_2O$-*t*-Bu) in methanol and lithium aluminum hydride reduction (diastereomeric ratio ca. 95:5) give the *β*-*xylo*-furanoside **540**, a protected form of 5-amino-5-deoxyidose or 5-*epi*-nojirimycin.[672] The lithium aluminum hydride reduction takes place from the sterically less hindered bottom of **539**. A free hydroxyl at C^4 has, however, a strong syn-directing effect that overrides the anti-directing effect of the C^5-substituent. This is demonstrated by ozonization of **539** and acetalization with neopentylglycol to give **541**, which on reduction gives the aminopentose derivative **542** in a diastereomeric ratio of ca. 97:3. The acetonide **543** affords **544** with reversed configuration at C^4 in a diastereomeric ratio of ca. 97:3, eq. (5.286). The 4-amino-4-deoxypentose derivative **547** has been prepared in an analogous way from furoisoxazoline **545**,[674] eq. (5.287).

1,3-Dipolar additions of nitrile oxides, generated from the nitro derivatives **548–550**, to the vinyldioxolanes **551** and **552** give the isoxazolines **553**:**554** in a diastereomeric ratio of ca. 1:4, *threo*:*erythro*, eq. (5.288)[675,676]; cf. Section 1.7. Lithium aluminum hydride reduction proceeded with similar selectivity (ca. 1:4) to give *ribo*-aminopolyols as main products. The isoxazoline **554**, obtained from the nitro compound **550** and vinyldioxolane **551** is reduced to D,L-lividosamine acetal (**555**), which by deprotection with hydrochloric acid gives D,L-lividosamine (**556**). Natural D-lividosamine is prepared from (S)-vinyldioxolane **552** obtained from D-glyceraldehyde.

Optically active **552** was also used in a synthesis of 2-deoxy-D-ribose (**560**). Cycloaddition to carboethoxyformonitrile oxide affords a 4:1 diastereomeric mixture of isoxazolines **557** and **558**,[67] eq. (5.289). Decarboxylation, hydrolysis of the cyano alcohol, and reduction give 2-deoxy-D-ribose.

Novel methodology has been developed for a three-step synthesis of deoxyaldoses and deoxyketoses: (a) regioselective addition of silyl nitronate or nitrile oxide to a diene; (b) stereospecific hydroxylation of the double bond; (c) unmasking of the aldol moiety by catalytic reduction of the 2-isoxazoline.[60] The syntheses of D,L-deoxyribose (**560**), D,L-oleose, D,L-digitoxose (**561**), D,L-2-deoxygalactose, D,L-1,3-dideoxyfructose, and D,L-3-deoxyfructose have been achieved from the major isoxazoline diastereomer. The synthesis of D,L-digitoxose (**561**) is shown in eq. (5.290). The osmylation proceeds with fair stereoselectivity[60,677,678] (ca. 4:1) in the same way as adding nitrile oxide to 5-vinylisoxazolines,[59] vinyldioxolanes, and allyl ethers.[67,675,676,679] Isoxazolines of type **562** or **563** represent useful precursors for amino sugars. In principle, it is possible to prepare lividosamine from **564** by the alternative route shown in eq. (5.291).[678]

$$\underset{\mathbf{564}}{\text{[structure]}} \xrightarrow[\text{2. Separation}]{\substack{1.\,\text{OsO}_4,\text{H}_2\text{O}_2 \\ 3.\,\text{LiAlH}_4}} \underline{556} \qquad (5.291)$$

Another potential route to carbohydrates, not yet fully explored, is represented in eq. (5.292). The isoxazolines **565–567** are readily available from D-glyceraldoxime and allyl acetate, acrolein diethylacetal, or methyl acrylate,[146] respectively. Selective reduction and subsequent rearrangements[137] could conceivably afford 3-deoxyhexoses of type **568**.

$$\underset{\mathbf{565\ \ R=CH_2OAc}\atop{\mathbf{566\ \ R=CH(OC_2H_5)_2}\atop \mathbf{567\ \ R=COOCH_3}}}{\text{[structure]}} \xrightarrow[\text{2. H}^+]{\text{1. Reduction}} \underset{\substack{\mathbf{568}\ x=\text{H, OH}\\ y=\text{OH, NH}_2}}{\text{[structure]}} \qquad (5.292)$$

5.22 AMINO ACIDS

Isoxazoline methodology has been used to synthesize amino acids. In 1964 a simple synthesis of α-hydroxy acids (**570**) was devised,[131] eq. (5.293), and the principle has since been utilized by several workers. Ethoxycarbonylformonitrile oxide is added to monosubstituted olefins to give 5-substituted 3-ethoxy-

$$|\text{C}_2\text{H}_5\text{OCCCNO}| + \underset{\substack{R=H,\,CH_3Ph,\,CH_2Cl,\\ COOCH_3}}{\overset{R}{\diagup\!\!\!\diagdown}} \longrightarrow \underset{\mathbf{569}}{\text{[structure]}} \xrightarrow[\text{2. OH}^-]{\text{1. Pd/C, H}_2\text{ or Na/Hg}} \underset{\mathbf{570}}{\text{[structure]}} \qquad (5.293)$$

$$\underset{\mathbf{571}}{\text{[structure]}} \xleftarrow[R=CH_2Cl]{\text{OH}^-}$$

carbonylisoxazolines, which are reductively cleaved by catalytic hydrogenation or by sodium amalgam to a diastereomeric mixture of the amino acids. It is not possible to stop the reduction at the α-keto acid stage, nor is there any significant asymmetric induction in the reductive cleavage. Ethylene affords homoserine (**570**) (R = H), and propylene, styrene, allyl chloride, and methyl acrylate give γ-methylhomoserine, γ-phenylhomoserine, 4-hydroxyproline,[131] (**571**), and γ-hydroxyglutamic acid[680] (**570**: R = COOH), respectively.

Cycloaddition of N-benzyl-α-ethoxycarbonyl nitrone to methyl acrylate yields the isoxazolidines **572** and **573** in a ratio of ca. 4:1, eq. (5.289). Hydrogenolysis of **572** over Pd and subsequent acid hydrolysis represent an alternative route to *threo*-γ-hydroxyglutamic acid (**577**).[681] By the same method **578** is prepared from isobutylene via **574**, and 4-hydroxyproline[682] (**571**) from acrolein via **575** or **576**.

$R^1 = H$, $R^2 = COOCH_3$
$R^1 = R^2 = CH_3$
$R^1 = H$, $R^2 = CHO$

572 $R^1 = COOCH_3$, $R^2 = H$
573 $R^1 = H$, $R^2 = COOCH_3$
574 $R^1 = R^2 = CH_3$
575 $R^1 = CHO$, $R^2 = H$
576 $R^1 = H$, $R^2 = CHO$

(5.294)

577

578

The amino acids **579** and **580**[683] are components in the nucleosidic nikkomycin antibiotics obtained from *Streptomyces tendae*. They are formed as major products in the nonstereoselective reduction,[683–685] eq. (5.295). The isoxazoline **581**, which is formed together with ca. 7% of the regioisomer **582**, is hydrolyzed, reduced, and demethylated to **579**. The amino acid **580** is prepared analogously from the pyridine derivative **583**.

The stereospecificity of the amino acid synthesis is improved by using lithium aluminum hydride as the reducing agent and a masked 3-carboxyl substituent, such as the anisyl or furyl group, which can be oxidized in a ruthenium-catalyzed reaction at a later stage,[673,686] eq. (5.296). 2-Hydroxycyclopentylglycine (**584**) is obtained as the only isomer in a yield of ca. 40%. N,O-Diacetyl-γ-phenylhomoserine (**585**) is obtained in an analogous fashion from styrene. However, the diastereomeric ratio (d.r.) of 92:8, which is obtained after the reduction, is not conserved during the acidic condition of deacetylation, eq. (5.297). At equilibrium, in acidic solution, the *cis-trans*-lactones occur in a ratio of 85:15, but pure *cis*-lactone **587** crystallizes out as a first crop from the solution. Since the *trans*-lactone **587** is less soluble in concentrated

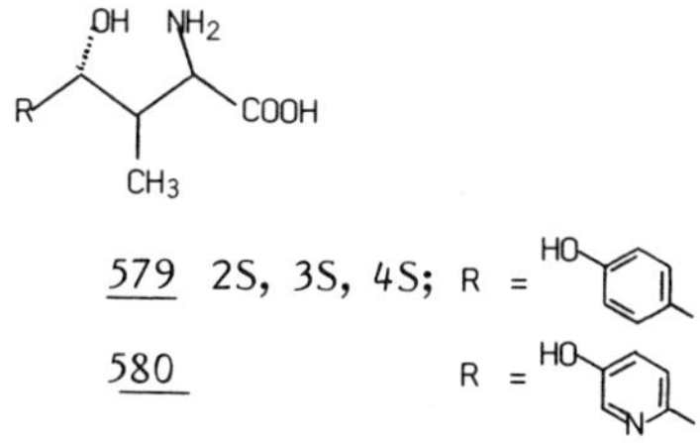

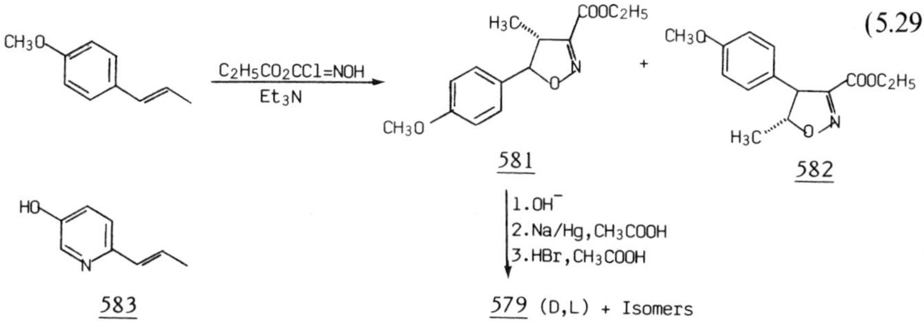

Amino Acids

579 2S, 3S, 4S; R = (4-hydroxyphenyl)

580 R = (6-hydroxypyridin-3-yl)

(5.295)

583

Ar = 4-methoxyphenyl, 2-furyl

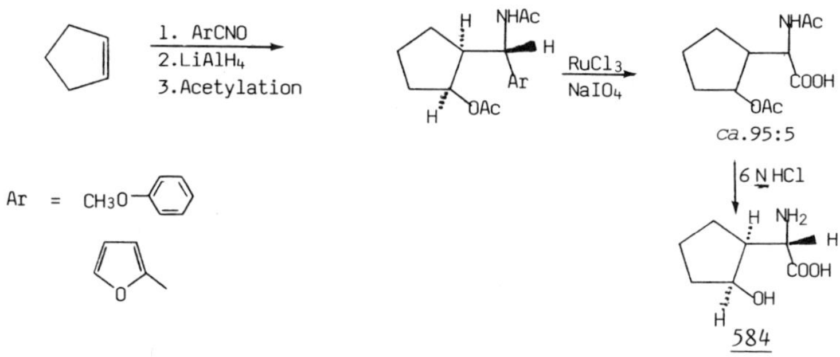

(5.296)

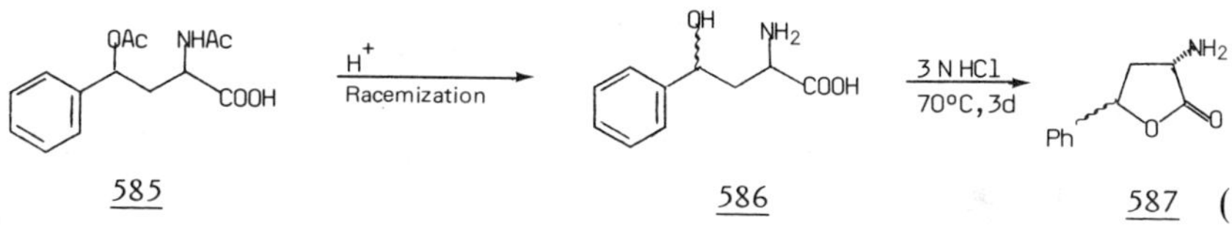

(5.297)

hydrochloric acid, a suspension of **587** is slowly enriched with respect to the *trans*-isomer in the solid phase, *trans*: *cis* ca. 90:10. By this procedure the product from the nonstereospecific reduction[131] can also be separated into its enantiomers. The procedure has been applied to the separation of the methyl ether of **579**. It is apparently possible to circumvent the protection-deprotection route, eq. (5.296), by direct stereoselective reduction (d.r. ca. 15–20:1) of the 3-*N*-*t*-butylcarboxamidoisoxazoline **588** with [(CH$_3$OCH$_2$CH$_2$O)$_2$AlH$_2$]Na, (Red-al), at $-78°C$, as shown in eq. (5.298).[687] The reported resistance of **589** towards racemization under acidic conditions[687] is, however, contradicted by a later report[673] and the d.r. value reported is actually the reverse of the anticipated value.

[Scheme showing compound 588 → 589 (major) + 590 (minor), reagents: 1. Red-al, THF −78°C; 2. PhCOCl, NEt$_3$] (5.298)

Intramolecular cycloaddition of α-nitrone esters has been reported to give geminally functionalized carbocyclic γ-hydroxy-α-aminoesters,[688a] eq. (5.299). The geminally functionalized heterocyclic amino acid **591** is obtained by cycloaddition of ethyl α,β-dehydroalaninate with benzonitrile oxide.[688b]

[Scheme for eq. (5.299): vinyl-substituted ketoester (R = H, CH$_3$) with COOEt group, 1. CH$_3$NHOH, 2. Rh/C, H$_2$ → cyclopentane with NHCH$_3$, COOCH$_3$, OH substituents] (5.299)

[Structure 591: isoxazoline bearing AcHN, COOC$_2$H$_5$, and Ph substituents]

The amino alcohol part (i.e., chiral hydroxyputresine) of the eucaryotic translation initiation factor eIF-4D (hypusine **592**) is synthesized by nitrone addition to allyl alcohol, eq. (5.300). Diastereomeric separation by preparative high performance liquid chromatography (HPLC), dimethysulfoxide oxidation, reductive amination with lysine, and reductive ring cleavage over Pd/C give the chiral amino acid **592**.[689] α-Amino acids **594** are formed by acylamination of cyclic nitrones **593** with N-phenylbenzimidoyl chloride in the presence of triethylamine, eq. (5.301),[690] similar to an earlier reported procedure for α-hydroxylation of nitrones.[154,691,692]

Several naturally occurring, biologically active isoxazoles have been isolated, such as cycloserine (**595**), muscimol (**596**), ibotenic acid (**597**) [from *Amanita* species], (αS,5S)-α-amino-3-chloro-4,5-dihydroisoxazole-5-acetic acid (AT-125), acivicin [from *Streptomyces sviceus*] (**598**), and tricholomic acid (**599**) from the mushroom *Tricholoma muscarium*, this has created interest in their syntheses and in the synthesis of analogues.

The obvious approach to the synthesis of AT-125 **598** is by adding chloronitrile oxide, generated in situ from phosgene oxime and base, to vinyl-glycine, eq. (5.302); however, none of the expected material was present in the

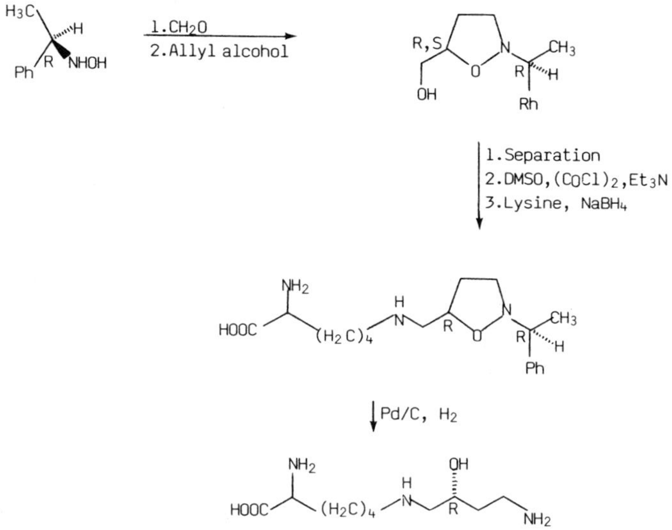

(5.300)

(5.301)

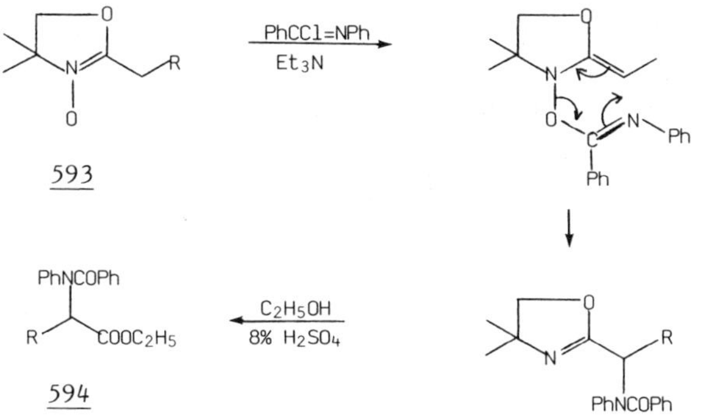

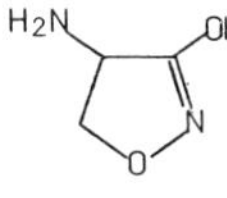

595 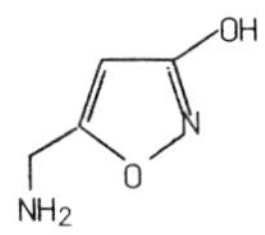 **596** 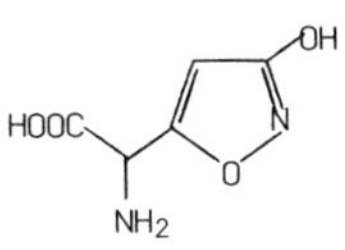 **597**

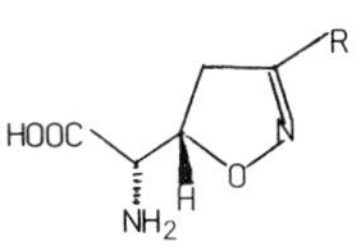

598 R=Cl
599 R=OH
600 R=Br

reaction product.[693] But chloronitrile oxide has been found to add to the more reactive nitronate salt **601** to give the adduct **602** as a mixture of diastereomers. Subsequent reduction of **602** with zinc amalgam gives **603**, a C^5 methyl homologue of **598**. Silver nitrate–generated chloronitrile oxide cycloadds to (S)-vinylglycine, protected by phthalylation, to give the desired **598** isomer after chromatographic separation of diastereoisomers (71:29) and deprotection.[694]

Generation of the more reactive bromonitrile oxide in the presence of vinylglycine leads to the corresponding 3-bromoisoxazoline **600**,[695] which gives **598** on treatment with refluxing methanolic hydrochloric acid.[696] Several other syntheses of **598**[697-700] and analogues[701] have been reported as well as the synthesis of the related tricholomic acid[702-706] (**599**) and of muscimol (**596**).[707] Paraformaldehyde and 5-hydroxypentane-1-oxime (**604**) react with alkenes to give isoxazolidines **605** and isoxazolines by subsequent oxidation with N-chlorosuccinimide,[708] eq. (5.303). The amino acid **606** is the dechloro derivative of AT-125 **598**. A synthesis of 3-chloroisoxazoles based on treatment of N-methyl-3-isoxazolinones with phosphorus oxychloride has been described,[709] eq. (5.304).

A most direct route to 3-isoxazolinones **608** seems to be the reaction of β-ketoesters with hydroxylamine. It turns out that this route preferentially leads to 5-isoxazolines **607**, eq. (5.305). However, it has been demonstrated that the pharmacologically interesting 3-isoxazolinones **608** become the major product, provided that the pH is kept at about 10 throughout the reaction and that the reaction mixture is quenched with an excess of strong acid.[710] Conjugate addition of N-substituted hydroxylamines to α,β-unsaturated esters followed by cyclization with lithium bis(trimethylsilyl)amide provides a procedure for synthesizing 5-isoxazolidinones **609**, which may be reduced to β-amino acids **610**,[711] eq. (5.306).

The γ-lactam analogue **613** of carbapenicillinic acid has been synthesized by the 1,3-dipolar addition of 5-methoxycarbonyl-1-pyrroline-1-oxide (**611**) to methyl acrylate. Reduction with Raney-Ni, cyclization, and separation of isomers afford the desired γ-lactam **612**, which is transformed to **613** by standard methods,[712] according to eq. (5.307). The compound is pharmacologically inactive.

The proline analogue **615**, 5-oxaproline, has been asymmetrically synthesized by the cycloaddition of ethylene to N-glycoside nitrones acting as chiral auxiliary. They are formed in situ by the condensation of protected D-mannose- or D-riboseoximes with glyoxalates,[713] eq. (5.308). The diastereoselectivity is in the range of 36–72%. The captopril analogue **616** is prepared by reacting **615** with methacryloyl chloride in pyridine and subsequent conjugate addition of thioacetic acid. Captopril functions as an inhibitor of certain carboxypeptidases.

L-Proline (**618**), L-homoserine (**619**), L-aspartic acid (**620**), and L-asparagine (**621**) analogues, in which a phosphate group replaces the carboxy group, have been synthesized according to eq. (5.309).[714]

5.23 MACROCYCLES, BIOTIN, MISCELLANEOUS NATURAL PRODUCTS

The 16-membered dilactone ring of pyrenophorin (**624**), produced by the pathogenic fungus *Pyrenophora avenae*, is constructed by the silyl nitronate method,[715] eq. (5.310). The silylated acryl ester **622** dimerizes rather than undergoes intramolecular cyclization, because of unfavorable orbital overlap and steric strain in the seven- or eight-membered transition states leading to the monomers. Elimination of silanol, reductive cleavage of the isoxazoline ring, and elimination of water give a mixture of ($\pm$)- and *meso*-**624** in a ratio of 1:1. The analogue ($-$)-vermiculine (**625**), isolated from *Penicillium vermiculatum*, has also been synthesized by using nitrile oxide cycloaddition in one of the key steps,[614] (eq. (5.241). The monomeric units **626** and **627**, corresponding to the dimeric macrocycles **624** and **625**, have been prepared by the nitrile oxide method, thus affording a formal synthesis of **624** and **625**.[716] Intramolecular cycloaddition of ω-nitroalkylacrylates leads to the macrocyclic antibiotic (±)-A26771B[717] **628** and 12- and 14–16-membered macrocarbocycles.[718] The 15-membered derivative has been converted into (±)-muscone (**629**), eqs. (5.311), (5.312).

The insecticidal milbemycins and avermectins produced by *Streptomyces* species contain a complex molecular framework presenting a challenge to the organic chemist. The spiroketal part **635** of milbemycin β_3 (**630**) is constructed by cycloaddition of nitrile oxide, obtained from the nitroacetal **633**, to the ketal **634** followed by reduction and elimination of the amino group,[719] eq. (5.313). The intermediary formed α,β-unsaturated aldehyde undergoes ring closure to **635**. Optically active hexahydrobenzofuran components **638**[720] and **644**[721] of avermectin A_{2a} (**632**) and milbemycin α_6 (**631**) have been synthesized from glucose diacetonide and glyceraldehyde acetonide, respectively, eqs. (5.314), (5.315). Glucose diacetonide is converted in several steps to the nitro compound **636**, which is cyclized intramolecularly via the nitrile oxide route, and reduced to **637**. Subsequent conversion to the appropriately functionalized hexahydrobenzofuran derivative **638** is carried out by standard procedures.

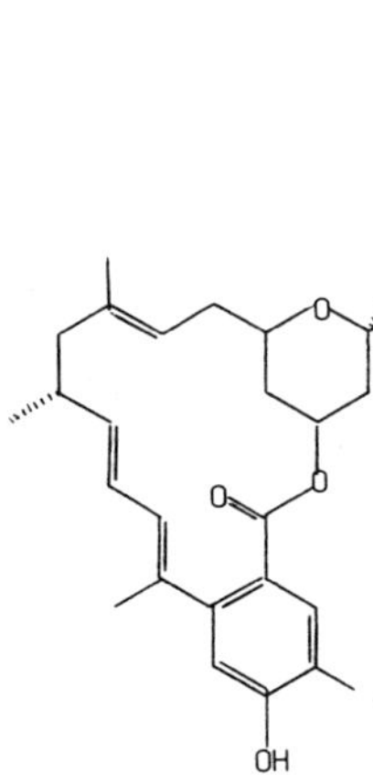

630

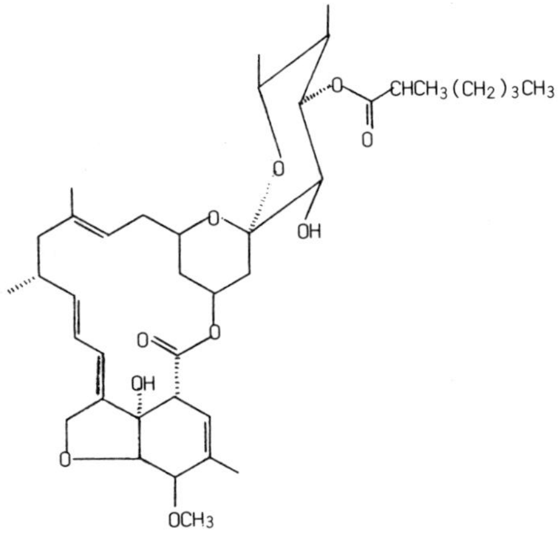

631

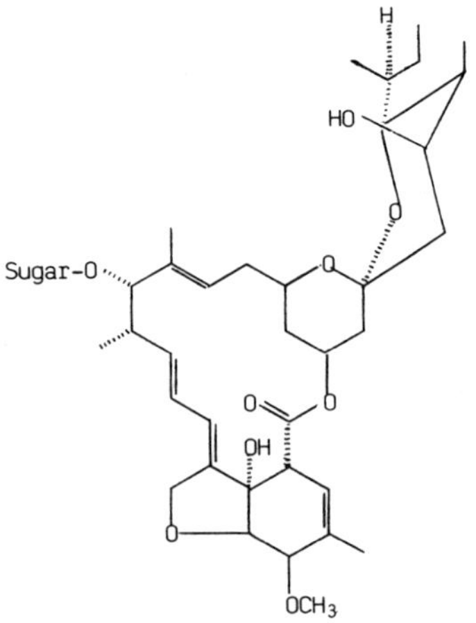

632

Glyceraldehyde acetonide has been treated with the sulfone **639** to give **640**, which in several steps is converted to the oxime **641**. Treating **641** with sodium hypochlorite generates the nitrile oxide, which undergoes intramolecular cyclization to a mixture of isomeric isoxazolines. It has been shown that the desired isoxazoline **642** is one of the major isomers. The methoxymethyl group (MOM) is removed by acid hydrolysis and replaced by a 2-bromo-2-propenyl group. Reductive cleavage with titanous ions, protective silylation, and butyl lithium copper–promoted cyclization give the desired reduced furan system **644**.

Maytansine (**645**) represents a group of antitumor active macrolides originally isolated from the East African plant *Maytenus serrata*. Model studies for the total synthesis have been carried out based on intramolecular nitrile oxide macrocycloaddition.[722] The isoxazoline moiety formed furnishes the 1,3-aminoalcohol functionality embedded in the 1,3-oxazin-2-one ring. The immediate targets are the macrocycle **646** and the functionalized macrocycle **647**, embodying the strained diene system. Compounds **648** and **649** are macrocyclized to **646** and **647**, respectively, in satisfactory yields on treatment with arylisocyanate and triethylamine, eqs. (5.316),(5.317).

The important metabolic cofactor biotin (**653**) has been synthesized by various routes involving dipolar cycloaddition chemistry. The appropriately functionalized thieno compound **650** is formed by intramolecular nitrile oxide cyclization,[723] eq. (5.318). It is reduced by the Zn/Ag couple to a mixture of **651** and minor amounts of the undesired thiophene **652**, which is not convertible into biotin. Catalytic hydrogenation of **650** gives **652** as the major product. Biotin is obtained by subsequent catalytic reduction of **651** and treatment with phosgene.

A second route for preparing **653** starts from cycloheptene, which is transformed to the nitro derivative **654**. Intramolecular cycloaddition and reduction with lithium aluminum hydride gives the aminoalcohol **655** with correct stereochemistry. Dimethylsulfoxide oxidation and Beckmann rearrangement of the oxime give the lactam **656**, which is converted to biotin **653**,[724] eq. (5.319).

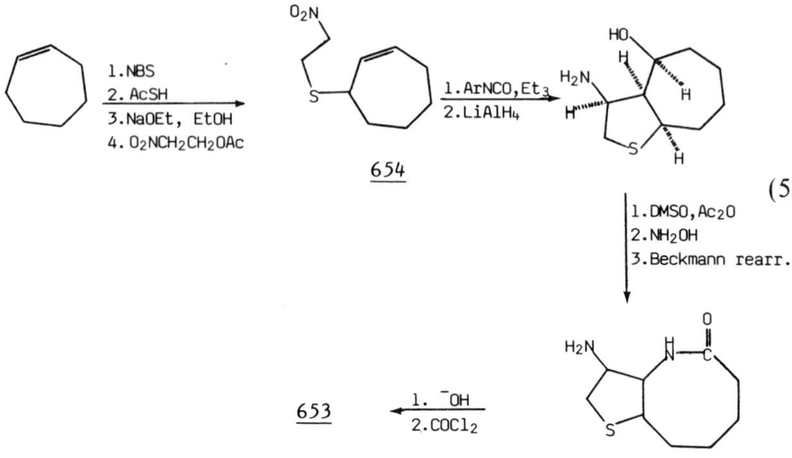

252   Applications in Synthesis

(5.319)

A third procedure starts from optically active L-cystein which is converted into the nitrone **657** with locked conformation by the 10-membered ring. Cycloaddition gives exclusively the desired chiral isomer,[725] eq. (5.320).

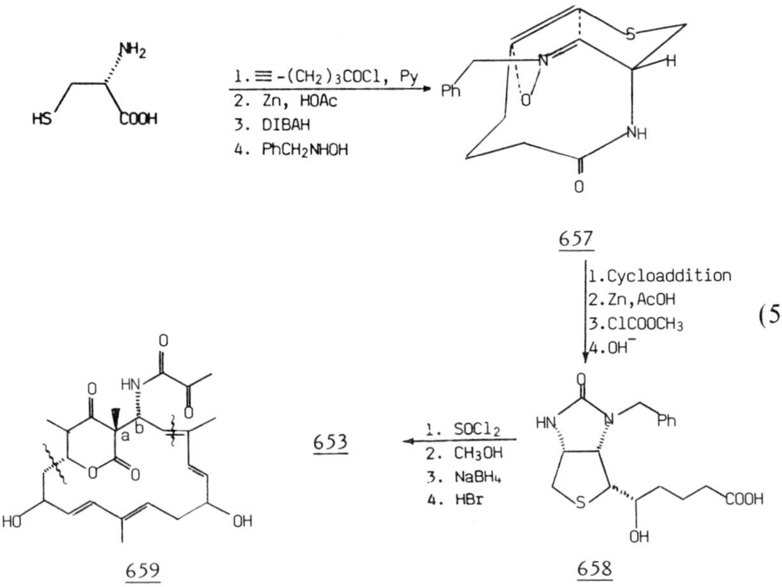

(5.320)

Lankacidin-C (**659**) represents another group of macrocyclic antibiotics that show antitumor activity. Attempts to synthesize the pyrone portion of the molecule have been reported.[726] 5-Substituted isoxazolines **660** are converted to cis-4-carbomethoxy-trans-4-methyl isoxazolines **661** by deprotonation using lithium diisopropylamide, followed by sequential addition of methyl chloroformate and methyl iodide. N-Methylation and reduction with sodium borohydride give modest yields of the isoxazolidine **662** of correct relative stereostructure at C^3 and C^4 (= a, b in **659**) for conversion into lankacidin analogues, eq. (5.321).

$$\text{660} \xrightarrow[\substack{1.\text{Li}(\text{iso Pr})_2\text{N} \\ 2.\text{ClCOOCH}_3 \\ 3.\text{CH}_3\text{I}}]{} \text{661} \xrightarrow[\substack{1.\overset{+}{\text{O}}(\text{CH}_3)_3 \\ 2.\text{NaBH}_4}]{} \text{662} \quad (5.321)$$

An early application of intramolecular nitrone cyclization has been reported in experiments on the synthesis of tetracyclins,[727] eq. (5.322). Coupling the B and D rings in **663** is done by treating the aldehyde with phenylhydroxylamine to yield the desired linear skeleton **664** via the nitrone.

$$\text{663} \xrightarrow{\text{PhNHOH}} \text{664} \quad (5.322)$$

The spiroketal toxic metabolite talaromycin B (**669**), isolated from the fungus *Talaromyces stipitatus*, is constructed by reacting the olefin **665** with the oxime **666** in the presence of sodium hypochlorite to give the isoxazoline **667**.[728] Reduction and acid-catalyzed intramolecular ketalization give **668**, which is converted to talaromycin B by methylation at C^{13}, eq. (5.323).

$$\text{665} + \text{666} \xrightarrow{\text{NaOCl}} \text{667} \xrightarrow[\substack{1.\text{RaNi},\text{H}_2 \\ 2.\text{H}^+}]{} \to \text{668} \to \text{669} \quad (5.323)$$

Since *C*-glucosides constitute a group of pharmacologically important compounds, extensive research has been devoted to the synthesis of analogues of the natural *C*-glucosides. From a number of sugar-based nitrones and nitrile oxides or alternatively unsaturated sugars, novel isoxazole derivatives have been prepared.[688,729-733] A few structures are exemplified in formulae **670–674**.

Blastmycinone (**674**),[734] a degradation product from the antibiotic antimycin A_3 is prepared by hydroxy carboxylation of the silyl ether of 3-buten-2-ol, eq. (5.324). The isoxazoline **675**, which is obtained as the major stereoisomer, is *trans*-alkylated with butyl iodide and reductively cleaved to give **676**. Elimination of the tetrahydropyranyl group, periodic cleavage, reductive debenzylation, lactonization, and esterification give **674**.[67]

(5.324)

CONCLUDING REMARKS

Thirteen years have elapsed since A. I. Meyer's "Heterocycles in Organic Synthesis" appeared. This monograph reviewed the use of heterocycles as an auxiliary in synthesis. A few pages were devoted to isoxazoles and their hydrogenated derivatives. The last decade has witnessed a tremendous development in the application of these heterocyclic systems to basic organic synthesis. This book, which is of comparable size to the cited monograph, reviews the isoxazole system and its precursors, namely, the nitrones, nitrile oxides, and nitronates.

It is evident that cycloadditions of nitrones, nitrile oxides, and nitronates to olefins and acetylenes leading to the isoxazoles, isoxazolines, and isoxazolidines, and subsequent reductive unmasking, is an excellent carbon-carbon coupling procedure. It represents a general and effective approach to a number of hydroxylated, carbonylated, aminated, and alkylated hydrocarbons, which are able to serve as building units in subsequent syntheses. In this connection should be mentioned, for example, the isoxazoline route to aldols, which supplements the classical Claisen condensation or the facile construction of condensed carbocyclic or heterocyclic ring systems. The method has already had celebrated triumphs in the field of synthesis of natural products, and its applications will be extended in this area. One can easily visualize a number of terpenoid and alkaloid skeletons, which are well suited to be assembled by the dipolar technique.

It may not be presumptious to anticipate that the methodology displayed in this book will belong to the basic arsenal of the synthetic chemist and, as such, be integrated in the synthetic planning in the same way as the Claisen, Diels-Alder, Wittig, Grignard, or Friedel-Crafts reactions function as cornerstones in synthesis.

REFERENCES

1. K. Torssell; O. Zeuthen, *Acta Chem. Scand. B32* (1978), 118.
2. E. W. Colwin; D. Seebach., *J. Chem. Soc., Chem. Commun. 1978*, 689.
3. E. W. Colwin; A. K. Beck; D. Seebach, *Helv. Chim. Acta 64* (1981), 2264.
4. D. Seebach; A. K. Beck; T. Mukhopadhyay; E. Thomas, *Helv. Chim. Acta 65* (1982), 1101.
5. V. Bertini; A. De Munno; P. Pino, *Chim. Ind. (Milan) 48* (1966), 491; *Gazz. Chim. Ital. 97* (1967), 173, 185.
6. A. De Munno; V. Bertini; F. Lucchesini, *J. Chem. Soc., Perkin II 1977*, 1121.
7. M. L. Casey; D. S. Kemp; K. G. Paul; D. D. Cox, *J. Org. Chem. 38* (1973), 2294.
8. L. Claisen; R. Stock, *Chem. Ber. 24* (1891), 130.
9. L. Claisen, *Chem. Ber. 25* (1892) 1776; *42* (1909), 59.
10. C. H. Eugster; L. Leichner; E. Jenny, *Helv. Chim. Acta, 46* (1963), 543.
11. S. Cusmano; S. Giambrone, *Gazz. Chim. Ital. 80* (1950), 702; F. J. Vinick; Y. Pan; H. W. Gschwend, *Tetrahedron Lett. 1978*, 4221.
12. P. Pino; R. Ercoli, *Rend. Ist. Lombardo Sci. Pt. I, 88* (1955), 378.
13. R. Justoni, *Rend. Ist. Lombardo Sci. Pt. I 71* (1938), 407.
14. N. K. Kochetkov, *Izv. Akad. Nauk SSSR, Ser. Khim. 1954*, 47.

15. L. Panizzi, *Gazz. Chim. Ital. 73* (1943), 13.
16. L. Panizzi, *Gazz. Chim. Ital. 72* (1942), 475.
17. A. Quilico; L. Panizzi, *Gazz. Chim. Ital. 72* (1942), 458.
18. L. Panizzi, *Gazz. Chim. Ital. 77* (1947), 206.
19. A. Quilico; G. Stagno d'Alcontres, *Gazz. Chim. Ital. 79* (1949), 654.
20. H. Yasuda, *Yakugaku Zasshi 79* (1959), 836.
21. S. Tagaki; H. Yasuda, *Yakugaku Zasshi 79* (1959), 467.
22. H. Yasuda, *Yakugaku Zasshi 79* (1959), 623.
23. A. Quilico; G. Gaudiano; L. Merlini, *Tetrahedron 2* (1958), 359.
24. H. Kano; Y. Makisumi; K. Ogata, *Chem. Pharm. Bull. (Japan) 6* (1958), 105.
25. A. Quilico; R. Justoni, *Rend. Ist. Lombardo Sci. Pt. I 69* (1936), 587.
26. N. K. Kochetkov; E. D. Khomutova, *Zh. Obshch. Khim. 29* (1959), 535.
27. H. B. Hill; W. J. Hale, *Am. Chem. J. 29* (1903), 253.
28. A. Quilico; R. Justoni, *Gazz. Chim. Ital. 70* (1940), 11.
29. A. Quilico; C. Musante, *Gazz. Chim. Ital. 72* (1942), 399.
30. E. J. Hessler, *J. Org. Chem. 41* (1976), 1828.
31. H. Lindemann; H. Cissée, *Liebigs Ann. 469* (1929), 44.
32. K.-H. Wünsch; A. J. Boulton, *Adv. Heterocycl. Chem. 8* (1967), 277.
33. R. K. Smalley, *Adv. Heterocycl. Chem. 29* (1981), 1.
34. K. von Auwers; Th. Bahr; E. Frese, *Liebigs Ann. 441* (1925), 54.
35. W. S. Johnson; W. E. Shelberg, *J. Am. Chem. Soc. 67* (1945), 1745, 1754.
36. W. S. Johnson; J. W. Petersen; C. D. Gutsche, *J. Am. Chem. Soc. 69* (1947), 2942.
37. D. K. Banerjee; K. M. Sivanandaiah, *Tetrahedron Lett. 5* (1960), 20.
38. M. E. Kuene, *J. Org. Chem. 35* (1970), 171.
39. G. V. Kondrateva; L. F. Kudryavtseva; S. I. Sav'yalov, *Zh. Obshch. Khim. 31* (1961), 3621.
40. S. Tagaki; H. Yasuda; A. Yokoyama, *Yakugaku Zasshi 81* (1961), 1639.
41. J. De Graw; L. Goodman; B. Weinstein; B. R. Baker, *J. Org. Chem. 27* (1962), 576.
42. K. von Auwers; A. E. Nold, *J. Prakt. Chem. 150* (1937), 57.
43. F. Winternitz; C. Menou; E. Arnal, *Bull. Soc. Chim. France 1960*, 505.
44. W. L. Meyer; G. B. Clemans; R. W. Huffman, *Tetrahedron Lett. 1966*, 4255.
45. L. Claisen, *Chem. Ber. 24* (1891), 3900.
46. A. Quilico; C. Musante, *Gazz. Chim. Ital. 70* (1940), 676.
47. S. Cusmano; T. Tiberio, *Gazz. Chim. Ital. 78* (1948), 896.
48. N. K. Kochetkov; S. D. Sokolov, *Adv. Heterocycl. Chem. 2* (1963), 365.
49. Z. Jerzmanowska; W. Basinski, *Rocz. Chem. 51* (1977), 2283.
50. W. Basinski; J. Jerzmanowska, *Pol. J. Chem. 53* (1979), 229.
51. W. Borsche, *Liebigs Ann. 390* (1912), 1.
52. R. Huisgen; M. Christl, *Angew. Chem. 79* (1967), 471; *Chem. Ber. 106* (1973), 3291.
53. A. Quilico; G. S. d'Alcontres, *Gazz. Chim. Ital. 79* (1949), 654.
54. G. S. d'Alcontres; G. Fenech, *Gazz. Chim. Ital. 82* (1952), 175.
55. G. Branchi; P. Grünanger, *Tetrahedron 21* (1965), 817.
56. A. Quilico; L. Panizzi, *Gazz. Chim. Ital. 72* (1942), 155.
57. K. Torssell; O. Zeuthen, *Acta Chem. Scand. B32* (1978), 118.
58. S. H. Andersen; N. B. Das; R. D. Jørgensen; G. Kjeldsen; J. S. Knudsen; S. C. Sharma; K. B. G. Torssell, *Acta Chem. Scand. B36* (1982), 1.
59. N. B. Das; K. B. G. Torssell, *Tetrahedron 39* (1983), 2247.
60. K. B. G. Torssell; A. C. Hazell; R. G. Hazell, *Tetrahedron 41* (1985), 5569.
61. T. P. Culbertson; G. W. Moersch; W. A. Neuklis, *J. Heterocycl. Chem. 1* (1964), 280.
62. U. Stache; W. Fritsch; H. Ruschig, *Liebigs. Ann. 685* (1965), 228.
63. G. W. Moersch; E. L. Wittle; W. A. Neuklis, *J. Org. Chem. 30* (1965), 1272; *32* (1967), 1387.
64. J. Kalvoda; H. Kaufmann, *J. Chem. Soc., Chem. Commun. 1976*, 209, 210.
65. G. Gerali; G. Sportoletti; G. Parini; A. Ius, *Farm. Sci. Ed. 24* (1969), 112, 231, 299.
66. A. P. Kozikowski; M. Adamczyk, *J. Org. Chem. 48* (1983), 366; A. P. Kozikowski; Y. Kitagawa; J. P. Springer, *J. Chem. Soc., Chem. Commun. 1983*, 1460.
67. A. P. Kozikowski; A. K. Ghosh, *J. Am. Chem. Soc. 104* (1982), 5788; *J. Org. Chem. 49* (1984), 2762.
68. A. P. Kozikowski, *Accts. Chem. Res. 17* (1984), 410.
69. D. P. Curran; S. A. Scanga; C. J. Fenk, *J. Org. Chem. 49* (1984), 3474.
70. P. A. Wade; H. R. Hinney, *J. Am. Chem. Soc. 101* (1979), 1319.
71. P. A. Wade; M. K. Pillay, *J. Org. Chem. 46* (1981), 5425; P. A. Wade; J. F. Bereznak, *J. Org. Chem. 52* (1987), 2973.

References

72. P. A. Wade; H.-K. Yen; S. A. Hardinger; M. K. Pillay; N. V. Amin; P. D. Vail; S. D. Morrow, *J. Org. Chem. 48* (1983), 1796.
73. P. A. Wade; N. V. Amin; H.-K. Yen; D. T. Price; G. F. Huhn, *J. Org. Chem. 49* (1984), 4595.
74. R. A. Whitney; E. S. Nicholas, *Tetrahedron Lett. 1981*, 3371.
75. C. Bellandi; M. De Amici; C. De Micheli; R. Gandolfi, *Heterocycles 22* (1984), 2187.
76. V. A. Tartakovskii; I. E. Chlenov; G. V. Lagodzinskaya; S. S. Novikov, *Doklady Akad. Nauk SSSR 161* (1965), 136; V. A. Tartakovskii; *Izv. Akad. Nauk SSSR, 1984* 165.
77. O. Mumm; G. Münchmeyer, *Chem. Ber. 43* (1910), 3335, 3345.
78. O. Mumm; C. Bergell, *Chem. Ber. 45* (1912), 3040, 3149.
79. A. Knust; O. Mumm, *Chem. Ber. 50* (1917), 563.
80. O. Mumm; H. Hornhardt, *Chem. Ber. 70* (1937), 1930.
81. R. B. Woodward; R. A. Olofson, *J. Am. Chem. Soc. 83* (1961), 1007; *Tetrahedron, Suppl. 7 1966*, 415.
82. R. B. Woodward; R. A. Olofson; H. Mayer, *J. Am. Chem. Soc. 83* (1961), 1010; *Tetrahedron, Suppl. 8 1966*, 321.
83. R. B. Woodward; D. J. Woodman, *J. Am. Chem. Soc. 88* (1966), 3169.
84. For further details, see *Specialist Periodical Reports, Amino Acids, Peptides and Proteins*, Vol. 1–14, Chem. Soc., London 1968–1985; D. J. Woodman In "Chemistry and Biochemistry of Amino Acids, Peptides and Proteins" (B. Weinstein, Ed.), Vol. 2, Dekker, New York, 1974, p. 207.
85. N. K. Kochetkov; E. D. Khomutova; M. V. Bazilevsky, *Zh. Obshch. Khim. 28* (1958), 2376.
86. J. E. McMurry; *Organic Synthesis 53* (1973), 59, 70; G. Stork; J. E. McMurry, *J. Am. Chem. Soc. 89* (1967), 5461.
87. G. Stork; S. Danishefsky; M. Ohashi, *J. Am. Chem. Soc. 89* (1967), 5459.
88. G. Stork; J. E. McMurry, *J. Am. Chem. Soc. 89* (1967), 5463.
89. G. Stork; J. E. McMurry, *J. Am. Chem. Soc. 89* (1967), 5464.
90. M. Ohashi, *Chem. Commun. 1969*, 893.
91. G. Stork; A. A. Hagedorn III, *J. Am. Chem. Soc. 100* (1978), 3609.
92. C. Kashima, *Heterocycles 12* (1979), 1343.
93. K. Bowden; G. Crank; W. J. Ross, *J. Chem. Soc. 1968*, 172.
94. R. G. Micetich, *Can. J. Chem. 46* (1970), 2006.
95. C. Kashima; S. Tobe; N. Sugiyama; M. Yamamoto, *Bull. Chem. Soc. (Japan) 46* (1973), 310.
96. C. Kashima; Y. Tsuda, *Bull. Chem. Soc. (Japan) 46* (1973), 3533.
97. C. Kashima; Y. Yamamoto; Y. Tsuda, *J. Org. Chem. 40* (1975), 526.
98. I. Iijima; N. Taya; M. Miyazaki; T. Tanaka; J. Uzawe, *J. Chem. Soc., Perkin I 1979*, 3190.
99. J. ApSimon; A. Holmes, *Heterocycles 6* (1977), 758.
100. C. Kashima; N. Mukai; Y. Tsuda, *Chem. Letters 1973*, 539.
101. C. Kashima; M. Uemori; Y. Tsuda; O. Omoto, *Bull. Chem. Soc. Japan 49* (1976), 2254.
102. S. Auricchio; S. Morrocchi; A. Ricca, *Tetrahedron Lett. 1974*, 2793.
103. S. Auricchio; R. Colle; S. Morrocchi; A. Ricca, *Gazz. Chim. Ital. 106* (1976), 823.
104. T. Tanaka; M. Miyazaki; I. Iijima, *Chem. Commun. 1973*, 233.
105. D. J. Brunelle, *Tetrahedron Lett. 1981*, 3699.
106. N. K. Kochetkov; S. D. Sokolov; V. M. Luboshnikova, *Zh. Obshch. Khim. 32* (1962), 1778.
107. V. Renzi; V. Dal Piaz; S. Pinzauti, *Gazz. Chim. Ital. 99* (1969), 753.
108. P. Bravo; G. Gavaraghi, *J. Heterocycl. Chem. 14* (1977), 37.
109. N. F. Haley, *J. Org. Chem. 43* (1978), 1233.
110. R. G. Micetich; C. C. Shaw; W. T. Hall; P. Spevak; R. A. Fortier; P. Wolfart; B. C. Foster; K. B. Bains, *Heterocycles 23* (1985), 571; R. G. Micetich; C. C. Shaw; W. T. Hall; P. Spevak; K. B. Bains, *Heterocycles 23* (1985), 585.
111. W. Lampe; J. Smolinska, *Bull. Akad. Polon. 6* (1958), 481.
112. C. Kashima; N. Mukai; Y. Yamamoto; Y. Tsuda; Y. Omote, *Heterocycles 7* (1977), 241.
113. S. A. Tischler; L. Weiler, *Tetrahedron Lett. 1979*, 4903.
114. M. Ohashi; T. Maruishi; H. Kakisawa, *Tetrahedron Lett. 1968*, 719.
115. G. Büchi; J. C. Vederas, *J. Am. Chem. Soc. 94* (1972), 9128.
116. A. Barco; S. Benetti; G. P. Pollini; P. G. Baraldi; M. Guarneri; C. B. Vicentini, *J. Org. Chem. 44* (1979), 105.
117. P. G. Baraldi; F. Moroder; G. P. Pollini; D. Simoni; A. Barco; S. Benetti, *J. Chem. Soc., Perkin I 1982*, 2983.
118. A. Barco; S. Benetti; P. G. Baraldi; M. Guarneri; G. P. Pollini; D. Simoni, *J. Chem. Soc., Chem. Commun. 1981*, 599.

119. A. Barco; S. Benetti; G. P. Pollini; P. G. Baraldi; M. Guarneri; D. Simoni; C. Gandolfi, *J. Org. Chem. 45* (1980), 3141; *46* (1981), 4518.
120. G. Casnati; A. Quilico; A. Ricca; P. VitaFinzi, *Tetrahedron Lett. 1966*, 233; *Chim. Ind. (Milan) 47* (1965), 993; *Gazz. Chim. Ital. 96* (1966), 1064, 1073.
121. S. Auricchio; A. Ricca, *Gazz. Chim. Ital. 103* (1973), 37.
122. S. Auricchio; A. Ricca; O. Vajna de Pava, *J. Heterocycl. Chem. 14* (1977), 159; *Chim. Ind. (Milan) 58* (1976), 1073.
123. A. Quilico; R. Fusco, *Gazz. Chim. Ital. 67* (1937), 589.
124. A. Quilico; G. Speroni, *Gazz. Chim. Ital. 76* (1946), 148.
125. C. Grundmann; P. Grünanger, "The Nitrile Oxides", Springer-Verlag, Berlin, 1971.
126. D. N. McGregor; U. Corbin; J. E. Swigor; L. C. Cheney, *Tetrahedron 25* (1969), 389.
127. A. A. Akhrem; F. A. Lakhvich; V. A. Khripach; I.-B. Klebanovich, *Tetrahedron Lett. 1976*, 3983; *Khim. Geterosikl. Soedin. 1975*, 329; *1979*, 230; *Dokl. Akad. Nauk SSSR 216* (1974), 1645.
128. Y. Takeuchi; F. Furusaki, *Adv. Heterocycl. Chem. 21* (1977), 207.
129. L. Panizzi, *Gazz. Chim. Ital. 76* (1946), 44.
130. G. H. Timms; E. Wildschmidt, *Tetrahedron Lett. 1971*, 195.
131. G. Drehfahl; H.-H. Hörhold, *Chem. Ber. 97* (1964), 159.
132. V. Jäger; H. Grund, *Angew. Chem. 88* (1976), 27; *Liebigs Ann. 1980*, 80.
133. A. A. Akhrem; F. A. Lakhvich; V. A. Khripach; I. B. Klebanovich; A. G. Pozdeev, *Khim. Geterosikl. Soedin. 1976*, 625.
134. A. A. Akhrem; F. A. Lakhvich; V. A. Khripach; I. B. Klebanovich, *Dokl. Akad. Nauk SSSR 244* (1979), 615.
135. A. A. Akhrem; F. A. Lakhvich; V. A. Khripach; I. I. Petrusevich, *Khim. Geterosikl. Soeden. 1976*, 891.
136. N. B. Das; K. B. G. Torssell, *Tetrahedron 39* (1983), 2227.
137. K. K. Sharma; K. B. G. Torssell, *Tetrahedron 40* (1984), 1085.
138. S. H. Andersen; K. K. Sharma; K. B. G. Torssell, *Tetrahedron 39* (1983), 2241.
139. A. P. Kozikowski; M. Adamczyk, *Tetrahedron Lett. 1982*, 3123.
140. D. P. Curran, *J. Am. Chem. Soc. 104* (1982), 4024; *105* (1983), 5826.
141. S. F. Martin; B. Dupre, *Tetrahedron Lett. 1983*, 1337.
142. R. Annunziata; M. Cinquini; F. Cozzi; A. Restelli, *J. Chem. Soc., Chem. Commun. 1984*, 1253; *J. Chem. Soc., Perkin I 1985*, 2293; R. Annunziata; M. Cinquini; F. Cozzi; A. Gilardi; A. Restelli, *J. Chem. Soc., Perkin I 1985*, 2289.
143. M. Cinquini; F. Cozzi; A. Gilardi, *J. Chem. Soc., Chem. Commun. 1984*, 551.
144. M. S. Schechter; N. Green; F. B. La Forge, *J. Am. Chem. Soc. 71* (1949), 3165.
145. S. K. Mukerji; K. K. Sharma; K. B. G. Torssell, *Tetrahedron 39* (1983), 2231.
146. K. E. Larsen; K. B. G. Torssell, *Tetrahedron 40* (1984), 2985.
147. D. P. Curran, *Tetrahedron Lett. 1983*, 3443.
148. S. K. Mukerji; K. B. G. Torssell, *Acta. Chem. Scand. B35* (1981), 643.
149. P. DeShong; J. M. Leginus, *J. Org. Chem. 49* (1984), 3421.
150. A. Liguori; G. Sindona; N. Uccella, *Tetrahedron 40* (1984), 1901.
151. D. C. Lathbury; P. J. Parsons, *J. Chem. Soc., Chem. Commun. 1982*, 291.
152. For a recent compilation of syntheses of 1,4-diones, see e.g. M. Miyashita; T. Yanami; A. Yoshikoshi, *J. Am. Chem. Soc. 98* (1976), 4679.
153. H. Emde; D. Domsch; H. Feger; U. Frick; A. Götz; H. H. Hergott; K. Hofmann; W. Kober; K. Krägeloh; T. Oesterle; W. Steppan; W. West; G. Simchen, *Synthesis 1982*, 1.
154. C. H. Cummins; R. M. Coates, *J. Org. Chem. 48* (1983), 2070.
155. G. Casnati; A. Ricca, *Tetrahedron Lett. 1967*, 327.
156. A. A. Akhrem; F. A. Lakhvich; V. A. Khripach, *Zh. Obshch. Khim. 45* (1975), 2572.
157. S. M. Kupchan; C. W. Sigel; M. J. Matz; C. J. Gilmore; R. F. Bryan, *J. Am. Chem. Soc. 98* (1976), 2295.
158. P. W. LeQuesne; S. B. Levery; M. D. Menacherry; T. F. Brennon; R. F. Raffaut, *J. Chem. Soc., Perkin I 1978*, 1572.
159. R. M. Carman; F. N. Lahey; J. K. MacLeod, *Austr. J. Chem. 20* (1967), 1957.
160. F. N. Lahey; J. K. MacLeod, *Austr. J. Chem. 20* (1967), 1943.
161. D. L. Dreyer; A. Lee, *Phytochem. 11* (1972), 763.
162. W. Parker; R. A. Raphael; D. J. Wilkinson, *J. Chem. Soc. 1958*, 3811.
163. G. Ohloff; I. Flament, *Prog. Chem. Org. Nat. Prod. 36* (1979), 231; for references to other methods, see H. Saïmoto; T. Hiyama; H. Nozaki, *Bull. Chem. Soc. Japan 56* (1983), 3078; H. Saimoto; M. Shinoda; S. Matsubara; K. Oshima; T. Hiyama; H. Nozaki, *Bull. Chem. Soc. Japan 56* (1983), 3088.

164. D. P. Curran; D. H. Singleton, *Tetrahedron Lett. 1983*, 2079.
165. (a) P. G. Baraldi; A. Barco; S. Benetti; S. Manfredini; G. P. Pollini; D. Simoni, *Tetrahedron Lett. 1984*, 4313.
    (b) P. G. Baraldi; A. Barco; S. Benetti; M. Guarneri; S. Manfredini; G. P. Pollini; D. Simoni, *Tetrahedron Lett. 1985*, 5319.
166. S. S. Ghabrial; I. Thomsen; K. B. G. Torssell, *Acta Chem. Scand. B41* (1987), 426.
167. V. Sprio; E. Ajello, *Ann. Chim. (Rome) 56* (1966), 858.
168. V. Sprio; S. Plescia; O. Migliara, *Ann. Chim. (Rome) 61* (1971), 271.
169. J. W. Cornforth, as reported by A. H. Jackson; K. H. Smith In "The Total Synthesis of Natural Products" (J. ApSimon, Ed.), Wiley-Interscience, New York, 1973, p. 261.
170. R. V. Stevens; M. Kaplan, *Chem. Commun. 1970*, 822.
171. R. V. Stevens; C. G. Christensen; W. L. Edmonson; M. Kaplan; E. B. Reid; M. P. Markland, *J. Am. Chem. Soc. 93* (1971), 6629.
172. R. V. Stevens; L. E. DuPree, Jr.; W. L. Edmonson; L. L. Magid; M. P. Wentland, *J. Am. Chem. Soc. 93* (1971), 6637.
173. R. V. Stevens; J. M. Fitzpatrick; P. B. Germeraad; B. L. Harrison; R. La Palme, *J. Am. Chem. Soc. 98* (1976), 6313.
174. R. V. Stevens; R. E. Cherpeck; B. L. Harrison; J. Lai; R. La Palme, *J. Am. Chem. Soc. 98* (1976), 6317.
175. R. V. Stevens, *Tetrahedron 32* (1976), 1599.
176. A. Eschenmoser, *Quart. Rev. Chem. Soc. 24* (1970), 366 and references cited therein.
177. H.-J. Wollweber; C. Wentrup, *J. Org. Chem. 50* (1985), 2041.
178. P. DeShong; N. E. Lowmaster; O. Baralt, *J. Org. Chem. 48* (1983), 1149.
179. J. P. Freeman, *Chem. Rev. 83* (1983), 241.
180. J. E. Baldwin; R. G. Pudussery; A. K. Qureshi; B. Sklarz, *J. Am. Chem. Soc. 90* (1968), 5325.
181. E. Winterfeldt; W. Krohn; H.-U. Stracke, *Chem. Ber. 102* (1969), 2346.
182. G. Schmidt; H.-U. Stracke; E. Winterfeldt, *Chem. Ber. 103* (1970), 3196.
183. I. Adachi; K. Harada; H. Kano, *Tetrahedron Lett. 1969*, 4875.
184. R. Grigg, *J. Chem. Soc., Chem. Commun. 1966*, 607.
185. R. M. Acheson; A. S. Bailey; I. A. Selby, *J. Chem. Soc., Chem. Commun. 1966*, 835.
186. N. Kahn; D. A. Wilson, *J. Chem. Res. (S) 1984*, 150.
187. H. Seidl; R. Huisgen; R. Knorr, *Chem. Ber. 102* (1969), 904.
188. I. Adachi; K. Harada; R. Miyazaki; H. Kano, *Chem. Pharm. Bull. Japan 22* (1974), 61.
189. M. C. Aversa; G. Cum; N. Uccella, *J. Chem. Soc., Chem. Commun. 1971*, 156.
190. G. Cum; G. Sindona; N. Uccella, *J. Chem. Soc., Perkin I 1976*, 719.
191. J. J. Tufariello; Sk. A. Ali; H. Klingele, *J. Org. Chem. 44* (1979), 4213.
192. C. H. Hassal; A. E. Lippman, *J. Chem. Soc. 1953*, 1059.
193. M. A. Abou-Gharbia; M. M. Joullie, *Heterocycles 12* (1979), 819; R. N. Pratt; D. P. Stokes; G. A. Taylor; P. C. Brookes, *J. Chem. Soc. (C) 1968*, 2086; D. P. Stokes; G. A. Taylor, *J. Chem. Soc. (C) 1971*, 2334.
194. D. P. Deltsova; N. P. Gambaryan; L. L. Knunyants, *Dokl. Akad. Nauk SSSR 206* (1972), 620.
195. A. D. Baker; D. Wong; S. Lo; M. Block; G. Horozoglu; N. L. Goldman; R. Engel; D. C. Liotta, *Tetrahedron Lett. 1975*, 25.
196. N. Murai; M. Komatsu; Y. Ohshiro; T. Agawa, *J. Org. Chem. 42* (1977), 448.
197. J. Motoyoshiya; I. Yamamoto; H. Gotoh, *J. Chem. Soc., Perkin I 1981*, 2727.
198. M. Hafiz; G. A. Taylor, *J. Chem. Soc., Perkin I 1980*, 1700.
199. O. Tsuge; H. Watanabe; K. Masuda; M. M. Yousif, *J. Org. Chem. 44* (1979), 4543.
200. E. Kaji; S. Zen, *Heterocycles 13* (1979), 187.
201. (a) H. G. Aurich; G. Blinne, *Chem. Ber. 107* (1974), 13.
    (b) T. Nishiwaki; K. Azechi; F. Fujiyama, *J. Chem. Soc., Perkin I 1974*, 1867.
202. (a) A. P. Kozikowski; X.-M. Cheng, *Tetrahedron Lett. 1985*, 4047, *1987*, 3189.
    (b) A. P. Kozikowski; C. S. Li; J. G. Scripco; X.-M. Cheng; cit. by A. P. Kozikowski, *Accts. Chem. Res. 17* (1984), 410.
203. G. W. Perold; F. V. K. von Reiche, *J. Am. Chem. Soc. 79* (1957), 465.
204. K. Kotera; Y. Takano; A. Matsuura; K. Kitahonoki, *Tetrahedron Lett. 1968*, 5759; *Tetrahedron 26* (1970), 539.
205. G. Chidichimo; G. Cum; F. Lelj; G. Sindona; N. Uccella, *J. Am. Chem. Soc. 102* (1980), 1372.
206. V. A. Tartakovskii; O. A. Lukyanov; S. S. Novikov, *Dokl. Akad. Nauk SSSR 178* (1968), 123.
207. R. Grée; R. Carrié, *J. Am. Chem. Soc. 99* (1977), 6667.
208. I. Adachi; R. Miyazaki; H. Kano, *Chem. Pharm. Bull. Japan 22* (1974), 70.
209. D. Döpp; A. M. Nour-el-din, *Tetrahedron Lett. 1978*, 1463.

210. K. Niklas, Dissertation, University of München, 1975, cited in Ref. 179.
211. T. Nishiwaki; T. Saito, *J. Chem. Soc. (C) 1971*, 3021.
212. J. M. J. Tronchet; E. Mihaly, *Helv. Chim. Acta 55* (1971), 1266.
213. B. Singh; E. F. Ullman, *J. Am. Chem. Soc. 89* (1967), 6911.
214. T. Nishiwaki, *Tetrahedron Lett. 1969*, 2049.
215. T. Nishiwaki; T. Kitamura; A. Nakano, *Tetrahedron 26* (1970), 453.
216. T. Nishiwaki; A. Nakano; H. Matsuoka, *J. Chem. Soc. (C) 1970*, 1825.
217. T. Nishiwaki; T. Saito; S. Onomura; K. Kondo, *J. Chem. Soc. (C) 1971*, 2644.
218. T. Nishiwaki; T. Saito, *J. Chem. Soc. (C) 1971*, 2648.
219. B. Singh; A. Zweig; J. B. Gallivan, *J. Am. Chem. Soc. 94* (1972), 1199.
220. T. Nishiwaki; F. Fujiyama, *J. Chem. Soc., Perkin I 1972*, 1456; *Synthesis 1972*, 569.
221. G. Szeimies; K. Mannhardt; W. Mickler, *Chem. Ber. 110* (1977), 2922.
222. S. Auricchio; O. Vajna de Pava, *J. Chem. Res. (S) 1983*, 132.
223. M. Maeda; M. Kojima, *J. Chem. Soc., Chem. Commun. 1973*, 539.
224. K. Dietliker; P. Gilgen; H. Heimgartner; H. Schmid, *Helv. Chim. Acta 59* (1976), 2074.
225. M. I. Komendatov; R. R. Bekmukhametov, *Khim. Geterosikl. Soedin.* (1975), 1292; M. I. Komendatov; R. R. Bekmukhametov; R. R. Kostikov, *Khim. Geterosikl. Soedin. 1978*, 1053.
226. G. Adembri; A. Camparini; F. Ponticelli; P. Tedeschi, *J. Chem. Soc., Perkin I 1977*, 971.
227. M. Maeda; M. Kojima, *J. Chem. Soc., Perkin I 1977*, 239; M. Kojima; M. Maeda, *Tetrahedron Lett. 1969*, 2379.
228. T. Sato; K. Yamamoto; K. Fukui, *Chem. Lett. 1973*, 111.
229. O. Tsuge; K. Sone; S. Urano; K. Matsuda, *J. Org. Chem. 47* (1982), 5171.
230. G. S. d'Alcontres, *Gazz. Chim. Ital. 80* (1950), 441.
231. A. Alberola; C. Andrés; A. G. Ortega; R. Pedrosa, *J. Heterocycl. Chem. 21* (1984), 1575.
232. V. Sprio; E. Ajello, *Ann. Chim. (Rome) 56* (1966), 1103.
233. W. Müller; U. Kraatz; F. Korte, *Chem. Ber. 106* (1973), 332.
234. L. Claisen, *Chem. Ber. 42* (1909), 59.
235. F. Bell, *J. Chem. Soc. 1941*, 285.
236. S. Cusmano, *Gazz. Chim. Ital. 69* (1939), 594, 621; *70* (1940), 86, 227, 235, 240.
237. H. Kano, *J. Pharm. Soc. (Japan) 73* (1953), 383, 387.
238. S. Cusmano; T. Tiberio, *Gazz. Chim. Ital. 78* (1948), 896; *80* (1950), 299.
239. T. S. Gardner; F. A. Smith; E. Wenis; J. Lee, *J. Org. Chem. 21* (1956), 530.
240. C. Musante; A. Stener, *Gazz. Chim. Ital. 89* (1959), 1579; C. Musante; S. Fatutta, *Gazz. Chim. Ital. 88* (1958), 879; M. Ruccia; N. Vivona; F. Piozzi, *Gazz. Chim. Ital. 97* (1967), 1494.
241. C. Musante, *Gazz. Chim. Ital. 72* (1942), 537; *73* (1943), 355.
242. A. Akhrem; A. M. Moiseenkov; M. B. Andaburskaya; A. J. Strakov, *J. Prakt. Chem. 314* (1972), 31.
243. T. Nishiwaki, *Synthesis 1975*, 20.
244. F. De Sarlo; G. Renzi, *Tetrahedron 22* (1966), 2995.
245. A. Mustafa; W. Asker; A. H. Harhash; N. A. L. Kassab, *Tetrahedron 19* (1963), 1577.
246. H. Wamhoff; D. Schramm; F. Korte, *Synthesis 1971*, 216.
247. G. Adembri; F. Ponticelli; P. Tedeschi, *J. Heterocycl. Chem. 9* (1972), 1219.
248. S. S. Joshi; I. R. Gambhir, *J. Org. Chem. 26* (1961), 3714.
249. A. N. Kost; I. I. Grandberg, *Adv. Heterocycl. Chem. 6* (1966), 347.
250. S. C. Sharma; K. Torssell, *Acta Chem. Acta B33* (1979), 379.
251. I. Adachi; H. Kano, *Chem. Pharm. Bull. Japan 17* (1969), 2201.
252. I. Adachi; K. Harada; R. Miyazaki; H. Kano, *Chem. Pharm. Bull. (Japan) 22* (1974), 61.
253. R. Faragher; T. L. Gildchrist, *J. Chem. Soc., Perkin I 1977*, 1196.
254. H. Kano, *J. Pharm. Soc. Japan 72* (1952), 150, 1118.
255. H. G. Aurich, *Liebigs Ann. 732* (1970), 195.
256. H. G. Aurich; G. Blinne, *Chem. Ber. 107* (1974), 13.
257. L. A. Reiter, *Tetrahedron Lett. 1985*, 3423; E. Ajello, *Ann. Chim. (Rome) 60* (1970), 343, 399.
258. H. Wamhoff, *Chem. Ber. 105* (1972), 748.
259. J. P. Ferris; F. R. Antonucci, *J. Am. Chem. Soc. 96* (1974), 2010.
260. T. Matsuura; Y. Ito, *Tetrahedron Lett. 1973*, 2283.
261. A. R. Gagneux; R. Göschke, *Tetrahedron Lett. 1966*, 5451.
262. E. F. Ullman; B. Singh, *J. Am. Chem. Soc. 88* (1966), 1844.
263. H. Göth; A. R. Gagneux; C. H. Eugster; H. Schmid, *Helv. Chim. Acta 50* (1967), 137; H. Göth; H. Schmid, *Chimia 20* (1966), 148.
264. A. Padwa; E. Chen; A. Ku, *J. Am. Chem. Soc. 97* (1975), 6484.

References

265. S. A. Lang, Jr.; Y. i. Lin, In "Comprehensive Heterocyclic Chemistry" (A. R. Katritzky and C. W. Rees, Eds.) Vol. 4, Pergamon, Oxford, 1984, p. 1.
266. B. J. Wakefield; D. J. Wright, *Adv. Heterocycl. Chem.* 25 (1979), 147.
267. A. Padwa; E. Chen, *J. Org. Chem.* 39 (1974), 1976.
268. A. Padwa, *Accts. Chem. Res.* 9 (1976), 371.
269. L. J. Darlage; T. H. Kinstle; C. L. McIntosh, *J. Org. Chem.* 36 (1971), 1088.
270. J. P. Ferris; F. R. Antonucci; R. W. Trimmer, *J. Am. Chem. Soc.* 95 (1973), 919.
271. J. P. Ferris; F. R. Antonucci, *J. Am. Chem. Soc.* 96 (1974), 2014.
272. J. P. Ferris; R. W. Trimmer, *J. Org. Chem.* 41 (1976), 13.
273. T. H. Kinstle; L. J. Darlage, *J. Heterocycl. Chem.* 6 (1969), 123.
274. H. Böshagen; W. Geiger, *Chem. Ber.* 103 (1970), 123.
275. T. Doppler; H. Schmid; H.-J. Hansen, *Helv. Chim. Acta* 62 (1979), 314.
276. K. H. Grellman; E. Tauer, *J. Photochem.* 6 (1977), 365.
277. M. A. Abou-Gharbia; M. M. Joullie, *Heterocycles* 9 (1978), 457.
278. A. F. Gettings; G. A. Taylor, *J. Chem. Soc., Chem. Commun.* 1972, 1146.
279. R. N. Pratt; D. P. Stokes; G. A. Taylor, *J. Chem. Soc., Perkin I* 1975, 498.
280. A. F. Gettings; D. P. Stokes; G. A. Taylor; C. B. Judge, *J. Chem. Soc., Perkin I* 1977, 1849.
281. A. R. Evans; M. Hafiz; G. A. Taylor, *J. Chem. Soc., Perkin I* 1984, 1241; C. P. Falshaw; N. A. Hashi; G. A. Taylor, *J. Chem. Soc., Perkin I* 1985, 1837.
282. M. Michalska; I. Orlich, *Bull. Acad. Pol. Sci., Ser. Chim.* 23 (1975), 655.
283. D. N. McGregor; U. Corbin; J. E. Swigor; L. C. Cheney, *Tetrahedron* 25 (1969), 389.
284. K.-y. Akiba; K. Kashiwagi; Y. Ohyama; Y. Yamamoto; K. Ohkata, *J. Am. Chem. Soc.* 107 (1985), 2721.
285. M. Ruccia; N. Vivona; D. Spinelli, *Adv. Heterocycl. Chem.* 29 (1981), 141.
286. H. C. van der Plas, "Ring Transformations of Heterocycles", Vols. 1, 2, Academic Press, New York, 1973.
287. T. Ajello, *Gazz. Chim. Ital.* 67 (1937), 55, 779.
288. (a) T. Ajello; S. Cusmano, *Gazz. Chim. Ital.* 68 (1938), 792; 69 (1939), 391.
    (b) F. de Sarlo; A. Guarna; A. Brandi; A. Goti, *Tetrahedron* 41 (1985), 5181.
289. T. Ajello; V. Sprio; J. Fabra, *Ric. Sci. Parte 2, Sez. B4 1964*, 575; A. Quilico; M. Freri, *Gazz. Chim. Ital.* 76 (1946), 1.
290. A. J. Boulton; A. R. Katritzky; A. M. Hamid, *J. Chem. Soc. (C) 1967*, 2005.
291. G. Nussberger, *Chem. Ber.* 25 (1892), 2142; A. Hantzsch; J. Heilbron, *Chem. Ber.* 43 (1910), 68; A. J. Boulton; I. J. Fletcher; A. R. Katritzky, *Chem. Commun. 1968*, 62; A. J. Boulton; R. C. Brown, *J. Org. Chem.* 35 (1970), 1662.
292. H. Lindemann; H. Cisseé, *Liebigs Ann.* 469 (1929), 44; *J. Prakt. Chem.* 122 (1929), 232.
293. K. Harzányi, *J. Heterocycl. Chem.* 10 (1973), 957.
294. H. Walser; T. Flynn; R. I. Fryer, *J. Heterocycl. Chem.* 11 (1974), 885.
295. H. Kano; E. Yamazaki, *Chem. Pharm. Bull. Japan* 10 (1962), 993; *Tetrahedron* 20 (1964), 461.
296. H. Kano; E. Yamazaki, *Tetrahedron* 20 (1964), 159.
297. H. Ulrich; J. N. Tilley; A. A. R. Sayigh, *J. Org. Chem.* 27 (1962), 2160.
298. R. B. Woodward; R. A. Olofson, *Tetrahedron Suppl. 7 1966*, 415.
299. T. Ajello; S. Cusmano, *Gazz. Chim. Ital.* 70 (1940), 770.
300. T. Ajello; B. Tornetta, *Gazz. Chim. Ital.* 77 (1947), 332.
301. T. Sasaki; T. Yoshioka; Y. Suzuki, *Bull. Chem. Soc. (Japan)* 44 (1971), 185.
302. N. Vivona; G. Macaluso; V. Frenna; M. Ruccia, *J. Heterocycl. Chem.* 20 (1983), 931.
303. N. Vivona; M. Ruccia; V. Frenna; D. Spinelli, *J. Heterocycl. Chem.* 17 (1980), 401.
304. G. L'abbé; F. Godts; S. Toppet, *Tetrahedron Lett. 1983*, 3149.
305. N. Vivona; G. Cusmano; G. Macaluso, *J. Chem. Soc., Perkin I* 1977, 1616.
306. M. Ohashi; H. Kamachi; H. Kakisawa; G. Stork, *J. Am. Chem. Soc.* 89 (1967), 5460.
307. G. Stork; M. Ohashi; H. Kamachi; H. Kakisawa, *J. Org. Chem.* 36 (1971), 2784.
308. P. Caramella; A. Querci, *Synthesis 1972*, 46.
309. P. Caramella; R. Metelli; P. Grünanger, *Tetrahedron* 27 (1971), 379.
310. I. Adachi, *Chem. Pharm. Bull. Japan* 17 (1969), 2209.
311. R. Grigg; R. Hayeo; J. L. Jackson; T. J. King, *Chem. Commun. 1973*, 349.
312. T. Kobayashi; M. Nitta, *Chem. Lett. 1983*, 1233.
313. M. Wilk; H. Schwab; J. Rochlitz, *Liebigs Ann.* 698 (1966), 149.
314. E. C. Taylor; D. R. Eckroth; J. Bartulin, *J. Org. Chem.* 32 (1967), 1899.
315. D. C. Campbell; C. W. Rees, *J. Chem. Soc. (C) 1969*, 748.
316. E. C. Taylor; J. Bartulin, *Tetrahedron Lett. 1967*, 2337.

317. G. Shaw; G. Sugowdz, *J. Chem. Soc. 1954*, 665.
318. W. J. Fanshawe; V. J. Bauer; S. R. Safir, *J. Org. Chem. 30* (1965), 2862.
319. H. Kashima; Y. Yamamoto; Y. Omote, *Heterocycles 4* (1976), 1387.
320. V. Sprio; E. Ajello, *Ann. Chim. (Rome) 56* (1966), 859, 1103; *57* (1967), 846.
321. M. Ruccia; N. Vivona; S. Plescia; V. Sprio, *J. Heterocycl. Chem. 8* (1971), 289.
322. V. Sprio; S. Plescia; O. Migliara, *Ann. Chim. (Rome) 61* (1971), 271.
323. G. Bianchi; R. Gandolfi; P. Grünanger, *J. Heterocycl. Chem. 5* (1968), 49.
324. G. Bianchi; A. Gamba-Invernizzi; R. Gandolfi, *J. Chem. Soc., Perkin I 1974*, 1757.
325. J. F. King; T. Durst, *Can. J. Chem. 40* (1962), 882.
326. J. L. Pinkus; H. A. Jessup; T. Cohen, *J. Chem. Soc. (C) 1970*, 242.
327. K.-H. Wunsche; A. J. Boulton, *Adv. Heterocycl. Chem. 8* (1967), 277 and references cited therein.
328. D. S. Kemp; R. B. Woodward, *Tetrahedron 21* (1965), 3019.
329. N. A. LeBel; T. A. Lajiness; D. B. Ledlie, *J. Am. Chem. Soc. 89* (1967), 3076.
330. J. P. Freeman; D. G. Pucci; G. Binsch, *J. Org. Chem. 37* (1972), 1894.
331. R. A. Olofson; R. K. Vander Meer; S. Stournas, *J. Am. Chem. Soc. 93* (1971), 1543.
332. M. Kinusaga; S. Hashimoto, *J. Chem. Soc., Chem. Commun. 1972*, 466.
333. L. K. Ding; W. J. Irwin, *J. Chem. Soc., Perkin I 1976*, 2382.
334. A. Padwa; K. F. Koehler; A. Rodriguez, *J. Am. Chem. Soc. 103* (1981), 4974; *J. Org. Chem. 49* (1984), 282.
335. J. J. Tufariello; G. E. Lee; P. A. Senaratne; M. Al-Nuri, *Tetrahedron Lett. 1979*, 4359.
336. R. V. Stevens; K. Albizati, *J. Chem. Soc., Chem. Commun. 1982*, 104.
337. T. Kametani; S.-P. Huang; A. Nakayama; T. Honda, *J. Org. Chem. 47* (1982), 2328.
338. T. Kametani; S.-P. Huang; S. Yokohama; Y. Suzuki; M. Ihara, *J. Am. Chem. Soc. 102* (1980), 2060.
339. T. Kametani; S.-P. Huang; T. Nagahara; M. Ihara, *J. Chem. Soc., Perkin I 1981*, 964, 2282.
340. T. Kametani; S.-P. Huang; M. Ihara, *Heterocycles 12* (1979), 1183, 1189; T. Kametani; A. Nakayama; Y. Nakayama; T. Ikuta; R. Kobo; E. Goto; T. Honda; K. Fukumoto, *Heterocycles 16* (1981), 53; M. Ihara; F. Konno; K. Fukumoto; T. Kametani, *Heterocycles 20* (1983), 2181.
341. T. Kametani; T. Nagahara; M. Ihara, *J. Chem. Soc., Perkin I 1981*, 3048.
342. T. Kametani; T. Nagahara; T. Honda, *J. Org. Chem. 50* (1985), 2327.
343. M. L. M. Pennings; D. N. Reinhoudt; S. Harkema; G. J. van Hummel, *J. Org. Chem. 48* (1983), 486.
344. U. Schöllkopf; I. Hoppe, *Angew. Chem. 87* (1975), 814.
345. E. Ajello, *J. Heterocycl. Chem. 8* (1971), 1035.
346. V. Sprio; S. Plescia, *J. Heterocycl. Chem. 9* (1972), 951.
347. E. Ajello; O. Migliara; V. Sprio, *J. Heterocycl. Chem. 9* (1972), 1169.
348. G. Dattolo; E. Ajello; S. Plescia; G. Cirrincione; G. Diadone, *J. Heterocycl. Chem. 14* (1977), 1021.
349. K. N. Houk; J. Sims; C. R. Watts; L. J. Luskus, *J. Am. Chem. Soc. 95* (1973), 7301.
350. S. Takahashi; H. Kano, *Chem. Pharm. Bull. (Japan) 12* (1964), 1290.
351. Ya. D. Samuilov; S. E. Solov'eva; A. I. Konovalov, *Zh. Obshch. Khim. 50* (1980), 138.
352. A. Dondoni; G. Barbaro, *Gazz. Chim. Ital. 105* (1975), 701.
353. K. Bast; M. Christl; R. Huisgen; W. Mack, *Chem. Ber. 105* (1972), 2825.
354. J. F. Barnes; M. L. Barrow; M. M. Harding; R. M. Paton; P. L. Ashcroft; J. Crosby; C. J. Joyce, *J. Chem. Res. (S) 1979*, 314.
355. K. Harada; E. Kaji; S. Zen, *Chem. Pharm. Bull. (Japan)* 28 (1980), 3296.
356. A. Corsaro; U. Chiacchio; A. Campagnini; G. Purello, *J. Chem. Soc., Perkin I 1977*, 2154; *1980*, 1635.
357. A. Corsaro; U. Chiacchio; G. Perrini; P. Caramella; G. Purello, *J. Heterocycl. Chem. 22* (1985), 797.
358. K. Kurabayashi; C. Grundmann, *Bull. Soc. Chem. Japan 51* (1978), 1484.
359. P. Beltrame; G. Gelli; A. Loi, *J. Chem. Res. (S) 1978*, 420.
360. S. Morrocchi; A. Ricca; L. Velo, *Tetrahedron Lett. 1967*, 331.
361. J. E. Franz; R. K. Howe; H. K. Pearl, *J. Org. Chem. 41* (1976), 620.
362. T. Sasaki; S. Eguchi; T. E. Sakai; T. Suzuki, *Tetrahedron 35* (1979), 1073.
363. G. Ferrara; A. Ius; C. Parini; G. Sportoletti; G. Vecchio, *Tetrahedron 28* (1972), 2461.
364. L. Fabbrini; G. Speroni, *Chim. Ind. (Milan) 43* (1961), 807.
365. G. D'Alò; L. Invernizzi; P. Grünanger, *Boll. Chim. Farm. 108* (1969), 792.

366. G. Rembarz; H. Brandner; E.-M. Bebenroth, *J. Prakt. Chem. 31* (1966), 221.
367. T. Sasaki; T. Yoshioka, *Bull. Chem. Soc, (Japan) 40* (1967), 2608.
368. T. Sasaki; T. Yoshioka, *Bull. Chem. Soc. (Japan) 41* (1968), 2206.
369. J. Plenkiurce; Z. Eckstein, *Bull. Acad. Pol. Sci. 15* (1967), 99.
370. T. O. Stevens, *J. Org. Chem. 33* (1968), 2660.
371. R. Eloy, *Bull. Soc. Chim. Belg. 73* (1964), 793.
372. M. S. Chang; J. U. Lowe, Jr., *J. Org. Chem. 32* (1967), 1577.
373. G. Leandri, *Boll. Sci. Chim. Ind. Bologna 14* (1956), 80.
374. G. Leandri; M. Palotti, *Ann. Chim (Rome) 47* (1957), 376.
375. D. Martin; H. J. Herrmann; S. Rackow; K. Nadolsky, *Angew. Chem. 77* (1965), 96.
376. D. Martin; A. Weise, *Chem. Ber. 99* (1966), 317.
377. M. Akiyama; Y. Iwakura; S. Shiraishi; Y. Imai, *Polymer. Lett. 4* (1966), 305.
378. R. Lenaers; F. Eloy, *Helv. Chim. Acta 46* (1963), 1067.
379. C. G. Overberger; S. Fujimoto, *Polymer Lett. 3* (1965), 735.
380. P. Grünanger; P. Vita Finzi, *Rend. Accad. Naz. Lincei 31* (1961), 277.
381. R. Huisgen; W. Mack; E. Anneser, *Tetrahedron Lett. 1961*, 587.
382. D. A. Klein; R. A. Fouty, *Macromolecules 1* (1968), 318.
383. P. Caramella; P. Grünanger In "1,3-Dipolar Cycoladdition Chemistry" (A. Padwa, Ed.), Vol. I, Wiley, New York, 1984, p. 291.
384. J. J. Tufariello In "1,3-Dipolar Cycloaddition Chemistry" (A. Padwa, Ed.), Vol. II, Wiley, New York, 1984, p. 83.
385. D. St. C. Black; R. F. Crozier; V. D. Davis, *Synthesis 7* (1975), 205.
386. T. Mukaiyama; T. Hoshino, *J. Am. Chem. Soc. 82* (1960), 5339.
387. L. Fabbrini; F. De Sarlo, *Chim. Ind. (Milan) 45* (1963), 242.
388. F. Lauria; V. Vecchietti; G. Tosolini, *Gazz. Chim. Ital. 94* (1964), 478.
389. S. Morrocchi; A. Ricca; L. Velo, *Chim. Ind. (Milan) 49* (1967), 168.
390. L. A. Simonyan; U. V. Zeifman; N. P. Gambaryan, *Izv. Akad. Nauk SSSR, Ser. Khim. 1968*, 1916.
391. R. Huisgen, *Angew. Chem. 75* (1963), 604.
392. F. M. Hershenson, *J. Heterocycl. Chem. 9* (1972), 739.
393. M. Rai; K. Krishan; A. Singh, *Ind. J. Chem. 15B* (1977), 848.
394. T. Sasaki; S. Eguchi; N. Toi, *J. Org. Chem. 44* (1979), 3711.
395. R. M. Srivastava; L. B. Clapp, *J. Heterocycl. Chem. 5* (1968), 61.
396. W. I. Awad; S. M. A. R. Omran; M. Sobhy, *J. Org. Chem. 31* (1966), 331.
397. W. I. Awad; M. Sobhy, *Can. J. Chem. 47* (1969), 1473.
398. A. Franke, *Liebigs Ann. 1978*, 717.
399. V. Nair, *Tetrahedron Lett. 1971*, 4831.
400. C. W. Rees; R. Somanathan; R. C. Storr; A. D. Woolhouse, *J. Chem. Soc., Chem. Commun. 1975*, 740.
401. K. Krishan; M. Rai; J. Singh; A. Singh, *Ind. J. Chem. 15B* (1977), 1041.
402. N. Singh; J. S. Sandhu; S. Mohan, *Tetrahedron Lett. 1968*, 4463.
403. N. Kornblum; R. A. Brown, *J. Am. Chem. Soc. 87* (1964), 2681.
404. V. A. Tartakovskii; S. S. Smagin; J. E. Chlenov; S. S. Novikov, *Izv. Acad. Nauk SSSR, Ser. Khim. 1965*, 552.
405. H. G. Aurich; K. Stork, *Chem. Ber. 108* (1975), 2764.
406. S. Morrochi; A. Ricca, *Chim. Ind. (Milan) 49* (1967), 629.
407. M. Fetizon; M. Golfier; R. Milcent; I. Papadakis, *Tetrahedron 31* (1975), 165.
408. G. Just; K. Dahl, *Tetrahedron 24* (1968), 5251.
409. W. M. Williams; W. R. Dolbier, *J. Org. Chem. 34* (1969), 155.
410. G. F. Bettinetti; A. Gamba, *Gazz. Chim. Ital. 100* (1970), 1144.
411. G. Bianchi; C. De Micheli; R. Gandolfi, *J. Chem. Soc., Perkin I 1972*, 1711.
412. P. Caramella; P. Frattini; P. Grünanger, *Tetrahedron Lett. 1971*, 3817.
413. C. Parini; S. Colombi; A. Ius; R. Longhi; G. Vecchio, *Gazz. Chim. Ital. 107* (1977), 559.
414. P. Caramella; G. Cellerino; A. C. Coda; A. G. Invernizzi; P. Grünanger; K. N. Houk; F. M. Albini, *J. Org. Chem. 41* (1976), 3349.
415. P. Caramella; G. Cellerino; P. Grünanger; F. M. Albini; M. R. Cellerino, *Tetrahedron 34* (1978), 3545.
416. G. Bailo; P. Caramella; G. Cellerino; A. G. Invernizzi; P. Grünanger, *Gazz. Chim. Ital. 103* (1973), 47.
417. J.-P. Gibert; R. Jacquier; C. Petrus, *Bull. Soc. Chim. France 1979*, II, 281.

418. P. Rajagopalan, *Tetrahedron Lett. 1969*, 311.
419. T. Sasaki; T. Yoshioka; Y. Suzuki, *Bull. Chem. Soc. (Japan) 42* (1969), 3335.
420. K. H. Magosch; R. Feinauer, *Angew. Chem. 83* (1971), 882.
421. F. Foti; G. Grassi; F. Risitano; F. Caruso, *J. Chem. Res. (S) 1983*, 230.
422. P. Caramella; E. Cereda, *Synthesis 1971*, 433.
423. R. Huisgen; J. Wulff, *Tetrahedron Lett. 1967*, 921; *Chem. Ber. 102* (1969), 1848.
424. T. Sasaki; T. Yoshioka, *Bull. Chem. Soc. (Japan) 42* (1969), 258.
425. C. Grundmann; R. Richter, *Tetrahedron Lett. 1968*, 963.
426. E. Beckmann, *Chem. Ber. 23* (1890), 1680, 3331; *27* (1894), 1957.
427. H. Goldschmidt, *Chem. Ber. 23* (1890), 2746.
428. H. Seidl; R. Huisgen; R. Grashey, *Chem. Ber. 102* (1969), 926.
429. J. Goerdeler; R. Schimpf, *Chem. Ber. 106* (1973), 1496.
430. O. Tsuge; M. Tashiro; R. Mizuguchi; S. Kanemasa, *Chem. Pharm. Bull. (Japan) 14* (1966), 1055.
431. D. St. C. Black; K. G. Watson, *Aust. J. Chem. 26* (1973), 2473; *Tetrahedron Lett. 1972*, 4191.
432. J. F. Elsworth; M. Lamchen, *J. Chem. Soc. (C) 1968*, 2423.
433. H. Staudinger; K. Miescher, *Helv. Chim. Acta 2* (1919), 554.
434. J. E. Baldwin; A. K. Qureshi; B. Sklarz, *J. Chem. Soc. (C) 1969*, 1073.
435. E. Beckmann; E. Fellrath, *Liebigs Ann. 273* (1893), 1.
436. R. Neidlein, *Arch. Pharm. 297* (1964), 623.
437. J. Thesing; W. Sirrenberg, *Chem. Ber. 92* (1959), 1748.
438. T. Sasaki; M. Ando, *Bull. Chem. Soc. (Japan) 41* (1968), 2960.
439. T. Takahashi; H. Kano, *Chem. Pharm. Bull. (Japan) 12* (1964), 1290; *Tetrahedron Lett. 1963*, 1687.
440. T. Takahashi; S. Hashimoto; H. Kano, *Chem. Pharm. Bull. Japan 18* (1970), 1176.
441. G. Zinner; U. Dybowski, *Arch. Pharm. 304* (1971), 877.
442. G. Zinner; N. P. Lupke; U. Dybowski, *Arch. Pharm. 305* (1972), 64.
443. G. Zinner; B. Geister, *Arch. Pharm. 306* (1973), 97, 898.
444. M. Masui; K. Suda; M. Yamaguchi; C. Yijima, *Chem. Pharm. Bull. Japan 21* (1973), 1605.
445. E. Hayashi, *J. Pharm. Soc. Japan 81* (1961), 1030.
446. C. Schenk; M. L. Beekes; J. A. M. van der Drift; Th. J. de Boer, *Rec. Trav. Chim. Pays. Bas 99* (1980), 278.
447. G. Zinner; U. Krueger, *Chem. Ztg. 101* (1977), 154.
448. P. Zápulský; A. Martvŏn; F. Považanec, *Coll. Chem. Chech., Chem. Commun. 41* (1976), 3799.
449. G. Zinner; E. Eghstessad, *Arch. Pharm. 312* (1979), 907.
450. E. Eghstessad; G. Zinner, *Arch. Pharm. 312* (1979), 1027; *313* (1980), 357.
451. M. Radau; K. Hartke, *Arch. Pharm. 305* (1972), 737.
452. M. Komatsu; Y. Ohshiro; T. Agawa, *J. Org. Chem. 37* (1972), 3192.
453. J. Streith; G. Wolff; H. Fritz, *Tetrahedron 33* (1977), 1349.
454. J.-P. Gibert; C. Petrus; F. Petrus, *J. Heterocycl. Chem. 16* (1979), 311.
455. J.-P. Gibert; R. Jacquier; C. Petrus; F. Petrus, *Tetrahedron Lett. 1974*, 755.
456. W. Kliegel, *Chem. Ztg. 100* (1976), 236.
457. R. Huisgen; W. Mack, *Tetrahedron Lett. 1961*, 583; *Chem. Ber. 105* (1962), 2805.
458. N. A. Genco; R. A. Partis; H. Alper, *J. Org. Chem. 38* (1973), 4365.
459. Y. Otsuji; Y. Tsujii, A. Yoshida; E. Imoto, *Bull. Chem. Soc. Japan 44* (1971), 223.
460. S. Morrocchi; A. Ricca; A. Zanarotti, *Chim. Ind. (Milan) 50* (1968), 352.
461. H. Matsukubo; M. Kato, *J. Chem. Soc., Chem. Commun. 1974*, 412.
462. N. G. Argyropoulos; N. E. Alexandrou; D. N. Nicolaides, *Tetrahedron Lett. 1976*, 83.
463. G. L'Abbé; G. Mathys, *J. Org. Chem. 39* (1974), 1221.
464. S. Morrocchi; A. Quilico; A. Ricca; A. Selva, *Gazz. Chim. Ital. 98* (1968), 891.
465. S. Morrocchi; A. Ricca; A. Selva; A. Zanarotti, *Gazz. Chim. Ital. 99* (1969), 565.
466. A. Quilico; G. Stagno d'Alcontres, *Gazz. Chim. Ital. 80* (1950), 140.
467. S. Shiraishi; S. Ikeuchi; M. Seno; T. Asahara, *Bull. Chem. Soc. Japan 50* (1977), 910.
468. F. De Sarlo; A. Guarna; A. Brandi, *J. Heterocycl. Chem. 20* (1983), 1505.
469. D. Mackey; L. H. Das; J. M. Dust, *J. Chem. Soc., Perkin I 1980*, 2408.
470. F. De Sarlo; A. Brandi, *J. Chem. Res. (S) 1980*, 122.
471. R. Huisgen; W. Mack; E. Anneser, *Angew. Chem. 73* (1961), 656.
472. R. Huisgen; W. Mack, *Chem. Ber. 105* (1972), 2815.
473. A. Battaglia; A. Dondoni; G. Mazzanti, *Synthesis 1971*, 378.
474. D. St. C. Black; K. G. Watson, *Austr. J. Chem. 26* (1973), 2491.
475. G. Mazzanti; G. Maccagnani; B. F. Bonini; P. Pedrini; B. Zwanenburg, *Gazz. Chim. Ital. 110* (1980), 163.

476. E. Vedejs; D. A. Perry, *J. Am. Chem. Soc. 105* (1983), 1683.
477. E. Vedejs; R. A. Buchanan, *J. Org. Chem. 49* (1984), 1840.
478. A. Battaglia; A. Dondoni; G. Maccagnani; G. Mazzanti, *J. Chem. Soc. (B) 1971*, 2096.
479. B. F. Bonini; G. Maccagnani; G. Mazzanti; L. Thijs; H. P. M. Ambrosius; B. Zwanenburg, *J. Chem. Soc., Perkin I 1977*, 1468.
480. B. F. Bonini; G. Maccagnani; G. Mazzanti; L. Thijs; G. E. Veenstra; B. Zwanenburg, *J. Chem. Soc., Perkin I 1978*, 1218.
481. E. Schaumann; G. Rühter, *Tetrahedron Lett. 1985*, 5265.
482. D. Noël; J. Vialle, *Bull. Soc. Chim. France 1967*, 2239.
483. J. Maignan; J. Vialle, *Bull. Soc. Chim. (France) 1973*, 1973.
484. S. Holm; J. A. Boerma; N. H. Nilsson; A. Senning, *Chem. Ber. 109* (1976), 1069.
485. K. Friedrich; M. Zamkanei, *Chem. Ber. 112* (1979), 1873.
486. T. Sasaki; T. Yoshioka, *Bull. Soc. Chim. (Japan) 41* (1968), 2211.
487. K. Dickoré; R. Wegler, *Angew. Chem. 78* (1966), 1023.
488. M. S. Raasch, *J. Org. Chem. 35* (1970), 3470.
489. A. Corsaro; U. Chiacchio; G. Alberghina; G. Purrello, *J. Chem. Res. (S) 1984*, 370.
490. E. Schaumann; U. Behrens, *Angew. Chem. 89* (1977), 750.
491. J.-M. Borsus; G. L'Abbé, G. Smets, *Tetrahedron 31* (1975), 1537.
492. C. Musante, *Gazz. Chim. Ital. 68* (1938), 331.
493. W. O. Foye; J. M. Kauffman, *J. Org. Chem. 31* (1966), 2417.
494. D. St. C. Black; K. G. Watson, *Aust. J. Chem. 26* (1973), 2177; *Angew. Chem. 84* (1972), 34.
495. M. Hamana; B. Umezawa; S. Nakashima, *Chem. Pharm. Bull. (Japan) 10* (1962), 969.
496. R. Grashey; G. Schroll; M. Weidner, *Chem. Ztg. 100* (1976), 496; R. Grashey; M. Weidner; C. Knorn; H. Bauer, *Chem. Ztg. 100* (1976), 496; R. Grashey; M. Weidner; G. Schroll, *Chem. Ztg. 100* (1976), 497.
497. H. J. Bestmann; R. Kunstmann, *Angew. Chem. 78* (1966), 1059; *Chem. Ber. 102* (1969), 1816.
498. H. J. Bestmann; T. Denzel; R. Kunstmann; J. Lengyel, *Tetrahedron Lett. 1968*, 2895.
499. G. Gaudiano; R. Mondelli; P. P. Ponti; C. Ticozzi; A. Umani-Ronchi, *J. Org. Chem. 33* (1968), 4431.
500. A. Umani-Ronchi; M. Acampora; G. Gaudiano; A. Selva, *Chim. Ind. (Milan) 49* (1967), 388.
501. M. I. Shevchuk; S. T. Shpak; A. V. Dombrovskii, *Zh. Org. Khim. 45* (1975), 2609.
502. G. L'Abbé; J. M. Bursus; P. Ykman; G. Smets, *Chem. Ind. (London) 1971*, 1491.
503. R. Huisgen; J. Wulff, *Tetrahedron Lett. 1967*, 917; *Angew. Chem. 79* (1967), 472; *Chem. Ber. 102* (1969), 746, 1833.
504. S. Zbaida; E. Breuer, *Tetrahedron 34* (1978), 1241; *J. Org. Chem. 47* (1982), 1073; *Experientia 35* (1979), 851; *J. Chem. Soc., Chem. Commun. 1978*, 6.
505. Y. Y. C. Yeung Lam Co; R. Carrié; A. Muench; G. Becker, *J. Chem. Soc. Chem. Commun. 1984*, 1634; Y. Y. C. Yeung Lam Co; R. Carrié, *J. Chem. Soc. Chem. Commun. 1984*, 1640.
506. T. Allspach; M. Regitz; G. Becker; W. Becker; *Synthesis 1986*, 31; W. Rösch; M. Regitz, *Synthesis 1987*, 689.
507. P. T. Meinke; G. A. Krafft; J. T. Spenser, *Tetrahedron Lett. 1987*, 3887.
508. P. Rajagopalan, *Tetrahedron Lett. 1964*, 887.
509. R. Huisgen; H. Blaschke; E. Brunn, *Tetrahedron Lett. 1966*, 405.
510. H. Blaschke; E. Brunn; R. Huisgen; W. Mack, *Chem. Ber. 105* (1972), 2841.
511. P. I. Paetzold; G. Stohr, *Chem. Ber. 101* (1968), 2874.
512. P. Beltrame; C. Vintani, *J. Chem. Soc. (B) 1970*, 873.
513. P. Beltrame; A. Comotti; C. Veglio, *Chem. Commun. 1967*, 996.
514. P. Rajagopalan; H. U. Daeniker, *Angew. Chem. 75* (1963), 91.
515. P. Rajagopalan; B. G. Advani, *J. Org. Chem. 30* (1965), 3369.
516. F. Eloy; R. Lenaers, *Bull. Soc. Chim. Belg. 74* (1965), 129.
517. F. Eloy, *Helv. Chim. Acta 48* (1965), 380.
518. E. S. Levchenko; B. N. Ugarov, *Zh. Org. Khim. 5* (1969), 148.
519. R. Albrecht; G. Kresze, *Chem. Ber. 98* (1965), 1205.
520. B. P. Stark; M. H. G. Ratcliffe, *J. Chem. Soc. 1964*, 2640.
521. O. Tsuge; M. Tashiro; S. Mataka, *Tetrahedron Lett. 1968*, 3877.
522. E. H. Burk; D. D. Carlos, *J. Heterocycl. Chem. 7* (1970), 177.
523. G. Trickes; H. P. Braun; H. Meier, *Liebigs Ann. 1977*, 1347.
524. I. V. Bodrikov; V. L. Krasnov; N. K. Tulegenova, *Zh. Org. Khim. 14* (1978), 2231.
525. H. Böshagen, *Chem. Ber. 100* (1967), 954.
526. C. Grundmann, *Synthesis 1970*, 344.

527. U. M. Kempe; T. K. Das Gupta; K. Blatt; P. Gygax; D. Felix; A. Eschenmoser, *Helv. Chim. Acta* 55 (1972), 2187.
528. T. K. Das Gupta; D. Felix; U. M. Kempe; A. Eschenmoser, *Helv. Chim. Acta* 55 (1972), 2198.
529. P. Gygax; T. K. Das Gupta; A. Eschenmoser, *Helv. Chim. Acta* 55 (1972), 2205.
530. M. Petrzilka; D. Felix; A. Eschenmoser, *Helv. Chim. Acta* 56 (1973), 2950.
531. S. Shatzmiller; P. Gygax; D. Hall; A. Eschenmoser, *Helv. Chim. Acta* 56 (1973), 2961.
532. S. Shatzmiller; A. Eschenmoser, *Helv. Chim. Acta* 56 (1973), 2975.
533. E. Shalom; J.-L. Zenou; S. Shatzmiller, *J. Org. Chem.* 26 (1977), 4213.
534. S. Levinger; S. Shatzmiller, *Tetrahedron* 34 (1978), 563.
535. M. Riediker; W. Graf, *Helv. Chim. Acta* 62 (1979), 205; *Angew. Chem.* 93 (1981), 491.
536. B. Hardegger; S. Shatzmiller, *Helv. Chim. Acta* 59 (1976), 2765; S. E. Denmark; C. J. Cramer; M. S. Dappen, *J. Org. Chem* 52 (1987) 877.
537. S. Shatzmiller; E. Shalom, *Liebigs Ann. 1983*, 897.
538. G. W. Kirby, *Chem. Soc. Rev.* 6 (1977), 1.
539. T. L. Gilchrist, *Chem. Soc. Rev.* 12 (1983), 53.
540. M. Ochiai; M. Obayashi; K. Morita, *Tetrahedron* 23 (1967), 2641.
541. A. Lablache-Combier; M. L. Villaume; R. Jacquesy, *Tetrahedron Lett. 1967*, 4959.
542. A. Lablache-Combier; M. L. Villaume, *Tetrahedron* 24 (1968), 6951.
543. E. Winterfeldt; W. Krohn, *Angew. Chem.* 79 (1967), 722.
544. H. Hjeds; B. Jerslev; K. J. Ross-Petersen, *Dansk Tidskr. Farm.* 46 (1972), 97.
545. N. K. A. Dalgård; K. E. Larsen; K. B. G. Torssell, *Acta Chem. Scand.* B38 (1984), 423.
546. R. Grigg; H. Q. Gunaratne; J. Kemp, *J. Chem. Soc., Perkin I 1984*, 41.
547. R. Grigg; M. Jordan; A. Tangthongkum; F. W. B. Einstein; T. Jones, *J. Chem. Soc., Perkin I 1984*, 47.
548. R. Grigg; S. Thianpantangul, *J. Chem. Soc., Perkin I 1984*, 653.
549. W. Oppolzer; K. Keller, *Tetrahedron Lett. 1970*, 1117.
550. T. Shimizu; Y. Hayashi; K. Teramura, *Bull. Soc. Chem. (Japan)* 58 (1985), 397.
551. "Cyclopentanoid Terpene Derivatives" (W. I. Taylor and A. R. Battersby, Eds.), Dekker, New York, 1969.
552. D. P. Curran, *Tetrahedron Lett. 1983*, 3443.
553. A. Barco; S. Benetti; G. P. Pollini; P. G. Baraldi; D. Simoni; C. B. Vicentini, *J. Org. Chem.* 44 (1979), 1734.
554. N. A. LeBel; J. J. Whang, *J. Am. Chem. Soc.* 81 (1959), 6334; A. C. Cope; N. A. LeBel, *J. Am. Chem. Soc.* 82 (1960), 4656.
555. A. Padwa, *Angew. Chem.* 88 (1976), 131; In "1,3-Dipolar Cycloaddition Chemistry" (A. Padwa, Ed.), Vol. II, Wiley, New York, 1984, p. 277 and references cited therein.
556. W. Oppolzer, *Angew. Chem.* 89 (1977), 10.
557. N. Balasubramanian, *Org. Prep. Proc.* 17 (1985), 25.
558. H. Iida; C. Kibayashi, *J. Synth. Org. Chem. (Japan)* 41 (1983), 652.
559. L. Garanti; A. Sala; G. Zecchi, *J. Org. Chem.* 40 (1975), 2403; *Synthesis 1975*, 666.
560. R. H. Wollenberg; J. E. Goldstein, *Synthesis 1980*, 757.
561. V. Jäger; H. J. Günther, *Angew. Chem.* 89 (1977), 253.
562. B. Bernet; A. Vasella, *Helv. Chim. Acta* 62 (1979), 1990, 2400.
563. R. J. Ferrier; P. Prasit, *J. Chem. Soc., Chem. Commun. 1981*, 983.
564. R. J. Ferrier; R. H. Furneaux; P. Prasit; P. C. Tyler; K. L. Brown; G.-J. Gainsford; J. W. Diehl, *J. Chem. Soc., Perkin I 1983*, 1621.
565. R. J. Ferrier; P. Prasit; G. J. Gainsford, *J. Chem. Soc., Perkin I 1983*, 1629.
566. F. J. Vinick; I. E. Fengler; H. W. Geschwend; R. K. Rudebaugh, *J. Org. Chem.* 42 (1977), 2936.
567. A. P. Kozikowski; P. D. Stein, *J. Am. Chem. Soc.* 104 (1982), 4023.
568. A. P. Kozikowski; P. D. Stein, *J. Org. Chem.* 49 (1984), 2301.
569. A. P. Kozikowski; Y. Y. Chen, *Tetrahedron Lett. 1982*, 2081.
570. N. A. LeBel; M. E. Post; J. J. Whang, *J. Am. Chem. Soc.* 86 (1964), 3759.
571. N. A. LeBel, *Trans. N.Y. Acad. Sci.* 27 (1965), 858.
572. N. A. LeBel; T. A. Lajiness, *Tetrahedron Lett. 1966*, 2173.
573. M. Raban; F. B. Jones; E. H. Carlson; E. Bannucci; N. A. LeBel, *J. Org Chem.* 35 (1970), 1496.
574. N. A. LeBel; E. G. Bannucci, *J. Org. Chem.* 36 (1971), 2440.
575. N. A. LeBel; D. Hwang, *Org. Synth.* 58 (1978), 106.
576. N. A. LeBel; G. M. J. Slusarczuk; L. A. Spurlock, *J. Am. Chem. Soc.* 84 (1962), 4360.
577. N. A. LeBel; N. D. Ojha; J. R. Menke; R. J. Newland, *J. Org. Chem.* 37 (1972), 2896.
578. T. Sasaki; S. Eguchi; T. Suzuki, *J. Chem. Soc., Chem. Commun. 1979*, 506.

579. T. Sasaki; S. Eguchi; T. Suzuki, *J. Org. Chem. 47* (1982), 5250.
580. T. Kusumi; S. Takahashi; Y. Sato; H. Kakisawa, *Heterocycles 10* (1978), 257.
581. S. Takahashi; T. Kusumi; Y. Sato; Y. Inouye; H. Kakisawa, *Bull. Chem. Soc. (Japan) 54* (1981), 1777.
582. N. A. LeBel; E. Banucci, *J. Am. Chem. Soc. 92* (1970), 5278.
583. M. L. Mihalović; L. Lorenc; Z. Maksimović; J. Kalvoda, *Tetrahedron 29* (1973), 2683.
584. N. A. LeBel; B. W. Caprathe, *J. Org. Chem. 50* (1985), 3490.
585. A. Padwa; H. Ku; A. Mazzu, *J. Org. Chem. 43* (1978), 381.
586. J. B. Bremner; L. V. Thuc, *Austr. J. Chem. 33* (1980), 379.
587. J. T. Bailey; I. Berger; R. Friary; M. S. Puar, *J. Org. Chem. 47* (1982), 857.
588. W. Oppolzer; H. P. Weber, *Tetrahedron Lett. 1970*, 1121.
589. W. Oppolzer; K. Keller, *Tetrahedron Lett. 1970*, 4313.
590. R. Brambilla; R. Friary; A. Ganguly; M. S. Puar; B. R. Sunday; J. J. Wright; K. D. Onan; A. T. McPhail, *Tetrahedron 37* (1981), 3615.
591. O. Tsuge; K. Ueno; S. Kanemasa, *Chem. Lett. 1984*, 285.
592. P. N. Confalone; E. M. Huie, *J. Org. Chem. 48* (1983), 2994; *52* (1987) 79.
593. R. L. Funk; L. H. M. Horcher, II; J. U. Daggett; M. M. Hansen, *J. Org. Chem. 48* (1983), 2632; R. L. Funk; G. L. Bolton; J. U. Daggett; M. M. Hansen; L. H. M. Horcher, II, *Tetrahedron 41* (1985), 3479.
594. R. L. Funk; G. Bolton, *J. Org. Chem. 49* (1984), 5021.
595. M. A. Schwartz; G. C. Swanson, *J. Org. Chem. 44* (1979), 953.
596. T. Iwashita; T. Kusumi; H. Kakisawa, *Chem. Lett. 1979*, 947.
597. M. A. Schwartz; A. M. Willbrand, *J. Org. Chem. 50* (1985), 1359.
598. P. M. Wovkulich; F. Barcelos; A. D. Batcho; J. F. Sereno; E. G. Baggiolini; B. M. Hennessy; M. R. Uskoković, *Tetrahedron 40* (1984), 2283.
599. K. Cooper; G. Pattenden, *J. Chem. Soc., Perkin I 1984*, 799.
600. D. P. Curran; P. B. Jacobs, *Tetrahedron Lett. 1985*, 2031; D. P. Curran; P. B. Jacobs; R. L. Elliot; B. H. Kim, *J. Am. Chem. Soc. 109* (1987), 5280.
601. A. P. Kozikowski; B. B. Mugrage; B. C. Wang; Z.-B. Xu, *Tetrahedron Lett. 1983*, 3705.
602. A. P. Kozikowski; K. Hiraga; J. P. Springer; B. C. Wang; Z.-B. Xu, *J. Am. Chem. Soc. 106* (1984), 1845.
603. A. P. Kozikowski; A. K. Ghosh, *Tetrahedron Lett. 1983*, 2623.
604. H. Nishiyama; H. Arai; T. Ohki; K. Itoh, *J. Am Chem. Soc. 107* (1985), 5310.
605. (a) R. Fusco; L. Garanti; G. Zecchi, *Chim. Ind. (Milan) 57* (1975), 16.
    (b) L. Garanti; G. Zecchi, *J. Heterocycl. Chem. 17* (1980), 609.
606. V. Jäger; V. Buss; W. Schwab, *Tetrahedron Lett. 1978*, 3133; *Liebigs Ann. 1980*, 122.
607. V. Jäger; V. Buss, *Liebigs Ann. 1980*, 101.
608. W. Stühmer; W. Heinrich, *Chem. Ber. 84* (1951), 224.
609. W. Stühmer; H. H. Frey, *Arch. Pharm. 286* (1953), 22.
610. M. V. Kashutina; S. L. Joffe; V. A. Tartakovskii, *Dokl. Akad. Nauk SSSR, Ser. Khim. 218* (1974), 109.
611. G. S. King; P. D. Magnus; H. S. Rzepa, *J. Chem. Soc., Perkin I 1972*, 437.
612. T. Kusumi; H. Kakisawa; S. Suzuki; K. Harada; C. Kashima, *Bull. Chem. Soc. Japan 51* (1978), 1261.
613. P. N. Confalone; E. D. Lollar; G. Pizzolato; M. R. Uskoković, *J. Am. Chem. Soc. 100* (1978), 6291.
614. K. F. Burri; R. A. Cardone; W. Y. Chen; P. Rosen, *J. Am. Chem. Soc. 100* (1978), 7069.
615. K. Narasaka; Y. Ukaji, *Chem. Lett. 1984*, 147.
616. V. Jäger; W. Schwab; V. Buss, *Angew. Chem. 93* (1981), 576; W. Schwab; V. Jäger, *Angew. Chem. 93* (1981), 578.
617. A. Hosomi; H. Shoji; H. Sakurai, *Chem. Lett. 1985*, 1049.
618. J. J. Tufariello; J. P. Tette, *Chem. Commun. 1971*, 469; *J. Org. Chem. 40* (1975), 3866.
619. J. J. Tufariello; G. E. Lee, *J. Am. Chem. Soc. 102* (1980), 373.
620. T. Iwashita; T. Kusumi; H. Kakisawa, *Chem. Lett. 1979*, 1337; *J. Org. Chem. 47* (1982), 230.
621. J. J. Tufariello; Sk. A. Ali, *J. Am. Chem. Soc. 101* (1979), 7116.
622. J. J. Tufariello; Sk. A. Ali, *Tetrahedron Lett. 1979*, 4445.
623. J. J. Tufariello; Sk. A. Ali, *Tetrahedron Lett. 1978*, 4647.
624. H. Otomasu; N. Takatsu; T. Honda; T. Kametani, *Tetrahedron 38* (1982), 2627.
625. T. Iwashita; M. Suzuki; T. Kusumi; H. Kakisawa, *Chem. Lett. 1980*, 383.
626. H. Iida; M. Tanaka; C. Kibayashi, *J. Chem. Soc., Chem. Commun. 1983*, 271.

627. H. Iida; Y. Watanabe; C. Kibayashi, *Chem. Lett. 1983*, 1195; *Chem. Pharm. Bull. Japan 33* (1985), 351.
628. H. Iida; Y. Watanabe; C. Kibayashi, *J. Chem. Soc., Perkin I 1985*, 261.
629. J. B. Bapat; D. St. C. Black; R. F. C. Brown; C. Ichlov, *Aust. J. Chem. 25* (1972), 2445.
630. J. J. Tufariello; E. J. Trybulski, *J. Chem. Soc., Chem. Commun. 1973*, 720.
631. J. J. Tufariello; G. B. Mullen, *J. Am. Chem. Soc. 100* (1978), 3638.
632. J. J. Tufariello; G. B. Mullen; J. J. Tegeler; E. J. Trybulski; S. C. Wong; Sk. A. Ali, *J. Am. Chem. Soc. 101* (1979), 2435.
633. J. J. Tufariello; J. J. Tegeler; S. C. Wong; Sk. A. Ali, *Tetrahedron Lett. 1978*, 1733.
634. J. J. Tufariello; H. Meckler; K. P. A. Senaratne, *Tetrahedron 41* (1985), 3447.
635. H. Iida; C. Kibayashi, *Tetrahedron Lett. 1981*, 1913.
636. J. J. Tufariello; J. J. Tegeler, *Tetrahedron Lett. 1976*, 4037.
637. J. J. Tufariello; R. G. Gatrone, *Tetrahedron Lett. 1978*, 2753.
638. S. Takano; K. Shishido, *J. Chem. Soc., Chem. Commun. 1981*, 940; *Heterocycles 19* (1982), 1439.
639. H. Iida; M. Tanaka; C. Kibayashi, *J. Chem. Soc., Chem. Commun. 1983*, 1143; *J. Org. Chem. 49* (1984), 1909.
640. K. Shishido; K. Tanaka; K. Fukumoto; T. Kametani, *Tetrahedron Lett. 1983*, 2783; *Chem. Pharm. Bull. (Japan) 33* (1985), 532.
641. E. Gössinger, *Tetrahedron Lett. 1980*, 2229; *Monatsh. Chem. 111* (1980), 143, 783.
642. E. Gössinger; B. Witkop, *Monatsh. Chem. 111* (1980), 803.
643. H. Oinuma; S. Dan; H. Kakisawa, *J. Chem. Soc., Chem. Commun. 1983*, 654.
644. J. W. Daly, *Progr. Chem. Org. Nat. Prod. 41* (1982), 205.
645. E. Gössinger; R. Imhof; H. Wehrli, *Helv. Chim. Acta 58* (1975), 96.
646. J. J. Tufariello; E. J. Trybulski, *J. Org. Chem. 39* (1974), 3378.
647. S. W. Baldwin; J. D. Wilson; J. Aubé, *J. Org. Chem. 50* (1985), 4432.
648. W. C. Lumma Jr., *J. Am. Chem. Soc. 91* (1969), 2820.
649. W. Oppolzer; S. Siles; R. L. Snowden; B. H. Bakker; M. Petrzilka, *Tetrahedron Lett. 1979*, 4391; *Tetrahedron 41* (1985), 3497.
650. W. Oppolzer; M. Petrzilka, *J. Am. Chem. Soc. 98* (1976), 6722; *Helv. Chim. Acta 61* (1978), 2755.
651. B. B. Snider; C. P. Cartaya-Marin, *J. Org. Chem. 49* (1984), 1688.
652. S. Eguchi; Y. Furukawa; T. Suzuki; K. Kondo; T. Sasaki; M. Honda; C. Katayama; J. Tanaka, *J. Org. Chem. 50* (1985), 1895.
653. A. P. Kozikowski; P.-W. Yuen, *J. Chem. Soc., Chem. Commun. 1985*, 847.
654. E. Dagne; N. Castagnoli, Jr., *J. Med. Chem. 15* (1972), 356.
655. M. Ménard; P. Rivest; L. Morris; J. Meunier; Y. G. Perron, *Can. J. Chem. 52* (1974), 2316.
656. A. E. Walts; W. R. Roush, *Tetrahedron 41* (1985), 3463.
657. W. Oppolzer; J. I. Grayson, *Helv. Chim. Acta 63* (1980), 1706; W. Oppolzer; J. I. Grayson; H. Wegmann; M. Urrea, *Tetrahedron 39* (1983), 3695.
658. A. P. Kozikowski; P. D. Stein, *J. Am. Chem. Soc. 107* (1985), 2569.
659. A. P. Kozikowski; H. Ishida, *J. Am. Chem. Soc. 102* (1980), 4265.
660. A. P. Kozikowski; Y.-Y. Chen, *J. Org. Chem. 46* (1981), 5248.
661. A. P. Kozikowski; Y.-Y. Chen; B. C. Wang; Z.-B. Xu, *Tetrahedron 40* (1984), 2345.
662. A. P. Kozikowski; P.-u. Park, *J. Org. Chem. 49* (1984), 1674; *J. Am. Chem. Soc. 107* (1985) 1763.
663. D. L. Comins; A. H. Abdullah, *Tetrahedron Lett. 1985*, 43.
664. P. M. Wovkulich; M. R. Uskoković, *J. Am. Chem. Soc. 103* (1981), 3956; *Tetrahedron 41* (1985), 3455.
665. P. DeShong; J. M. Leginus, *J. Am. Chem. Soc. 105* (1983), 1686.
666. P. DeShong; J. M. Leginus, *Tetrahedron Lett. 1984*, 5355.
667. A. Vasella; R. Voeffray, *Helv. Chim. Acta 65* (1982), 1134.
668. A. Vasella, *Helv. Chim. Acta 60* (1977), 426, 1273.
669. B. Bernet; E. Krawczyk; A. Vasella, *Helv. Chim. Acta 68* (1985), 2299.
670. R. Huber; A. Knierzinger; J.-P. Obrecht; A. Vasella, *Helv. Chim. Acta 68* (1985), 1730.
671. A. Corsico Coda; P. Grünanger; P. Veronesi, *Tetrahedron Lett. 1966*, 2911.
672. V. Jäger; I. Müller, *Tetrahedron Lett. 1982*, 4777; *Tetrahedron 41* (1985), 3519.
673. V. Jäger; H. Grund; V. Buss; W. Schwab; I. Müller; R. Schohe; R. Franz; R. Ehrler, *Bull. Soc. Chim. Belg. 92* (1983), 1039.
674. V. Jäger; I. Müller; E. F. Paulus, *Tetrahedron Lett. 1985*, 2997.
675. V. Jäger; R. Schohe, *Tetrahedron 40* (1984), 2199.
676. V. Jäger; R. Schohe; E. F. Paulus, *Tetrahedron Lett. 1983*, 5501.

677. R. Annunziata; M. Cinquini; F. Cozzi; L. Raimondi, *J. Chem. Soc., Chem. Commun. 1985*, 403.
678. R. Annunziata; M. Cinquini; F. Cozzi; L. Raimondi; A. Restelli, *Helv. Chim. Acta 68* (1985), 1217.
679. K. N. Houk; S. R. Moses; Y.-D. Wu; N. G. Rondan; V. Jäger; R. Schohe; F. R. Fronczek, *J. Am. Chem. Soc. 106* (1984), 3880.
680. T. Kusumi; H. Kakisawa; S. Suzuki; K. Harada; C. Kashima, *Bull. Chem. Soc. Japan 51* (1978), 1261.
681. Y. Inouye; Y. Watanabe; S. Takahashi; H. Kakisawa, *Bull. Chem. Soc. Japan 52* (1979), 3763.
682. J. Hara; Y. Inouye; H. Kakisawa, *Bull. Chem. Soc. Japan 54* (1981), 3871.
683. W. A. König; W. Hass; W. Dehler; H.-P. Fiedler; H. Zähner, *Liebigs Ann. 1980*, 622.
684. W. Hass; W. A. König, *Liebigs Ann. 1982*, 1615.
685. G. Zimmerman; W. Hass; H. Faasch; H. Schmalle; W. A. König, *Liebigs Ann. 1985*, 2165.
686. R. Franz, Thesis, Würzburg, 1984.
687. B. J. Banks; A. G. M. Barrett; M. A. Russell; D. J. Williams, *J. Chem. Soc., Chem. Commun. 1983*, 873.
688. (a) A. Toy; W. J. Thompson, *Tetrahedron Lett. 1984*, 3533.
    (b) H. Horikawa; T. Nishitani; T. Iwasaki; I. Inoue, *Tetrahedron Lett. 1983*, 2193.
689. C. M. Tice; B. Ganem, *J. Org. Chem. 48* (1983), 5043, 5048.
690. D. A. Abramovitch; R. A. Abramovitch; H. Benecke, *Heterocycles 23* (1985), 25.
691. N. J. A. Gutteridge; J. R. M. Dales, *J. Chem. Soc. (C) 1971*, 122.
692. J. P. Alazard; B. Khémis; X. Lusinchi, *Tetrahedron 31* (1975), 1427.
693. J. E. Baldwin; C. Hoskins; L. Kruse, *J. Chem. Soc., Chem. Commun. 1976*, 795.
694. P. A. Wade; M. K. Pillay; S. M. Singh, *Tetrahedron Lett. 1982*, 4563; *Tetrahedron 40* (1984), 601.
695. A. A. Hagedorn III; B. J. Miller; J. O. Nagy, *Tetrahedron Lett. 1980*, 229.
696. D. M. Vyas; Y. Chiang; T. W. Doyle, *Tetrahedron Lett. 1984*, 487.
697. J. E. Baldwin; L. I. Kruse; J. K. Cha, *J. Am. Chem. Soc. 103* (1981), 942.
698. J. E. Baldwin; J. K. Cha; L. I. Kruse, *Tetrahedron 41* (1985), 5241.
699. R. C. Kelly; I. Schletter; S. J. Stein; W. Wierenga, *J. Am. Chem. Soc. 101* (1979), 1054.
700. R. B. Silverman; M. W. Holladay, *J. Am. Chem. Soc. 103* (1981), 7357.
701. R. V. Stevens; R. P. Polniaszek, *Tetrahedron 39* (1983), 743.
702. H. Iwasaki; T. Kamiya; O. Oka; J. Ueyanagi, *Chem. Pharm. Bull. Japan 17* (1969), 866.
703. H. Iwasaki; T. Kamiya; C. Hatanaka; Y. Sunada; J. Ueyanagi, *Chem. Pharm. Bull. Japan 17* (1969), 873.
704. T. Kamiya, *Chem. Pharm. Bull. Japan 17* (1969), 879.
705. T. Kamiya, *Chem. Pharm. Bull. Japan 17* (1969), 886, 895.
706. R. M. Khomutov; E. S. Severin; G. V. Kovaleva, *Dokl. Akad. Nauk, SSSR 161* (1965), 1227.
707. For a compilation of recent syntheses of **596** and analogues, see M. Frey, Thesis, Würzburg, 1985; V. Jäger; M. Frey, *Liebigs Ann. 1982*, 817; *Synthesis 1985*, 1100 and T. A. Oster; T. M. Harris, *J. Org. Chem. 48* (1983), 4307.
708. S. Mzengeza; R. A. Whitney, *J. Chem. Soc., Chem. Commun. 1984*, 606.
709. G. Schlewer; P. Krogsgaard-Larsen, *Acta Chem. Scand. B38* (1984), 815.
710. N. Jacobsen; H. Kolind-Andersen; J. Christensen, *Can. J. Chem. 62* (1984), 1940.
711. J. E. Baldwin; L. M. Harwood; M. J. Lombard, *Tetrahedron 40* (1984), 4363.
712. J. E. Baldwin; M. F. Chan; G. Gallagher; P. Monk; K. Prout, *J. Chem. Soc., Chem. Commun. 1983*, 250; J. E. Baldwin; M. F. Chan; G. Gallagher; M. Otsuka; P. Monk; K. Prout, *Tetrahedron 40* (1984), 4513.
713. A. Vasella; R. Voeffray, *J. Chem. Soc., Chem. Commun. 1981*, 97; A. Vasella; R. Voeffray; J. Pless; R. Huguenin, *Helv. Chim. Acta 66* (1983), 1241.
714. A. Vasella; R. Voeffray, *Helv. Chim. Acta 65* (1982), 1953.
715. M. Asaoka; T. Mukuta; H. Takei, *Tetrahedron Lett. 1981*, 735.
716. P. G. Baraldi; A. Barco; S. Benetti; F. Moroder; C. P. Pollini; D. Simoni, *J. Org. Chem. 48* (1983), 1297.
717. M. Asaoka; M. Abe; T. Mukuta; H. Takei, *Chem. Lett. 1982*, 215.
718. M. Asaoka; M. Abe; H. Takei, *Bull. Soc. Chem. Japan 58* (1985), 2145.
719. A. B. Smith III; S. R. Schow; J. D. Bloom; A. S. Thompson; K. N. Winzenberg, *J. Am. Chem. Soc. 104* (1982), 4015.
720. M. Prasad; B. Frazer-Reid, *J. Org. Chem. 50* (1985), 1566.
721. A. P. Kozikowski; K. E. Maloney Huss, *Tetrahedron Lett. 1985*, 5759.
722. S. S. Ko; P. N. Confalone, *Tetrahedron Lett. 1984*, 947; *Tetrahedron 41* (1985), 3511.
723. M. Marx; F. Marti; J. Reisdorff; R. Sandmeier; S. Clark, *J. Am. Chem. Soc. 99* (1977), 6754.

724. P. N. Confalone; G. Pizzolato; D. L. Lollar-Confalone; M. R. Uskoković, *J. Am. Chem. Soc. 100* (1978), 6291; *102* (1980), 1954.
725. E. G. Baggiolini; H. L. Lee; G. Pizzolato; M. R. Uskoković, *Bull. Soc. Chim. Belg. 91* (1982), 967; *J. Am. Chem. Soc. 104* (1982), 6460.
726. M. J. Fray; E. J. Thomas, *Tetrahedron 40* (1984), 673.
727. J. E. Baldwin; D. H. R. Barton; N. J. A. Gutteridge; R. J. Martin, *J. Chem. Soc. (C) 1971*, 2184.
728. A. P. Kozikowski; J. G. Scripko, *J. Am. Chem. Soc. 106* (1984), 353.
729. J. M. J. Tronchet; A. Jotterand; N. Le-Hong, *Helv. Chim. Acta 52* (1969), 2569; J. M. J. Tronchet; B. Baehler; N. Le-Hong; P. F. Livio, *Helv. Chim. Acta 54* (1971), 921; J. M. J. Tronchet; E. Mihaly, *Helv. Chim. Acta 55* (1972), 1266; *Carbohydr. Res. 31* (1973), 159; *46* (1976), 127; J. M. J. Tronchet; F. Barbalat:Rey; N. Le-Hong; U. Burger, *Carbohydr. Res. 29* (1973), 297; J. M. J. Tronchet; N. Le-Hong, *Carbohydr. Res. 29* (1973), 311; J. M. J. Tronchet; S. Jaccard-Thorndahl; L. Faivre; R. Massard, *Helv. Chim. Acta 56* (1973), 1303; J. M. J. Tronchet; F. Perret, *Carbohydr. Res. 38* (1974), 169; J. M. J. Tronchet; J. Poncet, *Carbohydr. Res. 46* (1976), 119; J. M. J. Tronchet; A. P. Bonenfant; K. D. Pallie; F. Habashi, *Helv. Chim. Acta 62* (1979), 1622; J. M. J. Tronchet, *Biol. Medical 4* (1975), 81.
730. H. P. Albrecht; D. B. Repke; J. G. Moffat, *J. Org. Chem. 40* (1975), 2143.
731. G. Just; B. Chalard-Fauré, *Can. J. Chem. 54* (1976), 861.
732. D. Horton; J. H. Tsai, *Carbohydr. Res. 67* (1978), 357.
733. A. P. Kozikowski; S. Goldstein, *J. Org. Chem. 48* (1983), 1139.
734. M. Kinoshita; S. Aburaki; M. Wada; S. Umezawa, *Bull. Chem. Soc. Japan 46* (1973), 1279.
735. T. Rosen; C. H. Heathcock, *Tetrahedron 42* (1986) 4909; A. P. Kozikowski; C.-S. Li, *J. Org. Chem. 52* (1987) 3541.

Addendum

In order to minimize omissions of recent published work in the moving field of synthetic use of isoxazole derivatives, articles published in the period December 1985 to August 1987 are reviewed by chapter in this addendum. A few earlier relevant papers are also included.

CHAPTER 1

Self-consisting field (MCSCF) calculations of the mechanism for some 1,3-dipolar cycloadditions indicate a concerted path not involving radicals.[114] The direction of the attack of the nitrile oxide on a chiral alkene occurs from the sterically least-hindered side with the largest group in anti-conformation and the medium-sized group inside as in the preferred staggered transition-state (TS) structure 1.[1,2] This model rationalizes earlier stereoselectivities found for nitrile oxide additions to allyl ethers (M = alkoxy). An inside position for the hydrogen and antiposition for the alkoxy group (M) to avoid steric congestion in the TS has also been proposed.[91]

1

A few cycloaddition studies of cumulated olefinic bonds have been carried out. As predicted by FMO theory, nitrones add to phenylsulfonylpropadiene (2) across the activated double bond giving rise to the 5-exo-methylene isoxazolidine 3, which thermally rearranges to 4 rather than the 3-pyrrolidone 5, eq. (A.1). Base-catalyzed alkylation of 4 produces predominantly the γ-alkylated product 6.[3] In the hands of other investigators 3($R^1 = R^2 = $ Ph) rearranges to 3-pyrrolidone 5 and major amounts of benzazepinone and derived indoles.[92] Dipolar cycloaddition of nitrones to carbomethoxy substituted allenes gives 5-exo-methylene substituted isoazolidines, which upon thermolysis rearrange to 3-pyrrolidones.[93] Cycloadditions of nitrones with fluoroallene proceed regio- and stereospecifically to give 4-exo-fluoromethylene substituted isoxazolidines.[94]

N-Methyl-C-phenylnitrone and benzonitrile oxide add to 2,5-dimethyl-2,3,4-

hexatriene (**7**) according to eq. (A.2). FMO calculations of HOMO-LUMO interactions are in agreement with the experimental results.[4] The 6,7-double bond in the triene **8** reacts selectively with ethoxycarbonylformonitrile oxide to give the isoxazoline **9** and its fission product **10**, eq. (A.3).[5] In agreement with predictions, cycloheptatriene reacts regioselectively with nitrile oxides and nitrones to afford functionalized cycloheptadiene derivatives, eq. (A.4), but the iron tricarbonyl complex **11** was unreactive under the same conditions.[6] The iron tricarbonyl complex of tropone reacts with 3,4-dihydroisoquinoline-N-oxide to give the isoxazolidine **12**, which spontaneously cycloreverts to its components on oxidative removal of the iron tricarbonyl group, eq. (A.5).[7] Aromatic nitrile oxides cycloadd with iron tricarbonyl complex giving 3-aryl-4H-hepta[d]isoxazol-4-ones after oxidative decomplexation with trimethylamine oxide.[95] Reversed regioselectivity is observed in the cycloaddition of the 1,4-diazobutadiene nitrone **13** to allyl alcohol.[96] The allyl halides show common regioselectivity for the nitrone function and yield the expected 5-substituted isoxazolines. The unusual behavior of allyl alcohol is explained by an N,O-acetal formation preceding the cycloaddition, eq. (A.6). The Z-6-ethylidene olivanic acid derivative **14** undergoes cycloaddition with acetonitrile oxide to furnish **15** as the major regioisomer.[8]

The site and face selectivity of arylnitrile oxide cycloadditions to the bullvalene system **16**, as well as the valence isomerization, have been investigated, eq. (A.7).[9] The cyclobutene ring reacts selectively in an anti-fashion to the bullvalene residue to give predominantly isomer **17a**.

Double dehydrobromination of dibromosulfolane in the presence of N-methyl-C-phenylnitrone gives the isoxazolidine **18**, which by N-oxidation undergoes intramolecular nitrone cycloaddition to the tricyclic isoxazolidine **19**, eq. (A.8).[10] Nonbonded interaction exerts regiochemical control in the intramolecular nitrone cycloaddition, eq. (A.9), and only isomer **20a** is observed.[11]

(A.5)

(A.6)

(A.7)

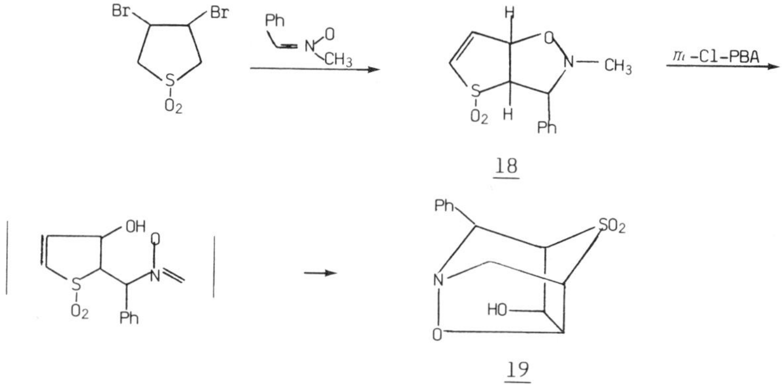

(A.8)

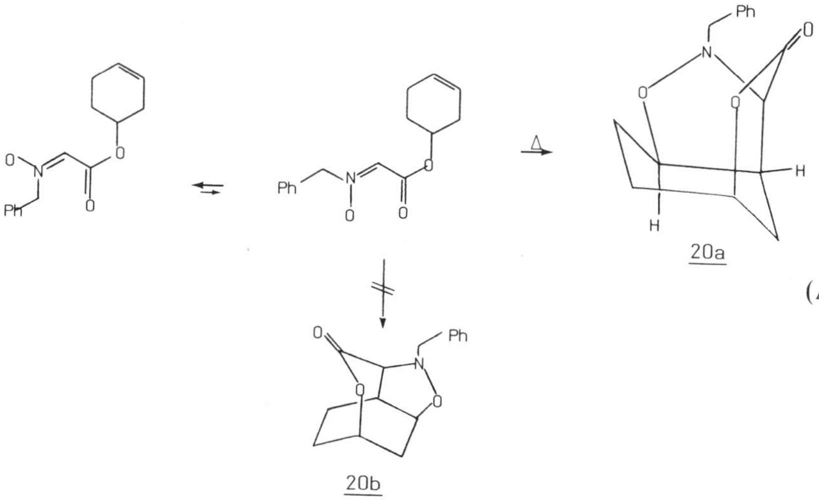

(A.9)

The ratio of endo- to exo-alkylated products of 3-methyl-2-isoxazolines increases by the use of a bulky base, eq. (A.10a). Thus the bulky lithium 2,2,6,6-tetramethylpiperidide gives **23:21 + 22** ~ 10:1, whereas lithium diethylamide gives **23:21 + 22** ~ 1:5–10. The trans to cis ratio **21:22** is often better than 25:1. Addition of HMPA improves the endo to exo ratio.[12] 3-Ethyl-4-

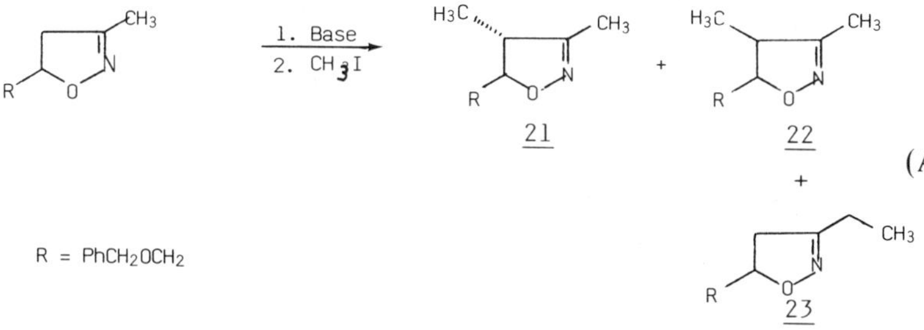

R = PhCH$_2$OCH$_2$

(A.10a)

methylisoxazoline undergoes base-catalyzed regio- and stereocontrolled exo-alkylation at $-80°C$ on the phase opposite from the 4-substituent with the ethyl group in a conformation avoiding 1,3-interactions,[97] eq. (A10b).

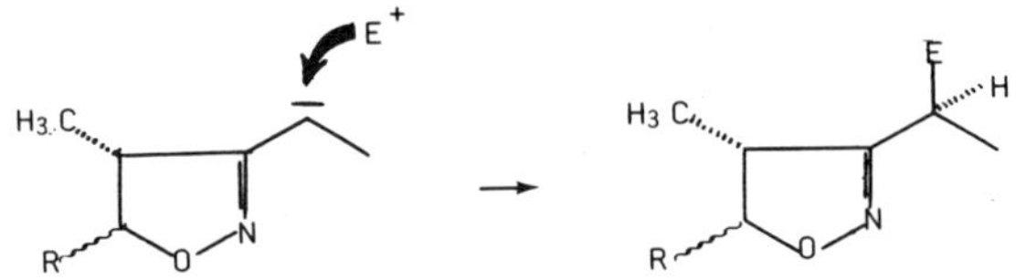

$$(A.10b)$$

2-Isoxazolines with electron-withdrawing groups on the 3-aryl moiety (—CN, —COOCH$_3$) undergo regiospecific $2+2$ photocycloaddition with indene, eq. (A.11).[13] Diels-Alder reactions have been carried out with 5-vinyl substituted isoxazoles[14] and with ethyl 4-nitro-3-phenylisoxazole-5-carboxylate.[15] This compound undergoes spiroannelation at C^5 with nitrogen binuclophiles,[98] and gives cyclopropane derivatives with diazoalkanes.[115]

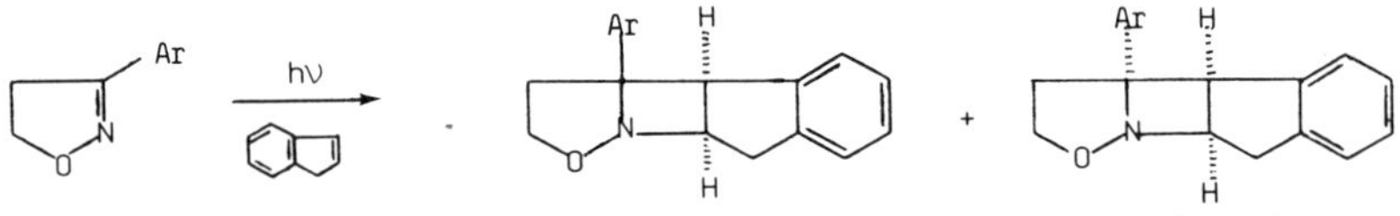

$$(A.11)$$

(major)   (minor)

5-Unsubstituted isoxazoles are prepared by borohydride reduction of 5-chloroisoxazoles.[16] Molybdenum hexacarbonyl has been applied to reductive N—O—bond cleavage of 2-isoxazolines.[17] The yield of amino alcohol from lithium aluminum hydride cleavage of isoxazolidine is improved by adding anhydrous cobalt chloride.[18] Modest asymmetric induction has been observed in nitrile oxide cycloadditions to chiral acrylates. The glucose derivative **24a** gives a 63:37 mixture of isomers.[85] Presently maximal selectivity has been obtained with the camphor derivative **24b**, 77:23.[99] Menthyl acrylate, cf. eq. (1.78), and the Cinchona- alkaloid derivative **24c**[6] gives negligible asymmetric induction. Variation of solvent and nitrile oxide has marginal effect. The corresponding methacrylates give lower diastereomeric ratios.

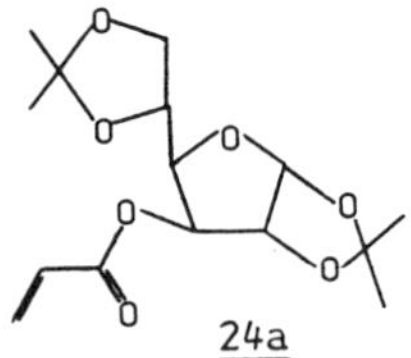

24a

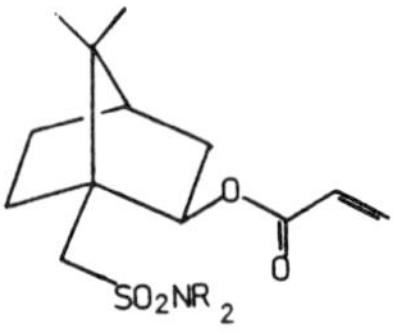

24b

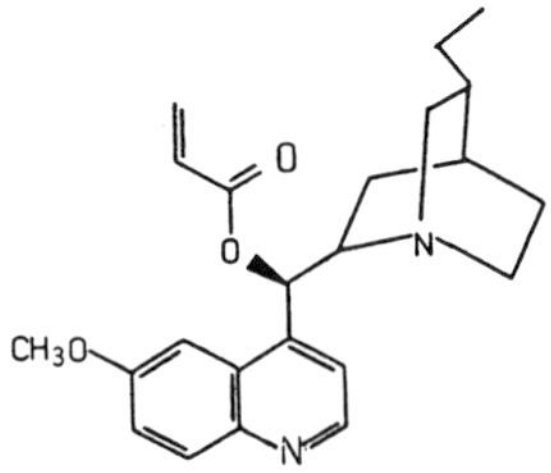

24c

CHAPTER 2

The phosphorous functionalized nitrile oxide eq. (A.12) gives isoxazolines with a variety of olefins. They can be used as Wittig reagents for introducing the isoxazoline moiety.[19]

$$\text{(A.12)}$$

Thermolysis of α-methoxycarbonyl-α-nitroacetanilide (**25**) in the presence of olefins or acetylenes leads to isoxazolines and isoxazoles, respectively, probably via the formanilide-derived nitrile oxide **26**, eq. (A.13).[20,21] No cycloaddition products were observed from the thermolysis of **27** in the presence of a dipolarophile. Nitrile oxide **26** is also formed in the reaction of nitromethane with phenylisocyanate in the presence of triethylamine. It forms 1,3-diphenyl-5-(hydroxyimino)-imidazolidine-2,4-dione (**28**) with aryl isocyanate.

$$\text{(A.13)}$$

The heteroaromatic bases, pyridine, quinoline, and isoquinoline, induce abnormal dimerization of benzonitrile oxide to the 1,4,2,5-dioxadiazine dimer **29**.[22] These bases also act as dipolarophiles giving rise to both mono- and bis-adducts, e.g., structures **30–34**. The rearrangement of nitrile oxides into isocyanates is catalyzed by UV irradiation.[100]

CHAPTER 3

A number of aldonitrones (**35**) have been prepared from *t*-butylhydroxylamine and aliphatic aldehydes, and are spectroscopically characterized by ^{1}H and ^{13}C NMR. They are unstable and tautomerize to the enamine form **36** with which they cycloadd to isoxazolidines **37** according to eqs. (A.14),(A.15). The formation of the dimers **38** and **39** have been observed in a few cases. The stabilities of the nitrones **35** and the derived dimerization products are discussed with reference to electronic and spatial effects and solvent effects. PbO$_2$ oxidizes **36**,

$$R-NHOH + R^1R^2CHCHO \longrightarrow \underset{\underset{O^-}{|}}{RN=CHCHR^1R^2} \rightleftharpoons \underset{\underset{OH}{|}}{RN-CH=CR^1R^2} \quad (A.14)$$

R = alkyl, aryl
R^1R^2 = alkyl, aryl, H   **35**   **36**

$$\mathbf{35} + \mathbf{36} \rightleftharpoons \text{37} \quad (A.15)$$

R' = H

$$\mathbf{36} \xrightarrow{PbO_2} R-\underset{\underset{O\cdot}{|}}{N}-CH=CR^1R^2 \xrightarrow{\mathbf{35}} R-\underset{\underset{O\cdot}{|}}{N}-\overset{\underset{O}{\|}}{C}HCHR^1R^2 + \underset{R-\underset{\underset{O\cdot}{|}}{N}-CHCHR^1R^2}{\overset{\underset{O}{\|}}{R^1R^2CCH=NR}} \quad (A.16)$$

37, and **39** to the corresponding nitroxide radicals, which are trapped by the nitrone **35** as in eq. (A.16).[23,24] These studies have been extended to oxidations of enol nitrones **40**. ^{1}H and ^{13}C NMR spectra of several α-arylnitrones **41**

have been recorded.[25] For R = CH$_3$ the C$^\alpha$—H varies from $\delta = 7.31$ (R^1 = SCH$_3$, electron donating) to $\delta = 7.53$ (R^1 = NO$_2$, electron withdrawing) and for R = Ph, C$^\alpha$–H varies from $\delta = 7.77$ [R^1 = N(CH$_3$)$_2$] to $\delta = 8.03$ (R^1 = NO$_2$). The ^{13}C$^\alpha$ shift is in the range of $\delta = 132.3$–135.3 from internal TMS.

Nitrones are conveniently prepared by electrochemical oxidation of secondary N-hydroxy amines using iodine as mediator.[26] Photolysis of cyclic N-hydroxylamines in the presence of 1,4-dicyanonaphthalene gives nitrones in good yield,[101] and hydrogen peroxide converts efficiently secondary amines directly into nitrones in a selenium dioxide catalyzed reaction.[102] Aryl and acyl substituted acetylenes form nitrones **42** with methyl- and phenylhydroxylamine. The N-methyl nitrone **42b** gives 4-isoxazoline **43** with a second molecule of acetylene, and the N-phenyl nitrone gives the isoxazolidine **44** with **42a**, eq. (A.17), cf. eq. (A.15). The acetylene **46** gives rise to the isoxazolidine **47** by intramolecular cyclization, eq. (A.18). Compound **43** rearranges on heating to the pyrrole **45**.[27]

C^α-Phenylnitrones are conveniently blocked by cyanotrimethylsilane to give **48** and are regenerated by silver fluoride. Treatment of **48** with lithium diisopropylamide affords cyanoimines **49**, eq. (A.19).[28] The reaction of C^α-alkylnitrones with chlorotrimethylsilane and triethylamine gives the *O*-silylated *N*-hydroxyenamines.[29]

The four-membered cyclic nitrones **50** prepared from nitroalkenes and ynamines, isomerize to α,β-unsaturated oximes **51** and **52** on treatment with potassium *t*-butoxide; only the *cis*-isomer **51** can cyclize to 6*H*-1,2-oxazin-6-ones **53**, eq. (A.20).[30]

The 3*H*-pyrrol-*N*-oxides (**54**) rearrange to the corresponding lactams by treatment with tetrafluoroquinone.[31] The synthesis, reactions, and spectral properties of the little known nitrones of thio esters (thioimidate *N*-oxides) are reported.[103]

CHAPTER 4

The lithium salt of 2-phenylnitroethane reacts with acetic anhydride or acetyl chloride to give an intermediate nitrile oxide which gives furoxan, chlorooxime, and hydroxamic acid, or in the presence of a dipolarophile, the corresponding cycloadduct.[32] An x-ray crystal structure of the major nitronate product **55**, obtained from the cycloaddition of nitrocyclohexene with cyclohexene, establishes the exo-pathway, eq. (A.21). As a result the earlier stereostructure[33] for the diketone **56** has to be reversed.[34] The nitroalkene **57** undergoes intramolecular Diels-Alder addition to give isomeric nitronates **58** and **59** ($\sim 1:1$).[35] For $R = CH_3$ the Diels-Alder addition proceeds stereoselectively to give the *trans*-isomer only. The intermediate nitronates behave as dipoles[36] and can add to another alkene. When **58** is treated with base followed by acid, a lactone is formed as shown in eq. (A.22).

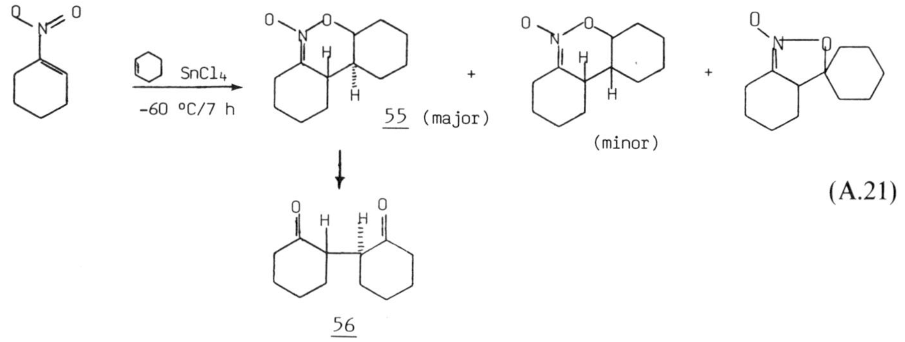

(A.21)

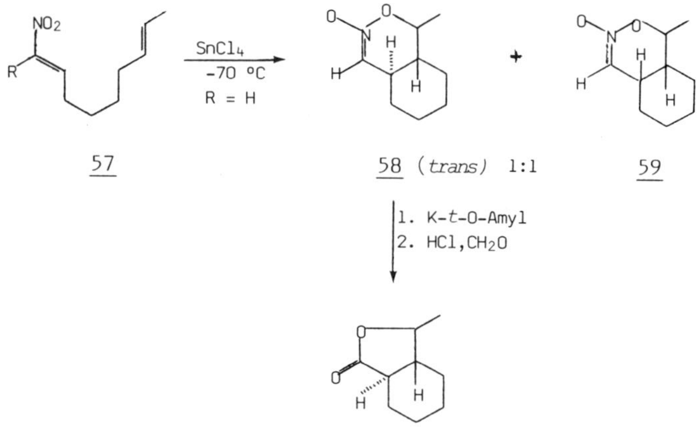

(A.22)

Nitroalkanes with C^β—H bonds react with twofold molar amounts of trialkylsilyl triflates **60** and triethylamine to give 2-(trialkylsilyloxy) oxime-*O*-trialkylsilyl ethers **63** via intermediate silyl nitronates **61** and *N,N*-disilyloxyenamines **62**. Reaction of **62** with amines gives 2-amino-oxime-*O*-trialkylsilyl ethers **64**, eq. (A.23).[37]

The azin-bisoxide structure **65a**[38] and the nitronate structure **65b**[39] have been suggested as being the dimers obtained by lead tetraacetate oxidation of benzaldoximes. The nitronate structure **65b** is unlikely, because nitronates usually are unstable compounds and the ^{13}C NMR spectrum of the phenyl derivative indicates that the two ArCH fragments are identical.[40] An x-ray study[41] shows that the phenyl derivative has the structure **65c**. It is centrosymmetric and planar, $M_r = 240.3$, monoclinic, $P2_1/c$, $a = 5.79(1)$, $b = 5.22(2)$, $c = 21.14(5)$ Å, $\beta = 112.1(1)°$, $v = 592.4$ Å^3, $z = 2$, $D_x = 1.35$ g cm^{-3}, $\lambda(CuK_\alpha) = 1.5418$ Å, $\mu = 7.11$ cm^{-1}, $F(000) = 252$, $T = 290$ K, $R = 0.22$ for 228 reflections. An X-ray crystal structure of the lithium salt of phenylnitromethane has been carried out.[104]

1-Nitroalkadienes are in a titanium tetrachloride-mediated reaction converted to the intermediate stannyl nitronates **66**, which cyclize on addition of triethylamine to a diastereomeric mixture of bicyclic isoxazolidines **67**, eq. (A.24).[42] Intramolecular cycloaddition of the nitroalkadiene system can also be initiated by conjugate addition of *t*-butylisonitrile.[105]

Alkyl nitrate nitration of 2-substituted 2-oxazolines leads to the corresponding α-nitro derivatives, which to a great extent exist in the nitronate form, eq. (A.25).[43] The C^α—H of the nitronate form is located at $\delta = 6.57$ ppm.

$$\text{(A.25)}$$

CHAPTER 5

Ring opening of 3-unsubstituted[44] or 3-ethoxycarbonyl substituted[5] isoxazolines has been applied for cyano-hydroxylation of alkenes. It has also been shown that 3-bromoisoxazolines, obtained by cycloaddition of bromonitrile oxide to alkenes, can be transformed into *cis*-β-hydroxy esters. This reaction has been applied to the synthesis of D-2-deoxy-ribose,[45] eq. (A.26), dihydromuscimol,[46] and to muscimol.[47] A general method is reported for the synthesis of Z-β-silyloxyacrylonitriles from isoxazoles, based on ring cleavage of 3-unsubstituted isoxazoles,[48] eq. (A.27). If the reaction is carried out above $-30°C$, mixtures of Z and E stereoisomers are obtained. These β-silyloxyacrylonitriles show only fair dienophilic reactivity towards homodienes, but give good yields with heterodienes. Flash vacuum pyrolysis of some 4-nitroisoxazoles gives α-nitro-α-cyanoketones.[116]

$$\text{(A.26)}$$

$$\text{(A.27)}$$

N-Ethyl-5-phenylisoxazolium-3¹-sulfonate, Woodwards reagent K, has been used as a reagent for identification of nucleophilic active sites in protein side chains, such as carboxylate, amino, thio, hydroxy, and imidazole groups.[49,50]

Allylic and homoallylic amines are prepared stereoselectively from isoxazolidines obtained by dipolar cycloaddition of nitrones with vinyl and allylsilanes,[51] eqs. (A.28),(A.29). The regioisomer **68** is the only one observed in the reaction, and the geometry of the olefins **69** and **70**, produced in the Peterson olefination step, is controlled by the choice of elimination conditions. The alkyl vinylsilane **71** displays diminished regioselectivity and gives a 7:3 mixture of 4- and 5-silylisoxazolidines.

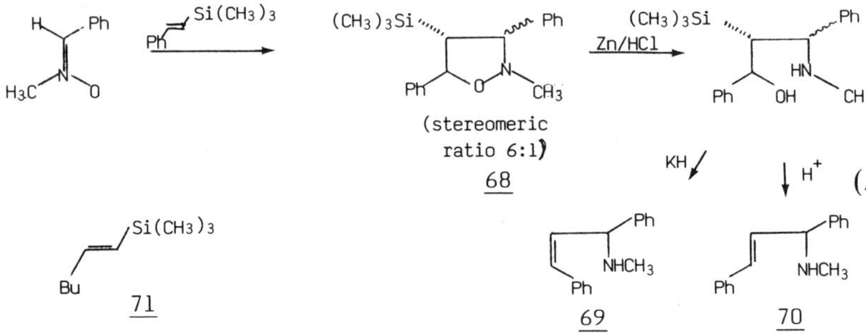

(A.28)

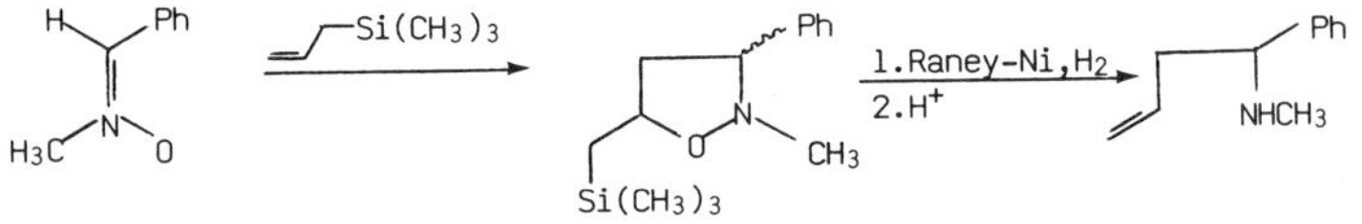

β,γ-Unsaturated ketones are similarly prepared from allylsilanes and nitrile oxides,[52] eq. (A.30). α,β-Unsaturated carbonyl compounds can be prepared by peracid oxidation of 4-isoxazolines derived from cycloaddition of nitrones with acetylenes or allenes.[106]

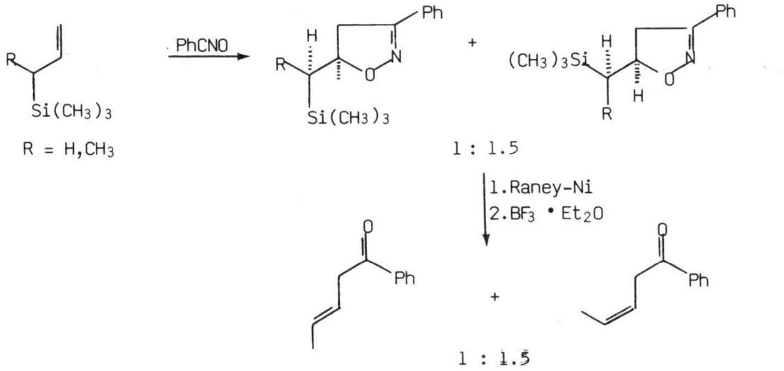

(A.30)

The acylation and [3,3] sigmatropic rearrangement of nitrones to α-acyloxyimines **71** has been adapted to synthesis of vicinal N-alkylamino alcohols by subsequent hydride reduction,[53] eq. (A.31).

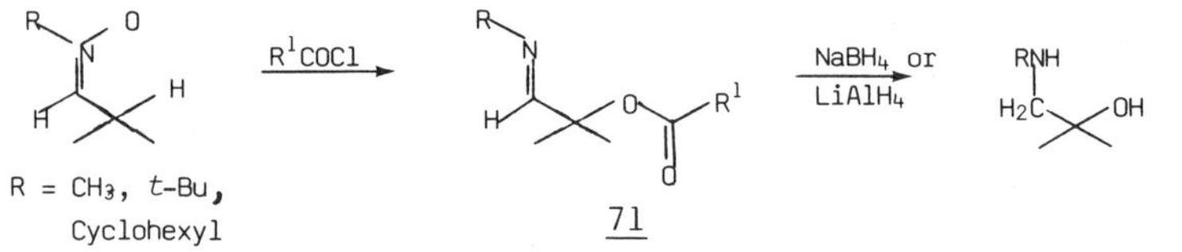

(A.31)

Isoxazoles are prone to undergo various ring transformations into other heterocycles or functionalities. Thus, the 3-oxopropionitrile formed by ring cleavage of isoxazole reacts in situ with an aromatic aldehyde to the arylidene derivative **72**,[54] eq. (A.32). Michael addition of malonodinitrile or cyanide gives the pyran and furan derivatives **73** and **74**.

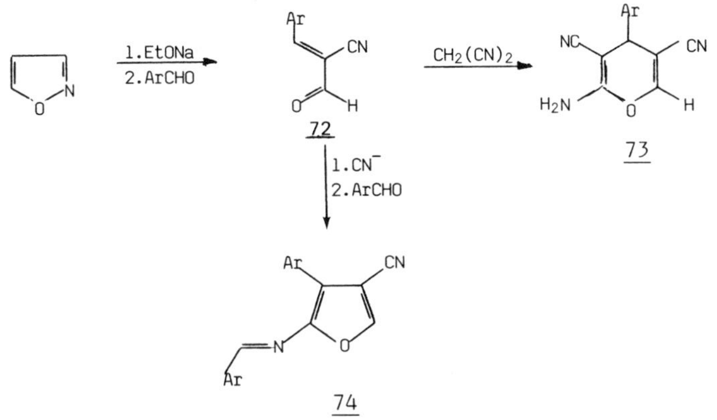

(A.32)

The isoxazole derivative **75** rearranges in refluxing xylene to the imidazole **76**,[55] eq. (A.33). 5-Aminoisoxazoles have been converted to acetylenes by diazotization in acetic acid:water mixtures,[56] eq. (A.34). Various imidazoles have been prepared from 4-aminoisoxazoles.[117]

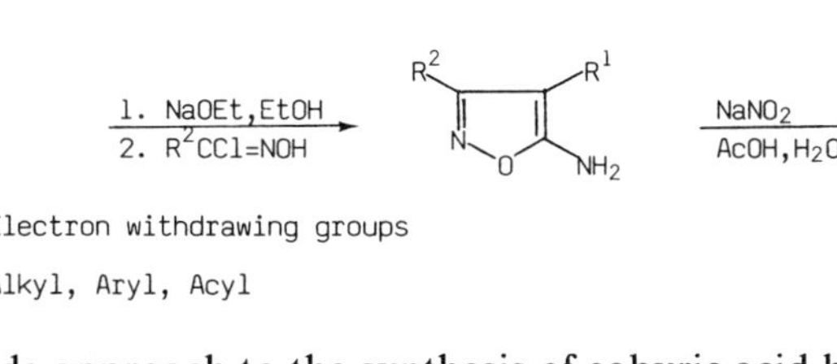

(A.33)

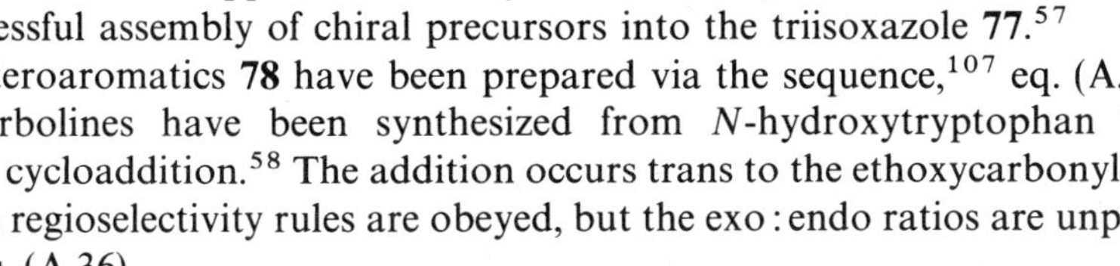

(A.34)

The nitrile oxide approach to the synthesis of cobyric acid has been extended to successful assembly of chiral precursors into the triisoxazole **77**.[57]

Biheteroaromatics **78** have been prepared via the sequence,[107] eq. (A.35).

β-Carbolines have been synthesized from N-hydroxytryptophan **79** by nitrone cycloaddition.[58] The addition occurs trans to the ethoxycarbonyl group and the regioselectivity rules are obeyed, but the exo:endo ratios are unpredictable, eq. (A.36).

Several pyrrolizidine, pyrrolidine, and piperidine alkaloids have been synthesized using nitrone-based methodology. As a key step the allene oxime **82** is cyclized with silver tetrafluoroborate to the nitrone **83**, which is trapped by methyl vinyl ketone as the isoxazolidine **84** and reduced in one step to **85**,[59] eq. (A.37). Jones oxidation, thioketalization, and Raney-Ni desulfuration give the pyrrolizidine alkaloid **86**, a venom constituent from *Solenopsis* ants. The dipolar cycloaddition occurs in a trans-relationship to the alkene side chain in a 1:1 exo-endo fashion. The *Solenopsis* alkaloid **88** is synthesized from pyrroline-1-oxide by a double nitrone cycloaddition sequence,[60] eq. (A.38). The pyrrolidine **87** is obtained in a trans-cis ratio of 87:13. By using butadiene in the second cycloaddition step the stereoselectivity is increased to 93:7. This trans-selectivity of the second cyloaddition step occurs predominantly—if not exclusively—in the corresponding piperidine series. The piperidine alkaloid solenopsine (**89**) is obtained by addition of 1-undecene to 6-methyl-Δ-1-piperdine-1-oxide followed by reductive ring cleavage and elimination of the hydroxy group,[61] eq. (A.39). The *Darlingia darlingiana* alkaloids **90–93** are synthesized with the same strategy from phenylbutadiene and pyrroline-1-oxide,[62] eq. (A.40). The exo-addition mode is favored. Compound **92** is prepared from **91**, **93** from **90**, and **95** from either **90** or **91**. The alkaloid **94** is synthesized from 6-phenyl-1-hexene and pyrroline-1-oxide.

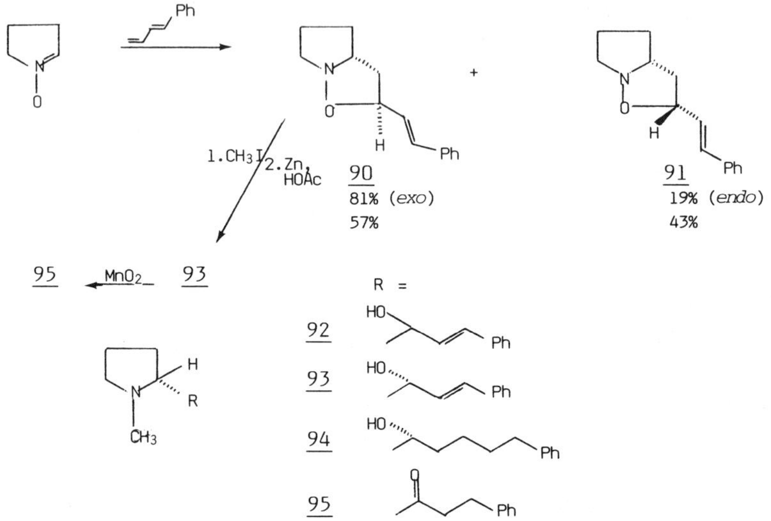

(A.40)

The pyrrolizidine **96**, obtainable from pyrroline-1-oxide-3-ketodimethylacetal and methyl-4-hydroxycrotonate,[63] has been transformed into *d,l*-croalbinecine (**98**),[64,65] *d,l*-loline (**101**), and *d,l*-norloline (**102**), according to eq. (A.41). The base

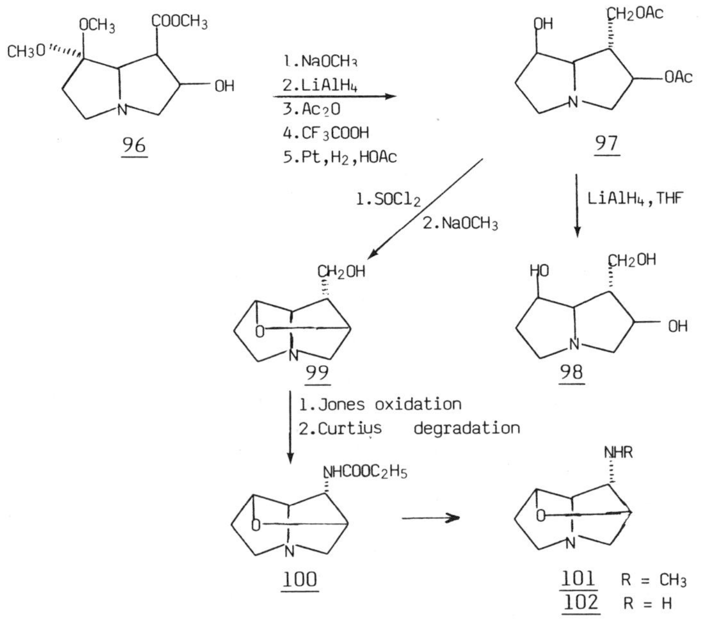

(A.41)

catalyzed inversion of the methoxycarbonyl group proceeds with ease as a result of unfavorable 1,3-interaction with the methoxy group. Reduction with lithium aluminum hydride, protection of the diol as acetate, hydrolysis of the ketal, catalytic hydrogenation to **97**, and finally, removal of the acetate groups with lithium aluminum hydride give *d,l*-croalbinecine (**98**), an alkaloid from *Crotalaria* species. The loline alkaloids **101** and **102** isolated from, e.g., *Lolium cuneatum* and *Festuca arundianacea*, are toxic to grazing cattle, and are also prepared from the intermediate (**97**). Chlorination with Vilsmeier's reagent gives the inverted chloride, which by treatment with alkoxide promotes the formation of the ether bridge in **99**. Oxidation of the alcohol to the carboxylic acid, Curtius' rearrangement, reduction, or hydrolysis give *d,l*-loline (**101**) or *d,l*-norloline (**102**). Nitrone based syntheses of new 3- and 5-hydroxypiperidine alkaloids from *Sedum acre*[108a] and the piperidine alkaloid dumetorine from *Discorea dumetorum Pax*, an African medical plant, are reported.[108b]

The spiroisoxazolidines **103**, obtained by cycloaddition of nitrones with methylenecyclopropane, undergo thermal rearrangement to piperidine-4-ones **104**,[66,67] eq. (A.42). This reaction gives an entry into indolizidine **105** and quinolizidine derivatives **106** by using cyclic nitrones. Enaminones are always produced as a side product at the thermolysis (FVT, 400°C/0.2 mmHg). The 5-spirocyclobutane isoxazolidines and 5-spirocyclobutane 2-isoxazolines give the corresponding azepin-4-ones by FVT at ca. 600°C.[68] A nitrone-based synthesis

(A.42)

of the *Nuphar* indolizidine 5-3(-furyl)-8-methyloctahydroindolizine was accomplished by a regiospecific cycloaddition of a nitrone with a 1,4-disubstituted butadiene.[118]

The ring skeleton of ptilocaulin (**107a**) and its 7-epimer (**107b**) is constructed by intramolecular nitrile oxide cycloaddition,[69] eq. (A.43). Cyclization and dehydration give a mixture of **108** and **109** which on Raney-Ni reduction gives a single isomer of the ketol **110**. Subsequent reduction, dehydration, and guanidination give **107a,b**.

Cycloaddition of ethoxycarbonylformonitrile oxide to vinyl- and propenylpyridines **111a–c** gives the isoxazolines **112a–c** regioselectively, which are reduced with the Zn/Cu couple in ammonia to the nikkomycin amino acids **113** as diastereoisomeric mixtures,[70] eq. (A.44). One pair of the isomers, **113a**, is identical with the natural amino acid (2S, 4R). None of the isomers **113b,c** is identical with the natural amino acid, implying that the *cis*-olefin **111** has to be used to give the correct relative stereostructure. However, it was observed that the *cis*-isomer of **112** spontaneously rearranges to the more stable *trans*-isomer. Therefore another route has to be chosen to achieve the synthesis of the natural amino acids.

The key intermediate **114** for the synthesis of detoxinine (**115**), the parent amino acid of the detoxin complex, has been synthesized via **116** by cycloaddition of the silyl nitronate of nitromethane to butadiene followed by reductive-ring cleavage of the isoxazoline with lithium aluminum hydride and protection of the hydroxy and amino groups. Expoxidation of **116** with *m*-chloroperbenzoic acid and magnesium triflate promoted cyclization to give a 1:1 diastereomeric mixture of **114**,[71] eq. (A.45).

A chiral synthesis of the amino acid negamycin (**116a**) and its unnatural isomer 3-epinegamycin (**116b**) starts with dipolar exo-cycloaddition of the chiral nitrone **117** with the protected allylamine **118** to produce a mixture of 3R, 5R, **119a**, and 3S, 5R, **119b**, isoxazolidines,[72] eq. (A.46). Nitrone **117** is generated as a mixture of E and Z isomers from 2,3:5,6-*di*-*O*-cyclohexylidene-D-gulo-furanose oxime and methyl glyoxalate. Hydrolysis of the chiral auxiliary followed by benzylation and reduction with lithium aluminum hydride give the isoxazolidines **120a** and **120b** in a 2:3 ratio, which in six steps are transformed into **116a,b**.

The side chain of 1α, 2,5-dihydroxyergocalciferol is constructed by a route involving a dipolar nitrone cycloaddition as a key step,[73] eq. (A.47). The cycloaddition gives exclusively the endo-products **121a,b** as a 1:1 mixture.

Intramolecular nitrile oxide cycloaddition is used for the construction of the decaline skeleton of forskolin (**122**), the major diterpene isolated from the Indian plant *Coleus forskolii*. The nitro compound **123**, obtained by Michael addition of nitromethane to α-damascone, cyclizes under Mukaiyama conditions to **124**. Acetalization followed by reductive-ring cleavage leads to the transfused hydroxyketone **125**,[74] eq. (A.48).

Further details of the synthesis of milbemycin β_3 have been given,[75] cf. Chapter 5, eq. (5.313). The lactone and octalin units of compactine (**368**), cf. Chapter 5, have been constructed by nitrile oxide methodology,[76] eqs. (A.49),(A.50).

The highly substituted α,β-unsaturated ketone **126** undergoes cycloaddition with furoxan **127** to isoxazoline **128**, which by Wittig olefination, hydroboration, and catalytic reduction gives **129** as a diastereomeric mixture, separable by chromatography. Compound **129a** was oxidized by periodic acid and hydrolyzed to crispatic acid (**130**),[77] eq. (A.51).

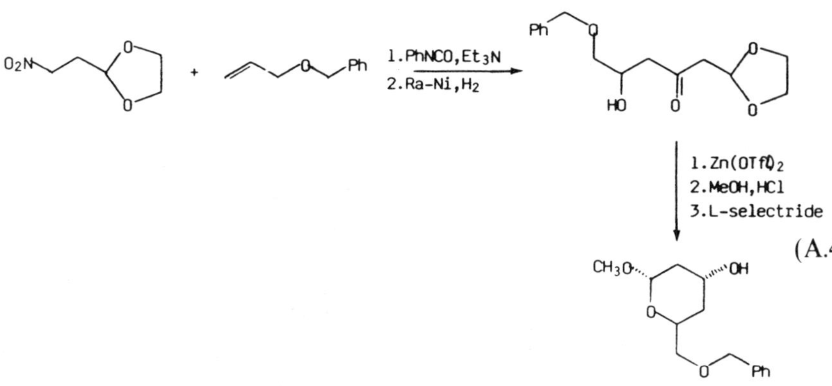

(A.49)

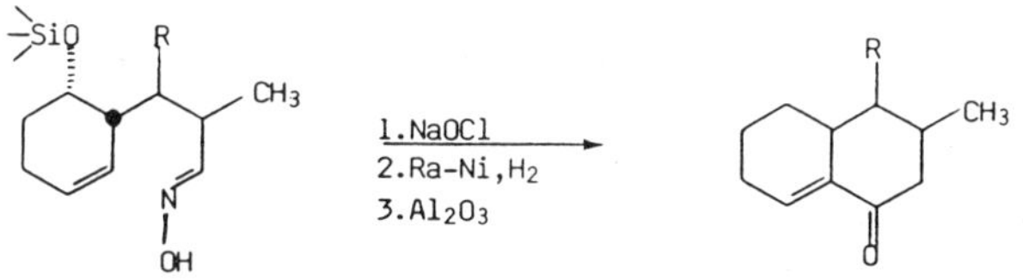

(A.50)

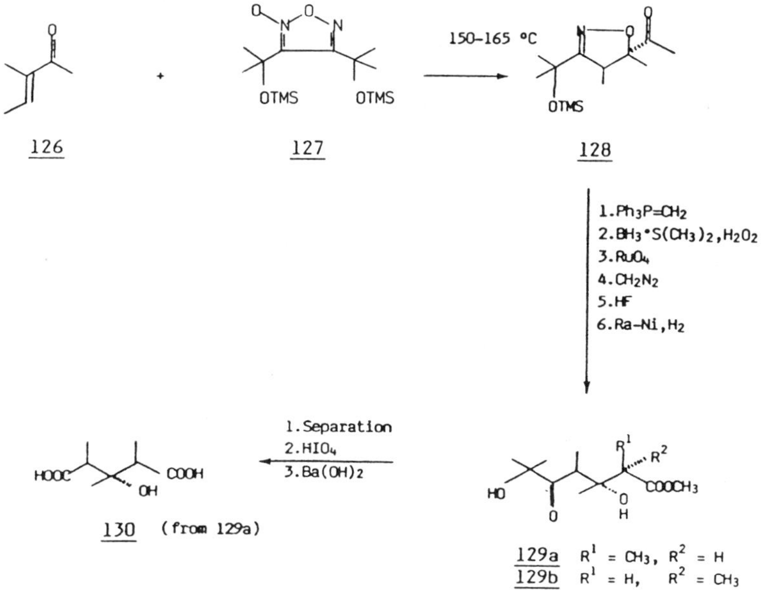

126   127   128

(A.51)

130 (from 129a)

129a $R^1 = CH_3$, $R^2 = H$
129b $R^1 = H$, $R^2 = CH_3$

Nitrile oxide methodology has been applied for the preparation of 1-acetyl-2,6-cyclohexanediol, an intermediate in the synthesis of bisabolangelone,[78] eq. (A.52).

3-α-Oxygenated isoxazolines are of interest in conjunction with the synthesis of carbohydrates. It has been shown that the 3-ethoxycarbonyl group can be reduced to the corresponding alcohol with sodium borohydride without ring

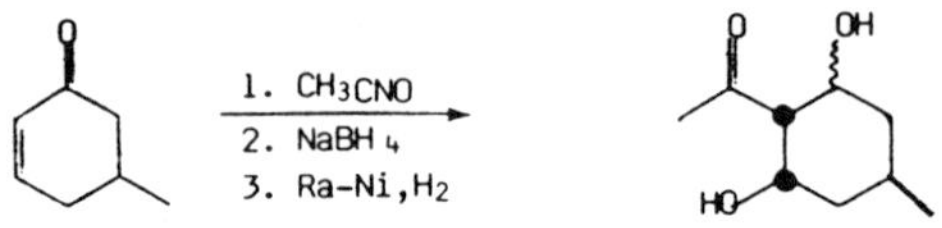

(A.52)

cleavage.[79-82] Swern oxidation of the alcohol leads to the isoxazoline-3-carboxaldehyde. The aldehyde can also be obtained directly by reducing the 3-ethoxycarbonyl group with diisobutylaluminum hydride.[83] These reactions supplement the synthesis via suitable protected 1,2-nitroalcohols,[84] nitroacetaldehyde acetals,[84,85] α-nitroketones,[81,82] silyl nitronates,[29,85] alkyl nitronates,[81] suitably protected α-hydroxyoximes,[84-88] and α-ketoximes.[86]

Novel methodology has been developed for the preparation of flavonoids based on nitrile oxide cycloaddition.[109] Salicylaldoxime is selectively chlorinated to hydroxylimino chloride with NCS without nuclear chlorination in the presence of pyridine and cycloadded to styrene or phenylacetylene. Reductive ring cleavage and cyclization give the flavanone or flavone, eq. (A.53). By reacting salicylnitrile oxide with the enamine of phenylacetaldehyde it is possible to synthesize the 4-phenylisoxazoline **132**, eq. (A.54), which serves as starting material for synthesizing isoflavonoids.

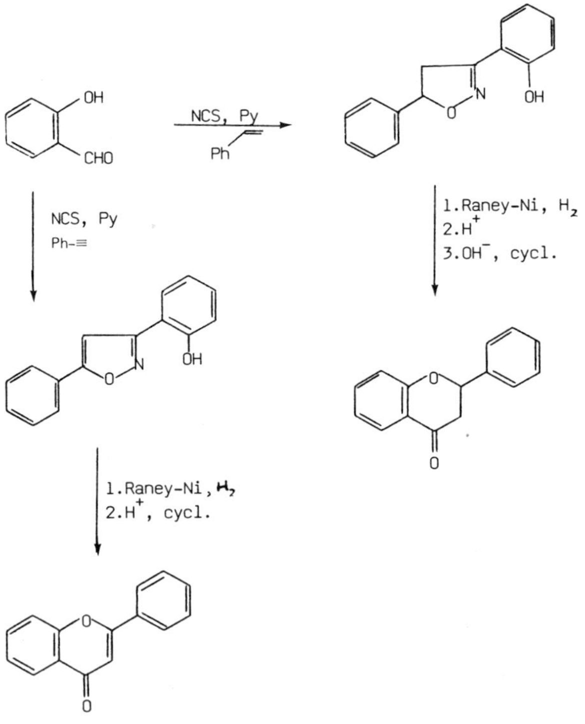

(A.53)

The quinolizidine alkaloid lasubine II (see section 5.20, **429**) has been synthesized by intramolecular nitrone cycloaddition with modest stereoselectivity, ca. 80%.[89]

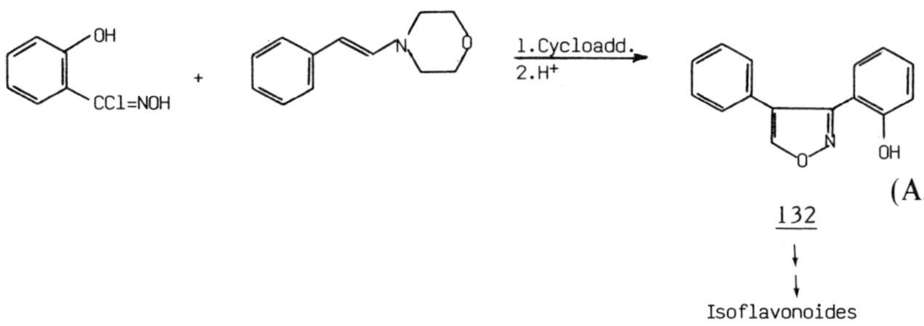

Chapter 5   295

(A.54)

As part of a synthesis of a leptospaerin analogue **130a** the lactam **131** was synthesized from the acetonide of R-glyceraldehyde by condensation with *N*-benzylhydroxylamine, cycloaddition with methyl acrylate in refluxing toluene and hydrogenation over Pd/C,[90] eq. (A.55).

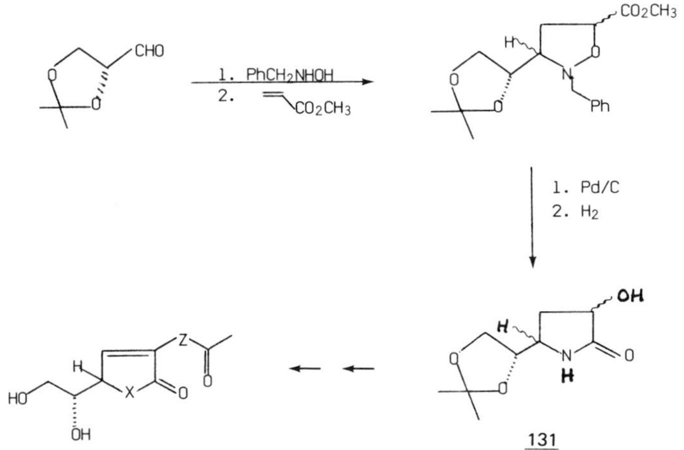

(A.55)

130a X = NH, Z = O
130b X = O, Z = NH (leptosphaerin)

The synthesis and reactions of the strained nitrones **133**, **134** have been described.[110] **133** has been transformed into **135**, which is an isomer of the pink bollworm moth sex pheromone, by a series of stereoselective nitrone cycloadditions and subsequent deaminations. This reaction constitutes a stereoselective E,E-1,5-diene synthesis. 3-Acyltetronic acids are prepared by cycloadding nitrile oxides with suitable acetylene derivatives, eq. (A.56). The corresponding 6-ring lactones are analogously obtained from 5-hydroxy-2-ynoates **136**.[111]

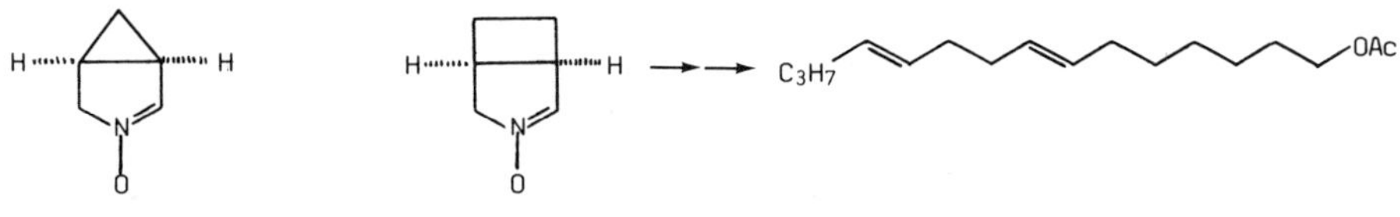

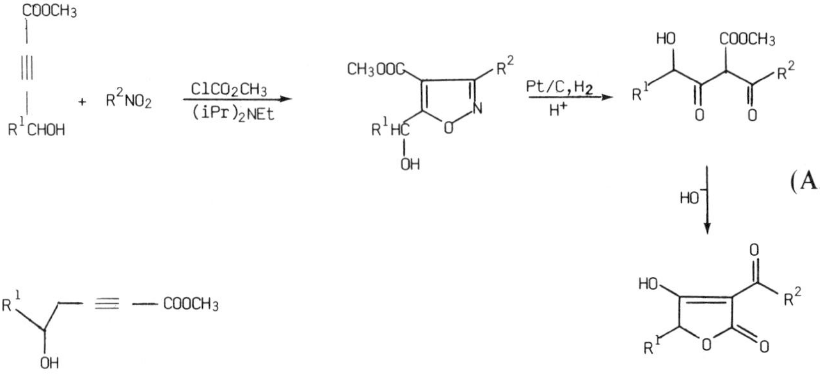

(A.56)

136

Aryl and heteroaryl C-glycosides are prepared by cycloadding nitro functionalized carbohydrates to allyl-substituted aromatics (furan, pyrrole, indole, and benzene), eq. (A.57).[112]

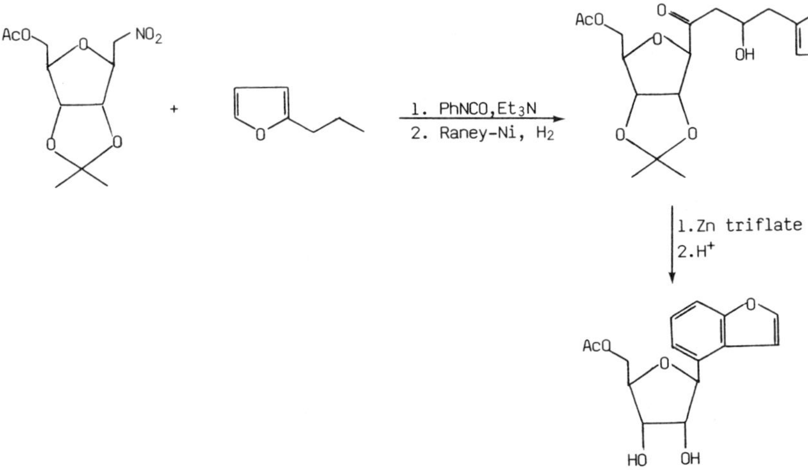

(A.57)

The chiral nitrone 137 condenses stereoselectively with ketene silyl acetate giving the O-silylated product 138 in a zinc iodide catalyzed reaction,[113] eq. (A.58). Hydrogenation over palladium, benzoylation and reduction with diisobutylaluminum hydride give N-benzoyl L-daunosamine (139).

α,β-Unsaturated aldoximes are efficiently chlorinated by NCS to hydroxamoyl chlorides. Senecioaldoxime gives with isoprene a mixture of isoxazolines 140 and 141 in a ratio of 3:1 and with 3-methyl-3-butenyl acetate the isoxazoline 142, which has a regular monoterpene skeleton,[6] eq. (A.59). Cleavage of the N–O bond is accomplished by N-methylation with dimethyl sulfate and subsequent electrochemical reduction under controlled potential. Compound 143 is a suitable intermediate for the synthesis of ocimenones.

REFERENCES

1. K. N. Houk; H.-Y. Duh; Y.-D. Wu; S. R. Moses, *J. Am. Chem. Soc. 108* (1986), 2754.
2. K. N. Houk; M. N. Paddon-Row; N. G. Rondan; Y.-D. Wu; F. K. Brown; D. C. Spellmeyer; J. T. Metz; Y. Li; R. J. Loncharich, *Science 231* (1986), 1108.
3. A. Padwa; S. P. Carter; U. Chiacchio; D. N. Kline, *Tetrahedron Lett. 1986*, 2683; A. Padwa; D. N. Kline; K. F. Koeler; M. Matzinger; M. K. Venkataraman, *J. Org. Chem. 52* (1987), 3909.
4. H. Gotthardt; R. Jung, *Chem. Ber. 119* (1986), 563.
5. J. P. Alazard; A. Leboff; C. Thal, *Tetrahedron 42* (1986), 1407.
6. I. Thomsen; K. B. G. Torssell, unpublished results.
7. R. Gandolfi; L. Toma; C. De Micheli, *Heterocycles 12* (1979), 5.
8. D. F. Corbett, *J. Chem. Soc., Perkin I 1986*, 421.
9. M. Burdisso; A. Gamba; R. Gandolfi; P. Pevarello, *Tetrahedron 42* (1986), 4355.

10. T.-M. Chan; R. Friary; B. Pramanik; M. S. Puar; V. Seidl; A. T. McPhail, *Tetrahedron 42* (1986), 4661.
11. D. M. Tschaen; R. R. Whittle; S. M. Weinreb, *J. Org. Chem. 51* (1986), 2604.
12. R. Annunziata; M. Cinquini; F. Cozzi; L. Raimondi, *Tetrahedron 42* (1986), 2129.
13. Y. Kawamura; T. Kumagai; T. Mukai, *Chem. Letters 1985*, 1937.
14. A. Brandi; F. De Sarlo; A. Guarna; A. Goti, *Heterocycles 23* (1985), 2019.
15. (a) R. Nesi; D. Giomi; S. Papaleo; L. Quartara, *J. Chem. Soc., Chem. Commun. 1986*, 1536.
    (b) R. Nesi; D. Giomi; P. Sarti-Fantoni; P. Tedeschi, *J. Chem. Soc. Perkin 1 1987*, 1005.
16. T. Ponticelli; P. Tedeschi, *Synthesis 1985*, 792.
17. P. G. Baraldi; A. Barco; S. Benetti; S. Manfredini; D. Simoni, *Synthesis 1987*, 276; T. Kobayachi; Y. Iionio; M. Nitta, *Nippon Kagaku Kaishi 1986*, 785.
18. J. J. Tufariello; H. Meckler; K. P. A. Senaratne, *Tetrahedron 41* (1985), 3447.
19. O. Tsuge; S. Kanemasa; H. Suga, *Chem. Letters 1986*, 183; O. Tsuge; S. Kanemasa; H. Suga; N. Nagagawa, *Bull. Chem. Soc. Japan 60* (1987), 2463.
20. T. Shimizu; Y. Hayashi; T. Ito; K. Teramura, *Synthesis 1986*, 488.
21. T. Shimizu; Y. Hayashi; K. Teramura, *Bull. Chem. Soc. Japan 59* (1986), 2038; T. Shimizu; Y. Hayashi; H. Shibafuchi; K. Teramura, *Bull. Chem. Soc. Japan 60* (1987), 1940.
22. A. Corsaro; G. Perrini; P. Caramella; F. M. Albini; T. Bandiera, *Tetrahedron Lett. 1986*, 1517.
23. H. G. Aurich; J. Eidel; M. Schmidt, *Chem. Ber. 119* (1986), 18, 36.
24. H. G. Aurich; O. Bubenheim; M. Schmidt, *Chem. Ber. 119* (1986), 2756.
25. C. Yijima; T. Tsujimoto; K. Suda; M. Yamauchi, *Bull. Chem. Soc. Japan 59* (1986), 2165.
26. T. Shono; Y. Matsumura; K. Inoue, *J. Org. Chem. 51* (1986), 549.
27. A. Padwa; G. S. K. Wong, *J. Org. Chem. 51* (1986), 3125.
28. A. Padwa; K. F. Koehler, *J. Chem. Soc., Chem. Commun. 1986*, 789.
29. K. Torssell; O. Zeuthen, *Acta Chem. Scand. B32* (1978), 118.
30. P. J. S. S. van Eijk; D. N. Reinhoudt; S. Harkema; R. Visser, *Rec. Trav. Chim. Pays-Bays 105* (1986), 103.
31. A. M. Nour El-Din, *Bull. Soc. Chem. Japan 59* (1986), 1239.
32. M. Cherest; X. Lusinchi, *Tetrahedron 42* (1986), 3825.
33. R. Criegee; H. G. Reinhardt, *Chem. Ber. 101* (1968), 102.
34. S. E. Denmark; C. J. Cramer; J. A. Sternberg, *Tetrahedron Lett. 1986*, 3693.
35. S. E. Denmark; C. J. Cramer; J. A. Sternberg, *Helv. Chim. Acta 69* (1986), 1971; S. E. Danmark; M. S. Dappen; C. J. Cramer; J. A. Sternberg, *J. Am. Chem. Soc. 108* (1986) 1306.
36. V. A. Tartakowskii, *Izv. Akad. Nauk SSSR, Ser. Khim. 1984*, 147.
37. H. Feger; G. Simchen, *Liebigs Ann. 1986*, 428, 1456.
38. L. Horner; L. Hockenberger; W. Kirmse, *Chem. Ber. 94* (1961), 290.
39. H. Kropf; R. Lambeck, *Liebigs Ann. 700* (1966) 18.
40. K. Torssell, unpublished results.
41. C. H. Haagensen; R. G. Hazell, unpublished results.
42. H. Uno; N. Watanabe; S. Fijiki; H. Suzuki, *Synthesis 1987*, 471.
43. H. Feuer; H. S. Bevinakatti; X.-G. Luo, *J. Heterocycl. Chem. 23* (1986), 825.
44. M. Christl; B. Mattauch; H. Irngartinger; A. Goldmann, *Chem. Ber. 119* (1986), 950.
45. P. Caldirola; M. Ciancaglione; M. DeAmici; C. De Micheli, *Tetrahedron Lett. 1986*, 4647.
46. P. Caldirola; M. De Amici; C. De Micheli, *Tetrahedron Lett. 1986*, 4651.
47. D. Chiarino; M. Napolitano; A. Sala, *Tetrahedron Lett. 1986*, 3181; *J. Heterocycl. Chem. 24* (1987) 43.
48. A. Alberola; A. M. Gonzalez; B. Gonzalez; M. A. Laguna; F. J. Pulido, *Tetrahedron Lett. 1986*, 2027.
49. P. Bodlaender; G. Feinstein; E. Shaw, *Biochemistry 8* (1969), 4941, 4949.
50. K. Llamas; M. Owens; R. L. Blakeley; B. Zerner, *J. Am. Chem. Soc. 108* (1986), 5543.
51. P. DeShong; J. M. Leginus; S. W. Lander, Jr., *J. Org. Chem. 51* (1986), 574.
52. D. P. Curran; B. H. Kim, *Synthesis 1986*, 312.
53. R. M. Coates; C. H. Cummins, *J. Org. Chem. 51* (1986), 1383.
54. J. A. Ciller; N. Martin; C. Seoane; J. L. Soto, *J. Chem. Soc., Perkin I 1985*, 2581.
55. A. C. Coda; G. Desimoni; A. Coda, *Heterocycles 23* (1985), 1893.
56. E. M. Becalli; A. Manfredi; A. Marchesini, *J. Org. Chem. 50* (1985), 2372.
57. R. V. Stevens; N. Beaulieu; W. H. Chan; A. R. Daniewski; T. Takeda; A. Waldner; P. G. Willard; U. Zutter, *J. Am. Chem. Soc. 108* (1986), 1039.
58. R. Plate; P. H. H. Hermkens; J. M. M. Smits; H. C. J. Ottenheijm, *J. Org. Chem. 51* (1986), 309.
59. D. Lathbury; T. Gallagher, *J. Chem. Soc., Chem. Commun. 1986*, 1017.
60. J. J. Tufariello; J. M. Puglis, *Tetrahedron Lett. 1986*, 1489.

61. W. Carruthers; M. J. Williams, *J. Chem. Soc., Chem. Commun. 1986*, 1287.
62. J. J. Tufariello; J. M. Puglis, *Tetrahedron Lett. 1986*, 1265.
63. J. J. Tufariello; G. E. Lee, *J. Am. Chem. Soc. 102* (1980), 373.
64. J. J. Tufariello; K. Winzenberg, *Tetrahedron Lett. 1986*, 1645.
65. J. J. Tufariello; H. Meckler; K. N. Winzenberg, *J. Org. Chem. 51* (1986), 3556.
66. A. Guarna; A. Brandi; A. Goti; F. De Sarlo, *J. Chem. Soc., Chem. Commun. 1985*, 1518.
67. A. Brandi; A. Guarna; A. Goti; F. De Sarlo, *Tetrahedron Lett. 1986*, 1727; See also A. Brandi; S. Carli; A. Guarna; F. De Sarlo, *Tetrahedron Lett. 1987*, 3845.
68. A. Goti; A. Brandi; F. De Sarlo; A. Guarna, *Tetrahedron Lett. 1986*, 5271.
69. A. Hassner; K. S. K. Murthy, *Tetrahedron Lett. 1986*, 1407.
70. W. A. König; H. Hahn; R. Rathmann; W. Hass; A. Keckeisen; H. Hagenmaier; C. Bormann; W. Dehler; R. Kurth; H. Zähner, *Liebigs Ann. 1986*, 407.
71. W. R. Ewing; B. D. Harris; K. L. Bhat; M. M. Joullié, *Tetrahedron 42* (1986), 2421.
72. H. Iida; K. Kasahara; C. Kibayashi, *J. Am. Chem. Soc. 108* (1986), 4647.
73. E. G. Baggiolini; J. A. Iacobelli; B. M. Hennessy; A. D. Batcho; J. F. Sereno; M. R. Uskokovič, *J. Org. Chem. 51* (1986), 3098.
74. P. G. Baraldi; A. Barco; S. Benetti; G. P. Pollini; E. Polo; D. Simoni, *J. Chem. Soc., Chem. Commun. 1986*, 757.
75. S. R. Schow; J. D. Bloom; A. S. Thompson; K. N. Winzenberg; A. B. Smith III, *J. Am. Chem. Soc. 108* (1986), 2662.
76. A. P. Kozikowski; C.-S. Li, *J. Org. Chem. 52* (1987) 3541. T. Rosen; C. H. Heathcock, *Tetrahedron 42* (1986), 4909.
77. D. P. Curran; C. J. Fenk, *Tetrahedron Lett. 1986*, 4865.
78. B. P. Riss; B. Muckensturm, *Tetrahedron Lett. 1986*, 4979.
79. P. G. Baraldi; D. Simoni; F. Moroder; S. Manfredini; L. Mucchi; F. D. Vecchia; P. Orsolini, *J. Heterocycl. Chem. 19* (1982), 557.
80. P. G. Baraldi; F. Moroder; G. P. Pollini, D. Simoni; A. Barco; S. Benetti, *J. Chem. Soc., Perkin I 1982*, 2983; P. G. Baraldi; A. Barco; S. Benetti; S. Manfredini; G. P. Pollini; D. Simoni, *Tetrahedron 43* (1987) 235.
81. P. A. Wade; N. V. Amin; H.-K. Yen; D. T. Price; G. F. Huhn, *J. Org. Chem. 49* (1984), 4595.
82. P. Caldirola; M. De. Amici; C. De Micheli; P. A. Wade; D. T. Price; J. F. Bereznak, *Tetrahedron 42* (1986), 5267.
83. R. Annunziata; M. Cinquini; F. Cozzi; L. Raimondi; A. Restelli, *Helv. Chim. Acta 68* (1985), 1217.
84. V. Jäger; I. Müller; R. Schohe; M. Frey; R. Ehrler; B. Häfele; R. Schröter, In *Lect. Heterocycl. Chem. 8* (1985), 79. Hetero Corp. Tampa USA. A review centered on work in Jäger's group.
85. K. B. G. Torssell; A. C. Hazell; R. G. Hazell, *Tetrahedron 41* (1985), 5569.
86. K. E. Larsen; K. B. G. Torssell, *Tetrahedron 40* (1984), 2985.
87. E. J. Thomas; R. H. Jones; C. G. Robinson, *Tetrahedron 40* (1984), 177.
88. B. Häfele; Diplom. Thesis, Würburg, 1983.
89. R. W. Hoffmann; A. Endesfelder, *Liebigs. Ann. 1986*, 1823.
90. G. A. Schiehser, J. D. White; G. Matsumoto; J. O. Pezzanite; J. Clardy, *Tetrahedron Lett. 1986*, 5587.
91. R. Annunziata; M. Cinquini; F. Cozzi; L. Raimondi; *J. Chem. Soc., Chem. Commun. 1987*, 529; R. Annunziata; M. Cinquini; F. Cozzi; G. Dondio; L. Raimondi, *Tetrahedron 43* (1987) 2369.
92. P. Parpani; G. Zecchi, *J. Org. Chem. 52* (1987) 1417.
93. A. Padwa; Y. Tomioka; M. K. Venkataraman, *Tetrahedron Lett. 1987*, 755.
94. W. R. Dolbier, Jr.; G. E. Wicks; C. R. Burkholder, *J. Org. Chem. 52* (1987) 2196.
95. M. Nitta; N. Kanomata, *Chem. Lett. 1986*, 1925.
96. J. Moskal; P. Milart, *Chem. Ber. 118* (1985) 4014.
97. D. P. Curran; J.-C. Chao, *J. Am. Chem. Soc. 109* (1987) 3036.
98. R. Nesi; S. Chimichi; D. Giomi; P. Sarti-Fantoni; P. Tedeshi, *J. Chem. Soc., Perkin I 1987*, 1005.
99. D. P. Curran; B. H. Kim; H. P. Piyasena; R. J. Loncharich; K. N. Houk, *J. Org. Chem. 52* (1987) 2137.
100. G. Maier; J. H. Teles, *Angew. Chem. 99* (1987) 152.
101. G. Pandey; G. Kumaraswamy; A. Krishna, *Tetrahedron Lett. 1987*, 2383.
102. S.-I. Murahashi; T. Shiota, *Tetrahedron Lett. 1987*, 2383.
103. R. M. Coates; S. J. Firsan, *J. Org. Chem. 51* (1986) 5198.
104. G. Klebe; K. H. Böhn; M. Marsch; G. Boche, *Angew. Chem. 99* (1987) 62.
105. J. Knight; P. J. Parsons, *J. Chem. Soc., Chem. Commun 1987*, 189.

106. A. Padwa; D. N. Kline; J. Perumattan, *Tetrahedron Lett. 1987*, 913.
107. S. S. Ghabrial; I. Thomsen; K. B. G. Torssell, *Acta Chem. Scand. B41* (1987), 426.
108. (a) W. Ibebeke-Bomangwa; C. Hootelé, *Tetrahedron 43* (1987) 935.
 (b) A. S. Amarasekara; A. Hassner, *Tetrahedron Lett. 1987*, 3151.
109. I. Thomsen; K. B. G. Torssell, *Acta Chem. Scand B42* (1988) in preparation.
110. J. J. Tufariello; A. S. Milowsky; M. Al-Nuri; S. Goldstein, *Tetrahedron Lett. 1987*, 263, 267.
111. H. Kawakami; S. Hirokawa; M. Asaoka; H. Takei, *Chem. Lett. 1987*, 85.
112. A. P. Kozikowski; X.-M. Cheng, *J. Chem. Soc., Chem. Commun. 1987*, 680.
113. Y. Kita; F. Itoh; O. Tamura; Y. Y. Ke; Y. Tamura, *Tetrahedron Lett. 1987*, 1431.
114. J. J. W. McDovall; M. A. Robb; U. Niazi; I. Bernardi; H. B. Schlegel, *J. Am. Chem. Soc. 109* (1987), 4642.
115. R. Nesi; D. Giomi; L. Quartara; S. Bracci; S. Papaleo, *J. Chem. Soc. Chem. Commun. 1987*, 1077.
116. J. D. Perez; D. A. Wunderlin, *J. Org. Chem. 52* (1987), 3637.
117. L. A. Reiter, *J. Org. Chem. 52* (1987), 2714.
118. J. J. Tufariello; A. D. Dyszlewski, *J. Chem. Soc. Chem. Commun. 1987*, 1138.

Author Index

Numbers set in roman type refer to page number, italicized to reference number.

Abbakumova, N. V. 88(*193*)
Abbott, R. 99(*45*)
Abdel-Wahab, A.-M. A. 102(*82*)
Abdullah, A. H. 232(*663*)
Abe, M. 113(*192*), 114(*192*), 247(*717, 718*)
Abou-Gharbia, M. 82(*80*), 86(*151*), 87(*80*)
Abou-Gharbia, M. A. 166(*193*), 171(*277*), 172(*193*)
Abramovitch, D. A. 242(*690*)
Abramovitch, R. A. 242(*690*)
Aburaki, S. 254(*734*)
Acampora, M. 188(*500*)
Acheson, R. M. 164(*185*)
Ackrell, J. 66(*120*)
Adachi, I. 18(*163, 167, 168*), 164(*183, 188*), 168(*208*), 171(*251, 252*), 177(*310*)
Adam, W. 27(*374*)
Adamczyk, M. 15(*126*), 16(*137*), 17(*137*), 134(*66*), 135(*66*), 136(*66*), 147(*139*), 148(*139*), 149(*139*), 150(*139*)
Adembri, G. 60(*46*), 168(*226*), 169(*226, 247*)
Adhikary, P. 104(*114*), 109(*114*)
Adolph, H. G. 96(*14*)
Advani, B. G. 27(*351*), 188(*515*)
Agawa, T. 166(*197*), 187(*452*)
Aiazzi-Mancini, M. 82(*109*)
Ajello, E. 160(*167*), 169(*232*), 171(*257*), 178(*320*), 183(*345*), 184(*347, 348*)
Ajello, T. 173(*287, 288a, 289*), 175(*299, 300*)
Akazawa, H. 59(*30*), 65(*94*)
Akbutina, F. A. 82(*97*)
Akhrem, A. A. 3, 13(*86*), 146(*127*), 147(*133–135*), 157(*156*), 169(*242*)
Akiba, K.-Y. 173(*284*)
Akimoto, T. 16(*124*)
Akimova, G. S. 27(*341*)
Akita, Y. 98(*41*)
Akiyama, M. 184(*377*)
Akmanova, N. A. 26(*353*), 47(*353*), 82(*97*)
Alazard, J. P. 76(*12*), 242(*692*), 272(*5*)
Alberghina, G. 187(*489*)
Alberola, A. 11(*71a*), 13(*90*), 19(*170*), 169(*231*), 283(*48*)
Albini, E. 27(*289, 343*), 28(*251*), 32(*251*)
Albini, F. M. 27(*284, 289, 343*), 28(*251*), 32(*251, 258*), 38(*302*), 55(*4*), 64(*4*), 185(*415*), 278(*22*)
Albizati, K. 27(*263*), 34(*263*), 181(*336*)
Albrecht, H. P. 254(*730*)
Albrecht, R. 188(*519*)
Alderson, G. W. 83(*130*)
Alemagna, A. 69(*148*)
Alessandri, L. 79(*49*), 82(*109*)
Alexandrou, N. E. 187(*462*)
Alford, E. J. 88(*168*)
Ali, Sk. A. 22(*195*), 23(*213*), 30(*274*), 31(*274*), 35(*265*), 38(*265*), 44(*265*), 165(*191*), 166(*191*), 214(*621, 622*), 215(*623*), 216(*632, 633*), 217(*623*)
Allspach, T. 188(*506*)
Al-Nuri, M. 27(*286*), 181(*335*), 295(*110*)
Alper, H. 187(*458*)
Altaf-ur-Rahman, M. 66(*120, 125*)
Alvarez, F. S. 98(*35*)
Amarasekara, A. S. 289(*108b*)
Ambrosius, H. P. M. 187(*479*)
Amin, N. V. 63(*72*), 102(*81*), 104(*113*), 105(*113*), 110(*113*), 135(*72, 73*), 294(*81*)
Andaburskaya, M. B. 169(*242*)
Andersen, S. H. 8(*25*), 15(*25*), 16(*25, 133*), 17(*25*), 21(*25*), 113(*188*), 114(*188*), 120(*188*), 133(*58*), 136(*58*), 147(*58, 138*), 148(*58, 138*), 151(*58, 138*), 155(*58*), 158(*58, 138*), 159(*58*), 178(*58*), 192(*58*)
Ando, M. 186(*438*), 187(*438*)
Andreeva, I. M. 75(*5*)
Andres, C. 169(*231*)
Angeli, A. 82(*109*)
Anneser, E. 64(*79*), 184(*381*), 187(*471, 472*)
Annunziata, R. 39(*368, 369*), 150(*142*), 239(*677, 678*), 271(*91*), 275(*12*), 294(*83*)
Antonucci, F. R. 171(*259*), 172(*270, 271*)
Apel, G. 98(*26*), 104(*26*), 119(*26*)
ApSimon, J. W. 100(*52*), 140(*99*)
Arai, H. 16(*124*), 209(*604*)
Araki, K. 27(*334, 340*)

301

Arbasino, M.   66(*106*)
Archibald, T. G.   99(*46*), 106(*128*)
Argyropoulus, N. G.   27(*344, 345*), 187(*462*)
Arimoto, M.   100(*51b*)
Arison, B. H.   26(*359b*)
Arlandini, E.   27(*361*)
Armand, J.   64(*85*), 66(*108, 109*), 98(*25*)
Arnal, E.   132(*43*)
Arndt, F.   104(*98*), 105(*98*), 110(*98*)
Asahara, T.   187(*467*)
Asaoka, M.   16(*139*), 113(*191, 192*), 114(*191, 192*), 246(*715*), 247(*717*), 295(*111*)
Ashburn, S. P.   87(*198*)
Ashcroft, P. L.   66(*122*), 184(*354*)
Asker, W.   169(*245*), 170(*245*)
Aube, J.   222(*647*)
Auricchio, S.   11(*68*), 12(*68*), 13(*83*), 140(*102, 103*), 145(*102, 103, 121, 122*), 168(*222*)
Aurich, H. G.   76(*21b*), 166(*201a*), 171(*255, 256*), 185(*405*), 279(*23, 24*)
Autenrieth, L.   27(*373*)
Autenrieth-Ansorge, L.   27(*373*)
Auwers, K.   66(*118*)
Aversa, M. C.   165(*189*)
Awad, W. I.   185(*396, 397*), 187(*396, 397*)
Azechi, K.   166(*201b*)

Bachetti, T.   69(*148*)
Bachmann, G. B.   68(*129*), 108(*156*)
Backer, H. J.   104(*117*)
Badovskaya, L. A.   27(*376*)
Baehler, B.   254(*729*)
Baggiolini, E. G.   205(*598*), 252(*725*), 292(*73*)
Bahr, Th.   132(*34*)
Bailey, A. S.   164(*185*)
Bailey, J. T.   200(*587*), 201(*587*)
Bailo, G.   27(*279*), 185(*416*)
Bains, K. B.   140(*110*), 141(*110*)
Baker, A. D.   76(*15*), 79(*67*), 166(*195*)
Baker, B. R.   132(*41*)
Baker, D. C.   108(*154*)
Bakker, B. H.   223(*649*), 225(*649*), 226(*649*)
Balasubramanian, N.   2, 88(*197*), 193(*557*), 223(*557*)
Baldwin, J. E.   18(*166*), 63(*66*), 76(*15*), 78(*36*), 82(*36*), 164(*180*), 168(*180*), 173(*180*), 186(*434*), 187(*434*), 243(*693*), 244(*697, 698*), 245(*711, 712*), 253(*727*)
Baldwin, S. W.   222(*647*)
Ballabio, M.   27(*361*), 30(*262*), 34(*262*)
Bamberger, E.   95(*6*), 97(*16*), 104(*116*), 109(*157*)
Bandiera, T.   27(*343*), 28(*251*), 32(*251, 258*), 278(*22*)
Banerjee, D. K.   132(*37*)
Banks, B. J.   241(*687*)
Bannucci, E. G.   22(*203*), 195(*573, 574*), 197(*573*), 199(*582*)
Bansal, R. K.   57(*14, 15*), 60(*14, 15*)

Bapat, J. B.   24(*235*), 216(*629*), 223(*629*)
Baraldi, P. G.   12(*76a*), 14(*109, 110*), 144(*116–118*), 145(*119*), 153(*116–119*), 158(*165*), 193(*119, 553*), 247(*716*), 276(*17*), 292(*74*), 294(*79, 80*)
Baralt, O.   163(*178*)
Baranski, A.   27(*337*), 104(*108*), 105(*108*), 110(*108*)
Barbalat-Rey, F.   254(*729*)
Barbaro, G.   26(*357*), 47(*357*), 66(*119*), 184(*352*)
Barbulescu, N.   16(*148*), 39(*309*)
Barcelos, F.   205(*598*)
Barco, A.   11(*56*), 12(*76a*), 14(*109, 110*), 144(*116–118*), 145(*119*), 153(*116–119*), 158(*165*), 193(*119, 553*), 247(*716*), 276(*17*), 292(*74*), 294(*80*)
Barlow, M. G.   39(*311*)
Barnes, J. F.   66(*122*), 184(*354*)
Barnes, R. P.   14(*105*)
Barrett, A. G. M.   241(*687*)
Barrett, G. R.   106(*126*)
Barrow, F.   75(*4*)
Barrow, M. J.   66(*122*)
Barrow, M. L.   184(*354*)
Barton, D. H. R.   76(*10*), 79(*63*), 84(*138b*), 88(*138b*), 253(*727*)
Bartulin, J.   177(*314, 316*)
Barzaghi, M.   28(*252*)
Basinski, W.   132(*50*)
Bast, K.   20(*174*), 26(*174, 259*), 27(*259*), 28(*250, 259*), 33(*250, 259*), 39(*259*), 44(*174*), 64(*80*), 184(*353*), 185(*353*), 186(*353*)
Bastani, B.   8(*26*), 113(*197*), 114(*197*), 115(*197*), 116(*197*), 117(*197*), 118(*197*), 119(*197*)
Bastide, J.   27(*270, 271*), 33(*270, 271*), 78(*48*)
Batcho, A. D.   205(*598*), 292(*73*)
Battaglia, A.   60(*39*), 66(*119*), 78(*46*), 187(*473, 478*)
Battioni, P.   27(*360*)
Bauer, H.   188(*496*)
Bauer, V. J.   178(*318*)
Baumann, A.   57(*12*), 60(*12*)
Bazilevsky, M. V.   10(*41*), 11(*41*), 138(*85*)
Beaulieu, N.   285(*57*)
Bebenroth, E.-M.   184(*366*)
Becalli, E. M.   285(*56*)
Beck, A. K.   113(*197–199*), 114(*197–199*), 115(*197*), 116(*197*), 117(*197*), 118(*197*), 119(*197, 198*), 120(*199, 216, 217*), 129(*4*), 130(*4*)
Beck, W.   8(*26*), 57(*21*), 60(*38*), 61(*54*)
Becker, G.   188(*505, 506*)
Becker, H.   82(*89*)
Becker, W.   188(*506*)
Beckmann, E.   60(*41*), 88(*187*), 186(*426, 435*)
Bedell, L.   67(*128a*)
Beekes, M. L.   86(*150*), 186(*446*)
Behrend, R.   80(*68*), 88(*166, 171*)

Behrens, U. 187(*490*)
Bekmukhametov, R. R. 168(*225*)
Belev, J. S. 96(*11*)
Bell, F. 169(*235*), 170(*235*)
Bellandi, C. 63(*73a*), 104(*118*), 105(*118*), 110(*118*), 135(*75*)
Bellavita, V. 82(*78*)
Bellino, A. 14(*112*)
Belly, A. 16(*127, 128*), 24(*225*)
Belokov, V. M. 96(*8*)
Beltrame, P. 39(*305*), 69(*143*), 184(*359*), 188(*512, 513*)
Beltrame, P. L. 28(*252*), 39(*305*)
Belzecki, C. 22(*326*), 24(*246*), 41(*325*), 42(*326*), 87(*199*)
Bereznak, J. F. 135(*71*), 294(*82*)
Bender, M. L. 102(*65*), 106(*121*)
Benderley, A. 77(*43*)
Benecke, H. 242(*690*)
Bened, A. 27(*293*)
Benetti, S. 11(*56*), 12(*76a*), 14(*109, 110*), 144(*116–118*), 145(*119*), 153(*116–119*), 158(*165*), 193(*119, 553*), 247(*716*), 276(*17*), 292(*74*), 294(*80*)
Benn, M. H. 69(*144, 145*), 70(*144, 145, 151*), 113(*201*), 114(*201*), 122(*201*)
Bergell, C. 137(*78*)
Berger, I. 200(*587*), 201(*587*)
Bergmann, E. D. 98(*26*), 104(*26*), 119(*26*)
Bernardi, F. 271(*114*)
Bernet, B. 42(*329*), 194(*562*), 234(*669*)
Berti, C. 82(*114*)
Bertini, V. 9(*32*), 60(*47*), 131(*5, 6*)
Berynkova, L. J. 96(*13*), 107(*13*)
Bestmann, H. J. 188(*497, 498*)
Bettinetti, G. F. 185(*410*)
Beutler, R. 57(*22*)
Bevinakatti, H. S. 283(*43*)
Bhat, K. L. 291(*71*)
Bianchi, G. 3, 14(*106, 113*), 15(*121, 122*), 16(*152*), 22(*207*), 23(*214*), 27(*278, 280*), 32(*275*), 38(*300*), 85(*140*), 179(*323, 324*), 185(*411*)
Biekert, E. 68(*135*)
Bihlmeyer, W. 26(*359a*)
Bimanand, A. Z. 29(*253*)
Binsch, G. 180(*330*)
Birkenbach, L. 56(*11*), 60(*11*), 62(*61*)
Bjørgo, J. 77(*32*), 78(*32*), 85(*32, 143*), 89(*32*)
Black, D. St. C. 2, 8(*22*), 22(*208*), 24(*235*), 27(*208*), 38(*299*), 76(*8*), 82(*105, 112, 113*), 83(*126, 127, 129, 130, 132*), 88(*132*), 185(*385*), 186(*431*), 187(*474, 494*), 188(*385*), 216(*629*), 223(*629*)
Blackman, N. A. 82(*113*), 83(*126, 127*)
Blakeley, R. L. 283(*50*)
Blaschke, H. 188(*509, 510*)
Blatt, A. H. 14(*102*), 16(*129*)
Blatt, K. 83(*128*), 189(*527*)
Blinne, G. 166(*201a*), 171(*256*)

Block, M. 166(*195*)
Bloom, J. D. 248(*719*), 292(*75*)
Blyumenfel'd, A. L. 117(*207*), 118(*207*), 119(*207, 210, 213*)
Boal, J. R. 57(*27a*), 58(*27a*)
Bocharova, L. A. 102(*80*)
Boche, G. 282(*104*)
Bodenseh, H. K. 60(*34, 35*)
Bodländer, P. 283(*49*)
Bodrikov, I. V. 188(*524*)
Boerma, J. A. 187(*484*)
Bolton, G. L. 202(*593*), 203(*593, 594*)
Bonenfant, A. P. 254(*729*)
Bonini, B. F. 187(*475, 479, 480*)
Bonnett, R. 75(*6*), 79(*57*), 82(*79*), 84(*79*), 87(*79*), 88(*79*)
Bordner, J. 27(*347*)
Borgardt, E. G. 101(*57*)
Bormann, C. 290(*70*)
Borsche, W. 132(*51*)
Borsus, J.-M. 187(*491*)
Borth, M. 11(*71c*)
Boscacci, A. B. 83(*129*)
Boswell Jr., G. A. 66(*113*)
Botwinnik, M. M. 11(*54*)
Boulton, A. J. 13(*94, 96*), 56(*10*), 66(*120, 125, 126*), 71(*126*), 132(*32*), 173(*290*), 174(*291*), 175(*291*), 179(*327*)
Bowden, K. 140(*93*)
Bowie, J. H. 77(*42*)
Boyd, D. R. 77(*32, 38*), 78(*32*), 79(*58*), 81(*74, 75*), 85(*32, 143–145*), 89(*32, 58*)
Böhn, K. H. 282(*104*)
Böshagen, H. 172(*274*), 188(*525*)
Bracci, S. 276(*115*)
Brady, O. L. 88(*184*)
Brambilla, R. 24(*230*), 202(*590*)
Branchi, C. 132(*55*)
Brandi, A. 22(*210*), 27(*277*), 55(*2*), 61(*53, 55*), 76(*19, 21a*), 173(*288b*), 187(*468, 470*), 276(*14*), 289(*66–68*)
Brandner, H. 184(*366*)
Brass, H. J. 24(*247*)
Braun, H. P. 188(*523*)
Bravo, P. 13(*80*), 140(*108*)
Bredereck, H. 57(*23*)
Bremner, J. B. 200(*586*)
Bren', V. A. 75(*5*)
Brenner, J. E. 16(*147*)
Brennon, T. F. 157(*158*)
Breuer, E. 2, 75(*2*), 79(*2*), 80(*2*), 82(*2, 102, 106–108, 117*), 86(*2*), 95(*2*), 96(*2*), 101(*2*), 188(*504*)
Briehl, H. 61(*57*)
Britteli, D. R. 66(*113*)
Brois, S. J. 20(*176*)
Brook, M. A. 99(*51a*)
Brookes, P. C. 166(*193*), 172(*193*)
Brown, C. W. 22(*197*)
Brown, F. K. 271(*2*)

Brown, G. B.   88(*186*)
Brown, K. L.   38(*298*), 194(*564*)
Brown, R. A.   98(*23*), 104(*95*), 105(*95*), 109(*95*), 111(*95*), 185(*403*)
Brown, R. C.   66(*120*), 174(*291*), 175(*291*)
Brown, R. F. C.   24(*235*), 82(*79*), 84(*79*), 87(*79*), 88(*79*), 216(*629*), 223(*629*)
Bruck, P.   110(*171*), 111(*171*)
Brunelle, D. J.   11(*71b*), 140(*105*)
Brunn, E.   188(*509, 510*)
Brück, B.   104(*114*), 109(*114*)
Brüning, I.   20(*174*), 26(*174*), 44(*174*), 110(*164*)
Brüntrup, G.   39(*310*)
Bryan, R. F.   157(*157*)
Bryson, T. A.   27(*347*)
Bubenheim, O.   279(*24*)
Buchanan, R. A.   110(*161*), 187(*477*)
Buchardt, O.   79(*52*)
Bucharev, V. N.   77(*41*)
Buczkowski, Z.   101(*62*)
Budagyants, M. I.   104(*99*)
Buehler, E.   88(*185, 186*)
Burdisso, M.   273(*9*)
Burger, M.   20(*175*), 27(*175*)
Burger, U.   254(*729*)
Burk, E. H.   188(*523*)
Burkholder, C. R.   271(*94*)
Burns, P.   24(*228*)
Burri, K. F.   211(*614*), 246(*614*)
Burrows, T. G.   40(*317*)
Bursus, J. M.   188(*502*)
Buss, H.   63(*65*), 64(*74*)
Buss, V.   16(*153, 154, 157*), 42(*153, 154*), 147(*673*), 210(*606, 607*), 211(*616*), 235(*673*), 240(*673*), 241(*673*)
Busti, G.   65(*89*)
Buttero, P. D.   28(*252*)
Büchi, G.   12(*72*), 142(*115*), 153(*115*)

Caballiera, N.   27(*374*)
Caldirola, P.   283(*45, 46*), 294(*82*)
Caldwell, R.   102(*68*)
Califano, S.   13(*95*), 59(*31*)
Calvin, M.   79(*53*), 85(*53*)
Camels, G.   69(*140*)
Campagnini, A.   184(*356*), 185(*356*)
Camparini, A.   168(*226*), 169(*226*)
Campbell, D. C.   177(*315*)
Caprathe, B. W.   199(*584*)
Caramella, P.   2, 8(*19*), 27(*279, 284, 289, 294, 342, 343*), 28(*251*), 32(*251, 258*), 38(*302*), 39(*303-307*), 55(*4*), 60(*39, 40*), 64(*4*), 78(*46*), 176(*308, 309*), 184(*357, 383*), 185(*357, 383, 412, 414-416*), 186(*422*), 188(*383*), 235(*414*), 278(*22*)
Cardone, R. A.   16(*156*), 42(*156*), 211(*614*), 246(*614*)
Carli, S.   289(*67*)
Carlos, D. D.   188(*522*)
Carlson, E. H.   195(*573*), 197(*573*)
Carman, R. M.   157(*159*)
Caroll, P. J.   42(*372*)
Carrié, R.   20(*178, 181*), 21(*181*), 22(*204*), 27(*181*), 104(*110-112*), 105(*112*), 110(*110-112*), 168(*207*), 188(*505, 506*)
Carruthers, W.   287(*61*)
Cartaya-Marin, C. P.   224(*651*), 225(*651*)
Carter, S. P.   271(*3*)
Caruso, F.   186(*421*), 187(*421*)
Casco, A.   66(*126*), 71(*126*)
Casey, M. L.   131(*7*)
Casnati, G.   64(*75*), 145(*120*), 157(*120, 155*), 178(*120*)
Castagnoli, Jr., N.   227(*654*)
Catsasoumis, A. G.   27(*281*)
Cawkill, E.   82(*81*)
Cellerino, G.   27(*279, 284*), 39(*303, 305*), 185(*414-416*), 235(*414*)
Cellerino, M. R.   27(*284*), 185(*415*)
Cereda, E.   27(*294*), 186(*422*)
Ceva, P.   27(*289*)
Cha, J. K.   244(*697, 698*)
Chalard-Fauré, B.   254(*731*)
Chalet, J. M.   40(*319*)
Challand, S. R.   111(*175*)
Chan, M. F.   245(*712*)
Chan, T.-M.   273(*10*)
Chan, W.-H.   285(*57*)
Chang, M. S.   184(*372*)
Chang, Y.-M.   60(*39*)
Chao, J.-C.   276(*97*)
Chapman, J. A.   66(*121*)
Chapman, O. L.   22(*194*)
Chen, E.   171(*264, 267*)
Chen, W. Y.   16(*156*), 42(*156*), 211(*614*), 246(*614*)
Chen, Y.-Y.   194(*569*), 230(*660, 661*)
Cheney, L. C.   145(*126*)
Cheng, X.-M.   166(*202*), 296(*112*)
Cherest, M.   76(*13*), 79(*13*), 281(*32*)
Cherpeck, R. E.   160(*174*)
Cherskaya, N. O.   113(*183*), 114(*183*), 115(*183*)
Chiacchio, U.   27(*342*), 184(*356, 357*), 185(*356, 357*), 187(*489*), 271(*3*)
Chiang, Y.   244(*696*)
Chiarino, D.   283(*47*)
Chidester, C. G.   63(*69*)
Chidichimo, G.   168(*205*)
Chigbo, F. E.   14(*105*)
Chimiak, A.   82(*85*)
Chimichi, S.   276(*98*)
Chin, C. G.   11(*57*)
Chistokletov, V. N.   24(*226, 227*), 27(*341*), 27(*366*)
Chlenov, I. E.   21(*184, 189, 191, 192*), 30(*192*), 104(*101, 102, 105, 106, 109*), 105(*101, 102, 105, 106, 109*), 110(*101, 102, 105, 106, 109, 159, 160*), 111(*166*), 135(*76*), 136(*76*), 147(*76*), 185(*404*)
Chmielewski, M.   27(*363*), 87(*199*)

Chong, Y.-M. 78(*46*)
Christensen, C. G. 160(*171*)
Christensen, J. 245(*710*)
Christl, M. 14(*107*), 15(*161*), 17(*161*), 20(*174*), 26(*174, 259*), 27(*259*), 28(*161, 250, 259*), 31(*161*), 33(*250, 259*), 39(*259, 310*), 44(*174*), 59(*32*), 60(*48, 51*), 61(*48*), 64(*80*), 133(*52*), 136(*52*), 184(*353*), 185(*353*), 186(*353*), 283(*44*)
Ciancaglione, M. 283(*45*)
Ciller, J. A. 285(*54*)
Cinquini, M. 39(*368, 369*), 42(*370, 371*), 150(*142, 143*), 239(*677, 678*), 271(*91*), 275(*12*), 294(*83*)
Cirrincione, G. 184(*348*)
Cisseé, H. 132(*31*), 174(*292*)
Claisen, L. 9(*34*), 11(*59, 62*), 12(*59, 74*), 131(*8, 9, 45*), 132(*8, 9*), 137(*9*), 169(*234*), 170(*234*)
Clapp, L. B. 14(*108*), 102(*71*), 185(*395*), 186(*395*)
Clardy, J. 295(*90*)
Clark, J. H. 98(*34*)
Clark, N. G. 82(*81*)
Clark, S. 251(*723*)
Clark, T. J. 106(*124b*)
Clark, V. M. 76(*8*), 79(*57*), 82(*79, 112*), 83(*130, 132*), 84(*79*), 87(*79*), 88(*79, 132*)
Clemans, G. B. 132(*44*)
Coates, R. M. 76(*9*), 87(*198*), 156(*154*), 280(*103*), 284(*53*)
Cocu, F. G. 39(*309*)
Coda, A. 285(*55*)
Coda, A. C. 28(*251*), 32(*251*), 185(*414*), 235(*414, 671*), 285(*55*)
Cogoli, A. 16(*152*)
Cohen, L. A. 104(*115*)
Cohen, T. 179(*326*)
Colle, R. 11(*68*), 12(*68*), 140(*103*), 145(*103*)
Colombi, S. 27(*348*), 185(*413*)
Colonna, M. 82(*114*)
Colvin, E. W. 8(*26*), 113(*196–198*), 114(*196–198*), 115(*197*), 116(*197*), 117(*197*), 118(*197*), 119(*196–198*), 120(*216, 217*), 122(*218*), 129(*2*), 129(*3*)
Comins, D. L. 232(*663*)
Comotti, A. 188(*513*)
Compagnini, A. 55(*4*), 64(*4*)
Confalone, P. N. 16(*155*), 42(*155*), 202(*592*), 210(*613*), 211(*613*), 250(*722*), 251(*724*)
Considine, W. J. 107(*149*), 108(*149*)
Cooley, J. H. 98(*24*)
Cooper, K. 206(*599*)
Cope, A. C. 80(*70*), 88(*70*), 193(*554*), 195(*554*)
Coppinger, G. M. 88(*163*)
Corbett, D. F. 273(*8*)
Corbin, U. 13(*84*), 145(*126*), 173(*283*)
Corezzi, S. 61(*55*)
Cornforth, J. W. 160(*169*)
Corsaro, A. 27(*342*), 55(*4*), 64(*4*), 184(*356, 357*), 185(*356, 357*), 187(*489*), 278(*22*)

Coulter, P. B. 79(*58*), 89(*58*)
Coutouli-Argyrapoulou, E. 27(*344, 345*), 66(*114*)
Coutts, R. T. 77(*43*)
Cox, D. D. 131(*7*)
Cozzi, F. 39(*368, 369*), 42(*370, 371*), 150(*142, 143*), 239(*677, 678*), 271(*91*), 275(*12*), 294(*83*)
Craig, P. N. 13(*89*)
Cramer, C. J. 189(*536*), 281(*34, 35*)
Crämer, E. 27(*374*)
Crank, G. 140(*93*)
Criegee, R. 281(*33*)
Croce, P. D. 28(*252*), 30(*262*), 34(*262*)
Crosby, J. 66(*121, 122*)
Crozier, R. F. 2, 8(*22*), 22(*208*), 27(*208*), 38(*299*), 185(*385*), 188(*385*)
Crumbie, R. L. 108(*155*)
Culbertson, T. P. 134(*61*)
Cum, G. 165(*189, 190*), 168(*205*)
Cummings, C. A. 66(*121*), 76(*9*)
Cummins, C. H. 156(*154*), 242(*154*), 284(*53*)
Cunico, R. F. 15(*141*), 67(*128a*)
Curran, D. P. 16(*135*), 17(*135*), 66(*124b*), 135(*69*), 136(*69*), 147(*140*), 149(*140*), 150(*140*), 152(*147*), 156(*147*), 158(*164*), 192(*552*), 193(*552*), 206(*600*), 276(*97, 99*), 284(*52*), 292(*77*)
Cusmano, S. 11(*60*), 14(*112*), 131(*11*), 132(*11, 47*), 169(*236, 238*), 173(*288*), 175(*299, 305*)

Daeniker, H. U. 188(*514*)
Daggett, J. U. 24(*232*), 38(*297*), 202(*593*), 203(*593*)
Dagne, E. 227(*654*)
Dahl, K. 66(*101*), 185(*408*)
Dales, J. R. M. 76(*11*), 242(*691*)
Dalgaard, N. K. A. 22(*206*), 24(*206*), 30(*206*), 86(*188*), 88(*188*), 191(*545*)
D'Alò, G. 184(*365*)
Dal Piaz, V. 11(*65*), 140(*107*)
Daly, J. W. 221(*644*)
Damavandy, J. A. 86(*156*)
Dan, S. 27(*365*), 221(*643*)
Danen, W.-C. 102(*89*)
Daniewski, A. R. 285(*57*)
Danishefsky, S. 4(*13*), 13(*13*), 139(*87*)
Dappen, M. S. 189(*536*), 281(*35*)
Da Prada, L. 27(*361*)
Darlage, L. J. 172(*269, 273*)
Das, L. H. 187(*469*)
Das, N. B. 8(*25*), 14(*114*), 15(*25, 114*), 16(*25, 114, 145*), 21(*25, 114, 145*), 30(*114*), 39(*312*), 40(*312*), 113(*188, 189*), 114(*188, 189*), 120(*188*), 133(*58, 59*), 136(*58*), 147(*58, 59, 136*), 148(*58, 136*), 150(*59*), 151(*58, 59*), 155(*58*), 158(*58, 59*), 159(*58, 59*), 178(*58*), 192(*58, 136*), 193(*136*), 239(*59*)
Das Gupta, T. K. 83(*128*), 189(*527–529*)

Datta, S. K.   65(*91*)
Dattolo, G.   184(*348*)
Davies, J. H.   69(*146*), 70(*146*)
Davis, R. H.   69(*146*), 70(*146*)
Davis, V. C.   2, 8(*22*), 82(*105*)
Davis, V. D.   185(*385*), 188(*385*)
Day, M. J.   79(*63*)
De Amici, M.   14(*113*), 27(*292*), 63(*73a*), 104(*118*), 105(*118*), 110(*118*), 135(*75*), 283(*45, 46*), 294(*82*)
Dean, J. M.   57(*25*), 64(*83, 87*), 65(*87*), 66(*87*)
De Benedetti, P. G.   27(*292*)
De Bernardis, J.   76(*15*)
De Boer, T. J.   86(*149, 150*), 186(*446*)
Declercq, J. P.   27(*293*), 64(*76*)
De Graw, J.   132(*41*)
Dehler, W.   240(*683*), 290(*70*)
De La Mare, H. E.   88(*163*)
Dell'Amico, V.   9(*32*)
Delpierre, G. R.   2, 22(*198, 199*), 75(*23*), 76(*23*), 79(*23*), 81(*23*), 86(*23*)
Deltsova, D. P.   166(*194*)
De Lucca, G.   39(*308*)
De Micheli, C.   3, 22(*207*), 27(*280, 292, 349*), 32(*275*), 38(*300, 301*), 63(*73a*), 104(*118*), 105(*118*), 110(*118*), 135(*75*), 185(*411*), 272(*7*), 283(*45, 46*), 294(*82*)
De Munno, A.   9(*32*), 60(*47*), 131(*5, 6*)
Denmark, S. E.   189(*536*), 281(*34, 35*)
Denzel, T.   188(*498*)
De Paolini, I.   62(*63*)
De Sarlo, F.   22(*210*), 27(*277*), 55(*2, 3*), 61(*53, 55*), 76(*19, 21a*), 169(*244*), 173(*288b*), 185(*387*), 186(*387*), 187(*468, 470*), 276(*14*), 289(*66–68*)
De Shong, P.   24(*231*), 27(*336*), 29(*256*), 30(*256*), 41(*323*), 43(*256*), 154(*149*), 163(*178*), 233(*665, 666*), 284(*51*)
Desimoni, G.   86(*154*), 285(*55*)
De Sio, F.   60(*46*)
Devi, P. S. U.   10(*43*)
Deville, C. G.   60(*40*)
De Ville, G.   102(*74*)
Dewald, M.   64(*88*)
Dewar, M.   55(*9*)
Diadone, G.   184(*348*)
Dicken, C. M.   27(*336*), 29(*256*), 30(*256*), 43(*256*)
Dickoré, K.   187(*487*)
Diehl, J. N.   38(*298*)
Diehl, J. W.   194(*564*)
Dietliker, K.   168(*224*), 171(*224*)
Ding, L. K.   180(*333*)
Dobashi, T. S.   77(*29, 34*), 80(*34, 71*), 85(*29*)
Dolbier, W. R.   185(*409*), 271(*94*)
Dombrovskii, A. V.   188(*501*)
Domelsmith, L. W.   78(*47*)
Domsch, D.   113(*195*), 114(*195*), 156(*153*)
Donaruma, L. G.   104(*97*)
Dondio, G.   271(*91*)

Dondoni, A.   26(*357*), 47(*357*), 66(*119*), 184(*352*), 187(*473, 478*)
Doppler, T.   172(*275*)
Dornov, A.   82(*110*)
Doyle, T. W.   244(*696*)
Döpp, D.   168(*209*)
Drehfahl, G.   16(*142*), 147(*131*), 239(*131*), 240(*131*), 241(*131*)
Dreyer, D. L.   157(*161*)
Dubey, S. K.   27(*339*)
Duchamp, D. J.   63(*69*)
Duden, P.   102(*72*)
Duh, H.-Y.   271(*1*)
Duke Jr., R. E.   3(*2*), 27(*2*), 33(*2*), 36(*2*)
Dull, M. F.   107(*142*)
Dunitz, J. D.   8(*26*), 113(*197*), 114(*197*), 115(*197*), 116(*197*), 117(*197*), 118(*197*), 119(*197*)
Dunn, F. B.   88(*184*)
Dupré, B.   16(*138*), 32(*138*), 33(*138*), 38(*138*), 150(*141*)
DuPree, L. E.   160(*172*)
Durand, R.   27(*293*)
Durst, T.   179(*325*)
Dust, J. M.   187(*469*)
Dybowski, U.   186(*441, 442*)
Dymowski, D.   63(*64*)
Dyslewski, A. D.   289(*118*)

Eckroth, D. R.   177(*314*)
Eckstein, Z.   184(*369*)
Edmonson, W. L.   160(*171, 172*)
Edwards, J. O.   24(*247*)
Eghstessad, E.   186(*449, 450*)
Eguchi, S.   184(*362*), 185(*394*), 198(*578, 579*), 226(*652*)
Ehrler, R.   2, 16(*157*), 147(*673*), 235(*673*), 240(*673*), 241(*673*), 294(*84*)
Ehrlich, P.   88(*178*)
Eidel, J.   279(*23*)
Einstein, F. W. B.   88(*191*), 191(*547*)
Elghandour, N.   27(*270*), 33(*270*)
Elguero, J.   13(*99*)
Elming, N.   4(*8*)
Eloy, F.   66(*104, 105*), 184(*371, 378*), 188(*516, 517*)
Elsworth, J. F.   83(*124*), 186(*432*)
Emde, H.   113(*195*), 114(*195*), 156(*153*)
Emmons, W. D.   79(*54*), 102(*70*)
Endesfelder, A.   294(*89*)
Endo, K.   24(*236*)
Endo, T.   12(*76b*), 24(*76b*)
Endres, A.   97(*17*)
Endres, G.   62(*62*)
Engel, R.   79(*67*), 166(*195*)
Erashko, V. I.   96(*13*), 101(*64*), 106(*64*), 107(*13*)
Ercoli, R.   132(*12*)
Erickson, R. E.   83(*122*)

Eschenmoser, A.   20(*177, 179*), 21(*179*), 83(*128*), 162(*176*), 189(*527-532*)
Ettlinger, M. G.   69(*150*)
Eugster, C. H.   130(*10*), 131(*10*), 171(*263*)
Evans, A. R.   172(*281*)
Ewing, W. R.   291(*71*)
Exner, O.   76(*17*), 77(*30*), 84(*137*)
Eyer, M.   102(*86b*), 108(*86b*)

Faasch, H.   240(*685*)
Fabbrini, L.   184(*364*), 185(*387*), 186(*387*)
Fabra, J.   173(*289*)
Fainzil'berg, A. A.   96(*10, 12, 13*), 101(*61, 64*), 106(*64*), 107(*13*)
Faivre, L.   40(*318, 319*), 254(*729*)
Falshaw, C. P.   172(*281*)
Fanshawe, W. J.   178(*318*)
Fanta, P. E.   104(*100*)
Fantechi, R.   39(*305*)
Faragher, R.   171(*253*)
Farina, V.   79(*66*)
Fatti, G.   9(*37*), 13(*79*)
Fatutta, S.   169(*240*)
Faulks, S.   40(*317*)
Feger, H.   113(*194, 195*), 114(*194, 195*), 156(*153*), 281(*37*)
Feinauer, R.   186(*420*)
Feinstein, G.   283(*49*)
Fekih, A.   84(*138b*), 88(*138b*)
Feldl, K.   57(*21*), 60(*38*)
Felix, D.   20(*177*), 83(*128*), 189(*527, 528, 530*)
Fellmann, J.-Y.   99(*47*)
Fellrath, E.   186(*435*)
Felsen, D.   79(*67*)
Fenech, G.   133(*54*)
Fengler, I. E.   194(*566*)
Fenk, C. J.   66(*124b*), 135(*69*), 136(*69*), 292(*77*)
Ferrara, G.   184(*363*)
Ferrier, R. J.   38(*298*), 194(*563-565*)
Ferris, J. P.   171(*259*), 172(*270-272*)
Fetizon, M.   185(*407*)
Feuer, H.   99(*43*), 101(*63*), 283(*43*)
Fiedler, H.-P.   240(*683*)
Fijiki, S.   283(*42*)
Fina, N. J.   24(*247*)
Finch, N.   4(*9*)
Finkbeiner, H. L.   108(*152, 153*)
Firestone, R. A.   26(*359b*)
Firsan, S. J.   280(*103*)
Fisera, L.   22(*212*), 27(*255, 376*)
Fitt, J. J.   4(*9*)
Fitzpatrick, J. M.   160(*173*)
Flament, I.   157(*163*)
Flanagan, P. W. K.   101(*58*)
Fletcher, I. J.   174(*291*), 175(*291*)
Flynn, T.   174(*294*)
Folting, K.   77(*39*)
Forrester, A. R.   83(*125*)

Forshult, S.   98(*31*)
Forster, M. O.   98(*29*)
Fortier, R. A.   140(*110*), 141(*110*)
Foster, B. C.   140(*110*), 141(*110*)
Foster, R.   76(*14*)
Foti, F.   186(*421*), 187(*421*)
Fouty, R. A.   184(*382*)
Fowler, J. S.   106(*124b*)
Foye, W. O.   187(*493*)
Föhlisch, B.   57(*23*)
Franchini, P. F.   60(*46*)
Frank, W.   57(*12*), 60(*12*)
Franke, A.   185(*398*)
Franz, J. E.   184(*361*)
Franz, R.   16(*157*), 147(*673*), 235(*673*), 240(*673, 686*), 241(*673*)
Fraser, R. R.   24(*239*)
Frattini, P.   185(*412*)
Fray, M. J.   15(*120*), 27(*288b*), 43(*288b*), 252(*726*)
Frazer-Reid, B.   248(*720*)
Freeman, J. P.   2, 18(*164*), 19(*164*), 164(*179*), 180(*330*)
Frenna, V.   175(*302, 303*)
Freri, M.   173(*289*)
Frese, E.   132(*34*)
Frey, H. H.   210(*609*)
Frey, J.   104(*11b*)
Frey, M.   2, 244(*707*), 294(*84*)
Freyer, A. J.   29(*256*), 30(*256*), 43(*256*)
Friary, R.   24(*230*), 200(*587*), 201(*587*), 202(*590*), 273(*10*)
Frick, U.   113(*195*), 114(*195*), 156(*153*)
Fridman, A. L.   111(*173*)
Friedrick, K.   187(*485*)
Fritsch, W.   134(*62*)
Fritz, G.   88(*170*)
Fritz, H.   187(*453*)
Frolkov, A. N.   27(*366*)
Frommeld, H.-D.   64(*83*), 69(*149*)
Fronczek, F. R.   40(*313*), 239(*679*)
Frost, J. W.   88(*165*)
Fryer, R. I.   174(*294*)
Fuga, V.   13(*91*)
Fujimoto, S.   184(*379*)
Fujita, E.   100(*51b*)
Fujiyama, F.   166(*201b*), 168(*220*), 169(*220*), 171(*220*)
Fujizawa, T.   107(*148*)
Fukui, K.   33(*268*)
Fukumoto, K.   182(*340*), 219(*640*)
Funk, R. L.   24(*232*), 38(*297*), 202(*593*), 203(*593, 594*)
Furneaux, R. H.   38(*298*), 194(*564*)
Furukawa, Y.   226(*652*)
Furusaki, F.   3, 20(*171*), 23(*171*), 71(*155*), 147(*128*)
Furuyama, H.   16(*140*)
Fusco, R.   8(*16*), 38(*295*), 66(*99*), 145(*123*), 209(*605a*)

Gabriel, S. 66(*117*)
Gadek, T. R. 88(*165*)
Gagneux, A. R. 171(*261, 263*)
Gainer, J. 13(*98*)
Gainsford, G.-J. 38(*298*), 194(*564, 565*)
Galeffi, E. 11(*50*)
Gallagher, G. 245(*712*)
Gallagher, T. 88(*200*), 286(*59*)
Gallivan, J. B. 168(*219*), 171(*219*)
Galucci, J. C. 39(*308*)
Gamba, A. 185(*410*), 273(*9*)
Gambaryan, N. P. 166(*194*), 185(*390*), 187(*390*)
Gambhir, I. R. 169(*248*)
Gambra, R. P. 111(*176*)
Gandolfi, C. 145(*119*), 153(*119*), 193(*119*)
Gandolfi, R. 3, 15(*121, 122*), 16(*152*), 22(*207*), 23(*214*), 27(*278, 280, 292, 349*), 32(*275*), 38(*300, 301*), 63(*73a*), 85(*140*), 104(*118*), 105(*118*), 135(*75*), 145(*323*), 179(*324*), 185(*411*), 272(*7*), 273(*9*)
Gandour, R. W. 60(*40*)
Ganem, B. 23(*219*), 24(*219*), 242(*689*)
Ganguly, A. 24(*230*), 202(*590*)
Garanti, L. 8(*16*), 38(*295*), 66(*99, 100*), 193(*559*), 209(*559, 605*)
Garcia, L. A. 26(*359b*)
Gardner, T. S. 169(*239*)
Gariboldi, P. 27(*292*)
Gasco, A. 56(*10*)
Gatrone, R. G. 218(*637*)
Gaudiano, G. 13(*80, 92*), 64(*82*), 132(*23*), 188(*499, 500*)
Gavaraghi, G. 140(*108*)
Gawrilow, N. I. 11(*54*)
Gazzaeva, R. A. 14(*100, 101*)
Gehrt, H. 82(*110*)
Geiger, W. 172(*274*)
Geister, B. 186(*443*)
Geittner, J. 26(*359a*)
Gelli, G. 69(*143*), 184(*359*)
Genco, N. A. 187(*458*)
Geneste, P. 27(*293*)
George, J. K. 3(*2*), 27(*2*), 33(*2*), 36(*2*)
Gerali, G. 134(*65*)
Gerecht, B. 61(*57*)
Germain, G. 27(*293*), 64(*76*)
Germeraad, P. B. 160(*173*)
Geschwend, H. W. 131(*11*), 132(*11*), 194(*566*)
Gettings, A. F. 172(*278, 280*)
Ghabrial, S. S. 65(*158*), 159(*166*), 227(*166*), 285(*107*)
Ghosh, A. K. 15(*120*), 40(*315*), 134(*67*), 135(*67*), 208(*603*), 237(*67*), 254(*67*)
Giambrone, S. 131(*11*), 132(*11*)
Gibert, J.-P. 185(*417*), 187(*454, 455*)
Gilardi, A. 42(*370, 371*), 150(*142, 143*)
Gilchrist, T. L. 171(*253*), 189(*539*)
Gilgen, P. 168(*224*), 171(*224*)
Gilmore, C. J. 157(*157*)

Ginsburg, D. 98(*26*), 104(*26*), 119(*26*)
Giomi, D. 276(*15, 98, 115*)
Giorgianni, P. 66(*119*)
Girutskaya, T. F. 26(*354*), 47(*354*)
Glower, D. J. 101(*59*)
Gloyer, S. E. 77(*26*), 80(*26*)
Gmelin, E. 70(*152*)
Gnichtel, H. 27(*373*), 88(*192*)
Godovikova, T. I. 106(*134*)
Godts, F. 175(*304*)
Goerdeler, J. 186(*429*)
Gohlke, R. S. 105(*120*), 106(*120*)
Goldmann, A. 283(*44*)
Goldman, N. L. 79(*67*), 166(*195*)
Goldschmidt, H. 186(*427*)
Goldstein, J. E. 193(*560*), 206(*560*)
Goldstein, R. F. 88(*184*)
Goldstein, S. 254(*733*), 295(*110*)
Golfier, M. 185(*407*)
Gonnermann, J. 102(*84*), 108(*84*)
Gonzales, A. M. 11(*71a*), 13(*90*), 19(*170*), 283(*48*)
Gonzales, B. 283(*48*)
Goodman, L. 132(*41*)
Goodrow, M. H. 77(*29*), 85(*29*)
Gore, P. H. 77(*28*)
Gorelik, V. P. 110(*168*), 111(*168*)
Goti, A. 61(*55*), 173(*288b*), 276(*14*), 289(*66–68*)
Goto, E. 182(*340*)
Gotoh, H. 166(*197*)
Gotthardt, H. 272(*4*)
Göschke, R. 171(*261*)
Gössinger, E. 22(*209*), 220(*641*), 221(*642, 645*)
Göth, H. 171(*263*)
Götz, A. 113(*195*), 114(*195*), 156(*153*)
Graf, W. 189(*535*)
Grandberg, I. I. 169(*249*)
Grant, R. D. 67(*128b*)
Grashey, R. 20(*173*), 24(*173, 223*), 25(*173*), 29(*173*), 34(*173*), 43(*173*), 186(*428*), 188(*428, 496*)
Grassi, G. 186(*421*), 187(*421*)
Grasso, L. 14(*103*), 16(*151*)
Graumann, J. 82(*91*)
Graybill, B. 102(*70*)
Grayson, J. I. 227(*657*)
Greci, L. 82(*114*)
Greé, R. 20(*178, 181*), 21(*181*), 27(*181, 282, 290*), 34(*248*), 104(*110–112*), 105(*112*), 110(*110–112*), 168(*207*)
Green, N. 151(*144*)
Grellman, K. H. 172(*276*)
Gribov, B. G. 106(*135–137*)
Griffin, C. E. 88(*177*)
Grigg, R. 18(*165*), 77(*40*), 88(*191*), 164(*184*), 177(*311*), 191(*546–548*)
Griller, D. 89(*194*)
Grob, J. 104(*116*)

Gronowitz, S. 4(*11*)
Gross, N. 16(*129*)
Grubbs, E. J. 77(*29, 34*), 80(*34, 71, 73*), 85(*29*)
Grund, H. 15(*115, 117*), 16(*157*), 17(*115, 117*), 42(*117*), 147(*132, 673*), 153(*132*), 235(*673*), 240(*673*), 241(*673*)
Grundmann, C. 2, 8(*14*), 55(*1*), 56(*1*), 57(*14, 15, 25, 26, 27a*), 58(*26, 27a*), 60(*14, 15*), 64(*83, 87*), 65(*87, 90, 91, 93*), 66(*87*), 69(*149*), 70(*1*), 145(*125*), 184(*125, 358*), 185(*125, 358*), 186(*125, 358, 425*), 188(*125, 526*)
Grünanger, P. 2, 8(*14, 19*), 11(*52*), 13(*52, 93*), 14(*103, 104, 106*), 15(*121*), 16(*149*, 151), 27(*278, 279, 284, 349*), 28(*249, 251*), 32(*251, 258, 275*), 55(*1*), 56(*1*), 70(*1*), 133(*55*), 145(*125*), 176(*309*), 179(*323*), 184(*125, 365, 380, 383*), 185(*125, 383, 412, 414–416*), 186(*125*), 187(*380*), 188(*125, 383*), 235(*414, 671*)
Gryszkiewiez-Trochimowski, E. 63(*64*)
Guarna, A. 27(*277*), 55(*2, 3*), 61(*53, 55*), 76(*19*), 173(*288b*), 187(*468*), 276(*14*), 289(*66–68*)
Guarneri, M. 12(*76a*), 14(*110*), 144(*116, 118*), 145(*119*), 153(*116, 118, 119*), 158(*165b*), 193(*119*)
Guette, J. P. 64(*85*), 66(*109*)
Guillovzo, G. P. 27(*293*)
Guimon, C. 27(*293*)
Gunaratne, H. Q. 191(*546*)
Gupta, B. G. B. 113(*202, 203*), 114(*202, 203*), 122(*203*)
Gurzynska, W. 107(*143*), 108(*143*)
Gutsche, C. D. 132(*36*)
Gutteridge, N. J. A. 76(*10, 11*), 242(*691*), 253(*727*)
Günther, H. J. 193(*561*)
Gygax, P. 83(*128*), 189(*527, 529, 531*)

Haagensen, C. H. 282(*41*)
Habashi, F. 254(*729*)
Häfele, B. 2, 294(*84, 88*)
Hafiz, M. 166(*198*), 172(*281*)
Hagedorn III, A. A. 40(*316*), 63(*67*), 140(*91*), 244(*695*)
Hagenmaier, H. 290(*70*)
Hahn, H. 290(*70*)
Hale, W. J. 132(*27*)
Haley, N. F. 140(*109*)
Hall, D. 189(*531*)
Hall, J. A. 60(*40*), 88(*168*)
Hall, W. T. 140(*110*), 141(*110*)
Hamana, M. 84(*133*), 187(*495*)
Hamelin, J. 22(*204*), 27(*266, 282, 290*), 34(*248*), 38(*266*)
Hamer, J. 2, 75(*22*), 76(*22*), 77(*22*), 79(*22*), 81(*22*), 86(*22*)
Hamid, A. M. 173(*290*)
Hammond, G. 102(*70*)
Hanamoto, T. 27(*364*)

Hansen, H.-J. 172(*275*)
Hansen, M. M. 24(*232*), 38(*297*), 202(*593*), 203(*593*)
Hantzsch, A. 95(*4*), 97(*4, 15*), 102(*67, 68*), 174(*291*), 175(*291*)
Hara, J. 85(*142*), 240(*682*)
Harada, K. 16(*143*), 18(*167*), 68(*130*), 107(*150*), 164(*183, 188*), 171(*252*), 184(*355*), 185(*355*), 187(*355*), 210(*612*), 240(*680*)
Hardegger, B. 189(*536*)
Harding, M. H. 66(*122*)
Harding, M. M. 184(*354*)
Hardinger, S. A. 63(*72*), 102(*81, 82*), 135(*72*)
Harhash, A. H. 169(*245*), 170(*245*)
Harikawa, H. 34(*264*)
Harkema, S. 77(*25*), 82(*25, 92, 93*), 84(*25*), 182(*343*), 280(*30*)
Harris, B. D. 291(*71*)
Harris, T. M. 244(*707*)
Harrison, B. L. 160(*173, 174*)
Hartke, K. 187(*451*)
Harwood, L. M. 245(*711*)
Harzanyi, K. 174(*293*)
Hashi, N. A. 172(*281*)
Hashimoto, S. 180(*332*), 186(*440*)
Hass, H. B. 102(*65*), 106(*121*)
Hass, W. 240(*683–685*), 290(*70*)
Hassal, C. H. 166(*192*)
Hassner, A. 27(*377*), 289(*69, 108b*)
Hatanaka, C. 244(*703*)
Hauck, H. 20(*173, 175*), 24(*173*), 25(*173*), 27(*175*), 29(*173*), 34(*173*), 43(*173*)
Hausmann, H. 57(*12*), 60(*12*)
Hausser, C. 40(*319*)
Haven Jr., A. C. 80(*70*), 88(*70*)
Hawkins, B. L. 59(*32*)
Haworth, G. A. 13(*98*)
Hawthorne, M. 102(*70*)
Hayakawa, T. 27(*334, 340*)
Hayashi, E. 186(*445*)
Hayashi, Y. 63(*73b*), 66(*123, 124a*), 68(*132*), 191(*550*), 277(*20, 21*)
Hayeo, R. 177(*310*)
Hazeldine, R. N. 39(*311*)
Hazell, A. C. 16(*132*), 39(*312*), 65(*98*), 113(*193*), 114(*193*), 133(*60*), 150(*60*), 238(*60*), 239(*60*), 276(*85*), 294(*85*)
Hazell, R. G. 16(*132*), 39(*312*), 65(*98*), 113(*193*), 114(*193*), 133(*60*), 150(*60*), 238(*60*), 239(*60*), 276(*85*), 282(*41*), 294(*85*)
Heathcock, C. H. 208(*735*), 292(*76*)
Hegarty, A. F. 64(*77*)
Heilbron, J. 174(*291*), 175(*291*)
Heimgartner, H. 168(*224*), 171(*224*)
Heine, H. W. 79(*65*)
Heinrich, W. 210(*608*)
Helmkamp, R. W. 102(*66*)
Hendrickson, J. B. 27(*267*), 38(*267*), 79(*59*), 110(*165*)
Hennessy, B. M. 205(*598*), 292(*73*)

Henning, R. 102(84, 85, 86a, 87), 108(84, 85, 86a, 87)
Henri-Rousseau, O. 27(270, 271), 33(270, 271)
Hepfinger, N. F. 88(177)
Hepler, L. G. 96(11)
Hepp, L. R. 27(347)
Herman, L. W. 100(52)
Hermkens, P. H. H. 285(58)
Herrgott, H. H. 113(195), 114(195), 156(153)
Herrmann, H. J. 184(375)
Hershberger, J. 102(77)
Hershenson, F. M. 185(392)
Hesse, R. H. 76(10), 79(63)
Hessler, E. J. 132(30)
Hill, H. B. 132(27)
Hinney, H. R. 63(70), 102(81), 135(70)
Hiraga, K. 207(602)
Hirai, H. 24(222), 41(324)
Hirao, I. 27(364)
Hirayama, S. 27(276, 338)
Hirokawa, S. 295(111)
Hiyama, T. 157(163)
Hjeds, H. 191(544)
Ho, T. L. 98(38, 40)
Hockenberger, L. 281(38)
Hodge, E. B. 97(19), 99(45)
Hoenicke, J. 16(158), 82(100, 115)
Hoffmann, R. 3(1)
Hoffmann, R. W. 294(89)
Hofmann, K. 113(195), 114(195), 156(153)
Hokama, T. 108(156)
Holla, B. S. 27(335)
Holladay, M. W. 244(700)
Holland, G. W. 98(32)
Hollemann, A. F. 69(141), 95(3)
Holm, S. 187(484)
Holmes, A. 140(99)
Honda, M. 226(652)
Honda, T. 16(140), 182(337, 340, 342), 215(624)
Hongu, T. 64(86), 65(86)
Hootelé, C. 289(108a)
Hope, H. 11(71c)
Hoppe, I. 11(63), 182(344)
Horcher, II, L. H. M. 24(232), 38(297), 202(593), 203(593)
Horikawa, H. 242(688b), 254(688b)
Horner, L. 66(115), 79(55), 281(38)
Hornhardt, H. 137(80)
Horozoglu, G. 166(195)
Horton, D. 254(732)
Hoshino, T. 61(58), 68(58), 107(151), 185(386)
Hoskins, C. 63(66), 243(693)
Hosomi, A. 211(617)
Houk, K. N. 3(2, 3), 25(358), 26(359b), 27(2, 272, 358), 29(253, 254), 30(254), 33(2, 272, 358), 36(2, 272), 38(302), 39(304, 306, 307), 40(313), 55(7), 60(39, 40), 78(46, 47), 183(349), 239(679), 271(1, 2), 276(89)
Howard Jr., E. 84(134)
Howe, R. K. 65(96), 184(361)
Hoyle, W. 13(98)
Höchtlen, A. 64(81), 69(81)
Hörhold, H.-H. 16(142), 147(131), 239(131), 240(131), 241(131)
Hsu, I. H. C. 4(9)
Huang, S.-P. 27(287), 182(337, 339, 340)
Huang, Y. 27(367)
Huber, R. 42(329), 234(670)
Huffman, R. W. 132(44)
Huguenin, R. 86(158), 245(713)
Huhn, G. F. 104(113), 105(113), 110(113), 135(73), 294(81)
Huie, E. M. 202(592)
Huisgen, R. 3, 3(4), 14(107), 15(161), 17(161), 20(172–175), 24(173, 223), 25(172, 173), 26(174, 259, 359a), 27(172, 175, 257, 259), 28(161, 250, 259), 29(173), 30(257), 31(161), 33(172, 250, 259), 34(173), 39(259), 43(173), 44(174), 45(352), 60(48, 51), 61(48), 64(78–80), 82(104), 110(164), 111(176), 133(52), 136(52), 164(187), 184(353, 381), 185(353, 391), 186(353, 423, 428), 187(428, 457, 471, 472), 188(423, 503, 509, 510)
Huldschinsky, I. 88(180), 136(52)
Hussain, S. A. 89(194)
Hwang, D. 23(216), 195(575), 197(575)

Iacobelli, J. A. 292(73)
Iball, J. 76(14)
Ibekeke-Bomangwa, W. 289(108a)
Ichlov, C. 24(235), 216(629), 223(629)
Igeta, H. 16(124)
Ihara, M. 27(287), 182(339–341)
Iida, H. 193(558), 216(626–628), 218(635), 219(639), 291(72)
Iijima, I. 11(69), 12(69), 140(98, 104), 145(104)
Iioni, Y. 276(17)
Iitaka, Y. 16(124), 106(133)
Ikeuchi, S. 187(467)
Ikuta, T. 182(340)
Illig, C. R. 35(265), 38(265), 43(265)
Imai, Y. 184(377)
Imamura, K. 27(335)
Imaye, K. 104(96)
Imhof, R. 22(209), 221(645)
Imoto, E. 187(459)
Inaba, M. 98(41)
Inoue, I. 34(264), 242(688b), 254(688b)
Inoue, K. 279(26)
Inouye, Y. 27(285), 27(365), 85(142), 86(146), 199(581), 240(681, 682)
Invernizzi, A. G. 15(122), 27(279), 38(300, 301), 179(324), 185(414, 416), 235(414)

Invernizzi, L. 184(*365*)
Irngartinger, H. 283(*44*)
Irwin, W. J. 180(*333*)
Isacescu, D. A. 97(*18*), 105(*119*), 106(*119*), 107(*119*)
Ische, F. 82(*110*)
Ishida, H. 230(*659*)
Ishidate, M. 24(*241*)
Ishikawa, H. 106(*130*)
Isler, O. 106(*123*)
Ismael, M. T. 102(*82*)
Isuge, O. 8(*17*)
Ito, S. 24(*236*)
Ito, T. 277(*20*)
Ito, Y. 16(*125*), 171(*260*)
Itoh, F. 296(*113*)
Itoh, K. 209(*604*)
Ius, A. 27(*348*), 184(*363*), 185(*413*)
Ivanov, A. I. 101(*61*), 111(*166*)
Ivanova, R. I. 102(*80*)
Ivshin, V. P. 111(*173*)
Iwakura, Y. 27(*283*), 64(*86*), 65(*86*), 184(*377*)
Iwasaki, H. 244(*702, 703*)
Iwasaki, T. 34(*264*), 242(*688b*), 254(*688b*)
Iwashita, T. 204(*596*), 213(*620*), 215(*625*), 218(*620*)
Iwatani, K. 59(*30*)

Jaccard-Thorndahl, S. 254(*729*)
Jackson, A. H. 160(*169*)
Jackson, J. L. 177(*311*)
Jackson, W. R. 40(*317*)
Jacobs, P. B. 206(*600*)
Jacobsen, N. 245(*710*)
Jacquesy, R. 88(*190*), 191(*541*)
Jacquier, R. 24(*225*), 185(*417, 455*)
Jakobsen, H. J. 78(*37*)
Jansen, F. 64(*88*), 84(*139a*)
Janzen, E. G. 82(*87*), 98(*42*)
Jawdosiuk, M. 102(*78*)
Jäger, V. 2, 15(*115–118*), 16(*153, 154, 157*), 17(*115, 117, 162*), 40(*313, 314*), 42(*116, 117, 153, 154*), 147(*132, 673*), 153(*132*), 193(*561*), 210(*606, 607*), 211(*616*), 235(*672–675*), 236(*674–676*), 239(*675, 676, 679*), 240(*673*), 241(*673*), 244(*707*), 294(*84*)
Jeffs, P. W. 79(*64*)
Jennings, W. B. 77(*32*), 78(*32*), 79(*58*), 85(*32, 144*), 89(*32, 58*)
Jenny, E. 131(*10*), 132(*10*)
Jensen, S. R. 69(*147*), 70(*147*)
Jerina, D. M. 77(*38*)
Jernov, J. L. 98(*32*)
Jerslev, B. 77(*39*), 191(*544*)
Jerzmanowska, Z. 132(*49, 50*)
Jessup, H. A. 179(*326*)
Joffe, S. L. 2, 8(*27, 159*), 16(*159*), 21(*186, 189*), 104(*102*), 105(*102*), 110(*102*), 113(*179, 181–185, 204*), 114(*183*), 115(*183, 185, 206*), 116(*205*), 117(*207*), 118(*205–210*), 119(*179, 207, 211–213*), 210(*610*)
Johnson, W. S. 132(*35, 36*)
Johnstone, L. M. 82(*113*)
Jonathan, N. 111(*169*)
Jones, D. S. 104(*100*)
Jones, F. B. 195(*573*), 197(*573*)
Jones, G. R. 77(*43*)
Jones, R. A. Y. 86(*156*)
Jones, R. H. 27(*288b*), 43(*288b*)
Jones, T. 191(*547*)
Jones, W. M. 104(*115*)
Jordan, M. 88(*191*), 191(*547*)
Joshi, S. S. 169(*248*)
Jotterand, A. 40(*319*), 254(*729*)
Joucla, M. 22(*204*), 27(*266, 282, 290*), 34(*248*), 38(*266*)
Joullie, M. M. 82(*80*), 86(*151*), 87(*80*), 166(*193*), 172(*193, 277*), 291(*71*)
Jovitschitsch, M. 68(*136*)
Joyce, C. J. 66(*122*)
Jørgensen, K. A. 111(*174*), 113(*174*)
Jørgensen, R. D. 8(*25*), 15(*25*), 16(*25*), 21(*25*), 113(*188*), 114(*188*), 120(*188*), 133(*58*), 136(*58*), 147(*58*), 148(*58*), 151(*58*), 155(*58*), 158(*58*), 159(*58*), 178(*58*), 192(*58*)
Judge, C. B. 172(*280*)
Junell, R. 96(*7*)
Jung, R. 272(*4*)
Just, G. 66(*101, 102*), 185(*408*), 254(*731*)
Justoni, R. 9(*33, 36*), 132(*13, 25, 28*)
Jürgens, B. 97(*21*), 107(*146*)
Jürgens, E. 79(*55*)

Kahn, N. 164(*186*)
Kahn, N. A. 68(*131*)
Kai, Y. 8(*26*), 113(*197*), 114(*197*), 115(*197*), 116(*197*), 117(*197*), 118(*197*), 119(*197*)
Kaji, E. 68(*130*), 102(*79*), 106(*130–133*), 107(*150*), 166(*200*), 184(*355*), 185(*355*), 187(*355*)
Kakisawa, H. 4(*13*), 13(*13*), 16(*143*), 27(*285, 365*), 38(*296*), 85(*142*), 86(*146*), 104(*96*), 142(*114*), 176(*306, 307*), 198(*580*), 199(*581*), 204(*596*), 210(*612*), 213(*620*), 215(*625*), 218(*620*), 221(*643*), 240(*680–682*)
Kallury, R. K. M. R. 10(*43*)
Kalvoda, J. 15(*126*), 134(*64*), 135(*64*), 199(*583*)
Kalyuzhnaya, V. G. 9(*31*)
Kamachi, H. 4(*13*), 13(*13*), 176(*306, 307*)
Kametani, T. 16(*140*), 27(*287*), 182(*337–342*), 215(*624*), 219(*640*)
Kaminski, L. S. 82(*116*)
Kamiya, T. 244(*702–705*)
Kamlet, M. J. 96(*14*), 101(*59*)
Kanemasa, S. 8(*17*), 186(*430*), 202(*591*), 235(*591*), 277(*19*)
Kano, H. 18(*163, 167, 168*), 132(*24*),

164(*183, 188*), 168(*208*), 169(*237*),
171(*251, 252, 254*), 175(*295, 296*), 184(*350*),
186(*439, 440*), 187(*439*)
Kanomata, N. 272(*95*)
Kaplan, L. A. 100(*53*)
Kaplan, M. 160(*170, 171*)
Karpeysky, M. Ya. 10(*40*)
Kashima, C. 11(*66, 70*), 12(*70*), 13(*88*),
16(*143*), 140(*92, 95–97, 100, 101*),
141(*95, 112*), 144(*97*), 153(*97*), 178(*319*),
179(*112*), 210(*612*), 240(*680*)
Kashiwagi, K. 173(*284*)
Kashmiri, M. A. 102(*76*)
Kashutina, M. V. 8(*27, 159*), 16(*159*), 21(*186*),
113(*181–183, 185*), 114(*183*), 115(*183, 185*),
116(*205*), 118(*205*), 210(*610*)
Kasahara, K. 291(*72*)
Kasparian, D. J. 68(*133*)
Kassab, N. A. L. 169(*245*), 170(*245*)
Katayama, C. 226(*652*)
Kato, M. 187(*461*)
Katritzky, A. R. 2, 13(*94, 96*), 16(*131*),
84(*135*), 89(*135*), 102(*74–76*), 173(*290*),
174(*291*), 175(*291*)
Kauffman, J. M. 187(*493*)
Kaufman, D. C. 60(*39*), 78(*46*)
Kaufmann, H. 15(*126*), 134(*64*), 135(*64*)
Kawakami, H. 295(*111*)
Kawamura, Y. 276(*13*)
Kawashima, A. 106(*124a*)
Kayen, A. H. M. 86(*149*)
Ke, Y. Y. 296(*113*)
Keana, J. F. W. 89(*196*)
Keckeisen, A. 290(*70*)
Keller, K. 24(*233*), 191(*549*), 201(*549*),
202(*589*)
Keller, T. H. 70(*151*), 113(*201*), 114(*201*),
122(*201*)
Kelly, D. P. 76(*15*)
Kelly, R. C. 244(*699*)
Kemp, D. S. 131(*7*), 179(*328*)
Kemp, J. 191(*546*)
Kempe, U. M. 83(*128*), 189(*527, 528*)
Kemula, W. 101(*60*)
Kerber, R. C. 102(*88, 92*), 104(*94*)
Kerr, D. A. 86(*160*)
Khasapov, B. N. 116(*205*), 118(*205, 208*),
119(*212*)
Khawaga, A. M. 102(*82*)
Khemis, B. 76(*12*), 242(*692*)
Kheruze, Yu. I. 27(*366*)
Khomutov, R. M. 244(*706*)
Khomutova, E. D. 9(*30, 38*), 10(*40, 41*),
11(*30, 41*), 132(*26*), 138(*85*)
Khovstenko, V. I. 77(*44*)
Khripach, V. A. 3, 13(*86*), 146(*127*),
147(*133–135*), 157(*156*)
Kibayashi, C. 193(*558*), 216(*626–628*),
218(*635*), 219(*639*), 292(*72*)
Kienzle, F. 98(*32*), 99(*47*)

Kihara, Y. 27(*283*)
Kijima, M. 12(*76b*), 24(*76b*)
Kim, B. H. 276(*99*), 284(*52*)
Kimura, J. 106(*124a*)
Kimura, K. 81(*77*)
King, G. S. 16(*160*), 210(*611*)
King, J. F. 179(*325*)
King, T. J. 177(*311*)
Kinoshita, M. 254(*734*)
Kinstle, T. H. 172(*269, 273*)
Kinusaga, M. 180(*332*)
Kirby, G. W. 189(*538*)
Kirby, P. 69(*146*), 70(*146*)
Kirchhoff, R. 98(*39*)
Kirmse, W. 281(*38*)
Kishida, M. 27(*375*)
Kissel, M. 57(*20*)
Kita, Y. 296(*113*)
Kitagawa, Y. 40(*322*), 134(*66*), 135(*66*),
136(*66*)
Kitahonoki, K. 167(*204*), 211(*204*)
Kitamura, T. 168(*215*), 171(*215*)
Kitasato, Z. 57(*13*), 60(*13*), 107(*147*)
Kjær, A. 69(*147, 150*), 70(*147*)
Kjeldsen, G. 8(*25*), 15(*25*), 16(*25*), 21(*25*),
113(*188*), 114(*188*), 120(*188*), 133(*58*),
136(*58*), 147(*58*), 148(*58*), 151(*58*), 155(*58*),
158(*58*), 159(*58*), 178(*58*), 192(*58*)
Klebanovich, I. B. 13(*86*), 146(*127*),
147(*133, 134*)
Klein, D. A. 184(*382*)
Klewe, B. 100(*54*)
Kliegel, W. 71(*157*), 76(*16*), 82(*84, 89–91*)
Klimova, V. A. 96(*8*), 98(*36*)
Kline, D. N. 271(*31*), 284(*106*)
Klingele, H. 165(*191*), 166(*191*)
Knaus, E. E. 27(*339*)
Knierzinger, A. 42(*329*), 234(*670*)
Knight, J. 283(*105*)
Knorn, C. 188(*496*)
Knorr, R. 27(*257*), 30(*257*)
Knudsen, J. S. 8(*25*), 15(*25*), 16(*25*), 21(*25*),
113(*188*), 114(*188*), 120(*188*), 133(*58*),
136(*58*), 147(*58*), 148(*58*), 151(*58*), 155(*58*),
158(*58*), 159(*58*), 178(*58*), 192(*58*)
Knunyants, L. L. 166(*194*)
Knust, A. 137(*79*)
Ko, S. S. 250(*722*)
Kobayashi, T. 12(*76a*), 177(*312*), 276(*17*)
Kober, W. 113(*195*), 114(*195*), 156(*153*)
Kobo, R. 182(*340*)
Kobylecki, R. J. 107(*144*)
Kochetkov, N. K. 2, 9(*28, 30, 38, 39*),
10(*40, 41, 42*), 11(*28, 30, 39, 41, 42, 67*),
132(*14, 48*), 138(*85*), 140(*106*), 147(*48*)
Kochs, P. 57(*26, 27a*), 58(*26, 27a*)
Koehler, K. F. 27(*255, 288a*), 43(*288a*),
180(*334*), 271(*3*), 280(*28*)
Kohler, E. P. 12(*75*), 106(*125, 126*)
Koizumi, T. 24(*222*), 41(*324*)

Kojima, M. 168(*223, 227*), 171(*223, 227*)
Kolind-Andersen, H. 245(*710*)
Komatsu, M. 166(*196*), 168(*223*), 187(*452*)
Komendatov, M. I. 168(*225*)
Kondo, K. 168(*217*), 226(*652*)
Kondrateva, G. V. 132(*39*)
Konnik, E. I. 16(*245*), 24(*245*)
Konno, F. 182(*340*)
Konovalov, A. I. 26(*354–356*), 27(*333*), 47(*354–356*), 184(*351*)
Konowalow, M. I. 95(*4*), 97(*4*)
Koppe, M. 66(*117*)
Korchemnaya, E. B. 96(*8*)
Korenevskii, V. A. 113(*183*), 114(*183*), 115(*183*)
Kornblum, N. 98(*23*), 102(*73, 88, 90–92*), 103(*93*), 104(*93, 95*), 105(*95*), 109(*95*), 111(*95*), 185(*403*)
Korobitsyna, J. K. 79(*60*), 85(*60*)
Korte, F. 169(*246*), 170(*246*)
Kost, A. N. 169(*249*)
Kostikov, R. R. 168(*225*)
Kostyanovskii, R. G. 77(*44, 45*), 88(*181*)
Kotera, K. 167(*204*), 211(*204*)
Kouwenhoven, C. G. 16(*123*)
Kovac, J. 22(*212*)
Kovaleva, G. V. 244(*706*)
Koyama, H. 59(*33*)
Koyano, K. 77(*27*), 78(*27, 35*), 85(*141*)
Kozikowski, A. P. 3, 8(*18*), 15(*120, 126*), 16(*136, 137*), 17(*137*), 40(*315, 322*), 134(*66–68*), 135(*66, 67*), 136(*66*), 147(*68, 139*), 148(*139*), 149(*139*), 150(*139*), 166(*202*), 194(*567–569*), 207(*601, 602*), 208(*68, 603, 735*), 224(*653*), 230(*658–661*), 232(*68, 662*), 237(*67*), 248(*721*), 253(*728*), 254(*67, 733*), 292(*76*), 296(*112*)
Kozina, N. D. 27(*376*)
Kozuka, S. 66(*116*), 67(*116*), 79(*61*)
König, E. 80(*68*), 88(*171*)
König, W. A. 240(*683–685*), 290(*70*)
Kössel, A. 68(*135*)
Krägeloh, K. 113(*195*), 114(*195*)
Kraatz, U. 169(*233*)
Krafft, G. A. 188(*507*)
Krasnaya, Zh. A. 106(*129*)
Krasnov, V. L. 188(*524*)
Krawczyk, E. 42(*329*), 234(*669*)
Kreher, R. 86(*153*)
Kresze, G. 188(*519*)
Kreuz, K. L. 99(*44*)
Krimm, H. 79(*56*)
Krishan, K. 185(*393, 401*)
Krishna, A. 279(*101*)
Krishnan, K. 82(*98*)
Krogsgaard-Larsen, P. 244(*709*)
Krohn, W. 164(*181*), 191(*543*)
Kröhnke, F. 79(*50*), 88(*179*)
Kropf, H. 66(*103*), 281(*39*)
Krueger, U. 186(*447*)

Kruse, L. 63(*66*), 243(*693*), 244(*697*)
Kruse, L. I. 244(*698*)
Ku, A. 171(*264*)
Ku, H. 200(*585*)
Kubota, T. 59(*30, 33*), 65(*94*), 78(*48*)
Kucherov, V. F. 106(*129*)
Kudryavtseva, L. F. 132(*39*)
Kuene, M. E. 132(*38*)
Kumagai, T. 276(*13*)
Kumaraswamy, G. 279(*101*)
Kumuzawa, T. 99(*49, 50*)
Kunstmann, R. 188(*497, 498*)
Kupchan, S. M. 157(*157*)
Kurabayashi, M. 184(*358*), 185(*358*), 186(*358*)
Kurita, Y. 107(*148*)
Kurkovskaya, L. N. 82(*103*)
Kurth, R. 290(*70*)
Kurtz, P. 60(*45*)
Kurutz, I. 79(*60*), 85(*60*)
Kusumi, T. 16(*143*), 38(*296*), 104(*96*), 198(*580*), 199(*581*), 204(*596*), 210(*612*), 213(*620*), 215(*625*), 218(*620*), 240(*680*)
Kvagina, L. 24(*227*)
Kwoh, S. 98(*32*)

L'Abbe, G. 27(*350*), 187(*463, 491*), 188(*502*)
Lablache-Combier, A. 24(*224*), 88(*190*), 191(*541, 542*)
La Forge, F. B. 151(*144*)
Lagercrantz, C. 84(*139b*), 98(*31*)
Lagodzinskaya, G. V. 20(*183*), 21(*189, 192*), 30(*192*), 104(*102, 106*), 105(*102, 106*), 106(*138*), 110(*102, 106*), 135(*76*), 136(*76*), 147(*76*)
Lagowski, J. M. 2, 84(*135*), 89(*135*)
Laguna, M. A. 13(*90*), 19(*170*), 283(*48*)
Lahey, F. N. 157(*159, 160*)
Lai, J. 160(*174*)
Lajiness, T. A. 22(*202*), 179(*329*), 195(*572*)
Lakhvich, F. A. 3, 13(*86*), 146(*127*), 147(*133–135*), 157(*156*)
Lambeck, R. 66(*103*), 281(*39*)
Lamberton, A. H. 110(*170, 171*), 111(*170–172*)
Lamchen, M. 2, 22(*198–200*), 75(*23*), 76(*23*), 79(*23*), 81(*23*), 82(*116*), 83(*124*), 86(*23*), 185(*432*)
Lampe, W. 11(*64*), 12(*64*), 141(*111*)
Lander Jr., S. W. 284(*51*)
Landers, H. 12(*73*)
Lang Jr., S. A. 3, 71(*154*), 171(*265*), 173(*265*)
Lange, G. 119(*214*)
Langella, M. R. 13(*93*), 16(*149*)
Lapalme, R. 160(*173, 174*)
Lappenberg, M. 82(*86*)
Lapshina, Z. Ya. 21(*185*), 104(*107*), 105(*107*), 110(*107*)
La Rosa, C. 30(*262*), 34(*262*)
Larsen, B. S. 77(*42*)

Larsen, K. E.   8(*15*), 22(*206*), 24(*206*), 30(*206*), 40(*15*), 64(*97*), 65(*97*), 86(*188*), 88(*188*), 152(*146*), 191(*545*), 239(*146*), 294(*86*)
Larson, H. O.   111(*167*)
Lathbury, D.   155(*151*), 286(*59*)
Lathbury, D. C.   88(*200*)
Lauria, F.   185(*388*)
Lawesson, S.-O.   77(*42*)
Lazar, R.   39(*309*)
Leandri, G.   184(*373, 374*)
LeBel, N. A.   8(*21*), 22(*193, 201–203*), 23(*193, 215, 216*), 24(*240*), 88(*197*), 179(*329*), 193(*554*), 195(*554, 570–575*), 197(*571, 573, 575–577*), 198(*571*), 199(*582, 584*), 216(*570*)
Leboff, A.   272(*5*)
Ledlie, D. B.   179(*329*)
Lee, A.   157(*161*)
Lee, G. A.   65(*92*)
Lee, G. E.   27(*286*), 181(*335*), 213(*619*), 288(*63*)
Lee, H. L.   252(*725*)
Lee, J.   169(*239*)
Lee, T. D.   89(*196*)
Leginus, J. M.   24(*231*), 41(*323*), 154(*149*), 233(*665, 666*), 284(*51*)
Lehn, J. M.   20(*182*)
Le-Hong, N.   40(*319, 320*), 254(*729*)
Lehr, F.   102(*83–85, 87*), 108(*83–85, 87*), 120(*216, 217*), 122(*218*)
Leibzon, V. N.   16(*245*), 24(*245*)
Leichner, L.   131(*10*), 132(*10*)
Leiterman, H.   24(*223*)
Lelj, F.   168(*205*)
Lenaers, R.   66(*104, 105*), 184(*378*), 188(*516*)
Lengyel, J.   188(*498*)
Leonteva, L. M.   2, 113(*179, 204*), 117(*207*), 118(*207*), 119(*179, 207, 211–213*)
LeQuesne, P. W.   157(*158*)
Leuchs, K.   88(*166*)
Levand, O.   111(*167*)
Levchenko, E. S.   188(*518*)
Levery, S. B.   157(*158*)
Levi, S.   82(*108*)
Levin, A. A.   8(*27*), 113(*181, 182*)
Levinger, S.   189(*534*)
Lewis, G. E.   77(*42*)
Ley, H.   57(*20*)
Leyshon, W. M.   81(*76*), 86(*161*)
Li, C. S.   166(*202b*), 208(*735*), 292(*76*)
Li, Y.   271(*2*)
Licandro, E.   28(*252*)
Lidor, R.   17(*119*)
Liebermann, S. V.   106(*122*)
Liguori, A.   22(*196*), 155(*150*)
Lin, Y.-i.   3, 171(*265*), 173(*265*)
Lin, Y. S.   24(*239*)
Linda, P.   13(*99*)
Lindemann, H.   88(*183*), 132(*31*), 174(*292*)
Lindlar, H.   106(*123*)

Liotta, D. C.   79(*67*)
Lippmaa, E. T.   116(*205*)
Lippman, A. E.   166(*192*)
Lipscomb, W. N.   77(*39*)
Liu, K. C.   65(*96*)
Livio, P. F.   254(*729*)
Llamas, K.   283(*50*)
Llewellyn, F. J.   100(*55*)
Lo, S.   166(*195*)
Lobry de Bruyn, C. A.   60(*42*)
Loi, A.   69(*143*), 184(*359*)
Lollar, E. D.   16(*155*), 42(*155*), 210(*613*)
Loncharich, R. J.   271(*2*), 276(*99*)
Lollar-Confalone, E. D.   251(*724*)
Lombard, M. J.   245(*711*)
Longhi, R.   27(*348*), 185(*413*)
Lorberth, J.   119(*214*)
Lorenc, L.   199(*583*)
Love, R. F.   99(*44*)
Lowe Jr., J. U.   184(*372*)
Lowmaster, N. E.   163(*178*)
Luboshnikova, V. M.   11(*67*), 140(*106*)
Lucchesini, F.   131(*6*)
Lucchini, U.   27(*374*)
Lue, X.-G.   283(*43*)
Luger, P.   27(*373*)
Luke, W. K. H.   111(*167*)
Lukyanov, O. A.   21(*187*), 104(*103*), 105(*103*), 110(*103, 163, 168*), 111(*168*), 168(*206*)
Lumma Jr., W. C.   24(*234*), 223(*648*), 225(*648*)
Lund, H.   12(*78*), 16(*78*), 24(*78*)
Lupke, N. P.   186(*442*)
Lusinchi, X.   76(*12, 13*), 79(*13*), 84(*136, 138b*), 88(*138b*), 242(*692*), 281(*32*)
Luskus, L. J.   3(*3*), 55(*7*), 183(*349*)
Lyapkin, N. M.   55(*8*)

MacLeod, J. K.   157(*160*)
Macaluso, A.   2, 75(*22*), 76(*22*), 77(*22*), 79(*22*), 81(*22*), 86(*22*)
Macaluso, G.   175(*302, 305*)
Maccagnani, G.   187(*475, 478, 480*)
Macdonald, J. G.   27(*291*)
Mack, W.   20(*174*), 26(*174, 259*), 27(*259*), 28(*250, 259*), 33(*250, 259*), 39(*259*), 44(*174*), 45(*352*), 64(*78–80*), 184(*353, 381*), 187(*457, 471*), 188(*510*)
Mackey, D.   187(*469*)
Maeda, M.   168(*223, 227*), 171(*223, 227*)
Maeno, S.   27(*375*)
Magdesieva, N. N.   27(*346*)
Magid, L. L.   160(*172*)
Magnus, P. D.   16(*160*), 210(*611*)
Magosch, K. H.   186(*420*)
Maier, G.   278(*100*)
Maignan, J.   187(*483*)
Maimind, V. I.   77(*41*)
Mairanovskii, S. G.   16(*245*), 24(*245*), 96(*8*)
Maiorana, S.   28(*252*)

Mak, A. L. C. 77(43)
Mak, S. 98(42)
Makarenkova, L. M. 104(109), 105(109), 110(109), 113(184, 204), 115(206), 117(207), 118(206–210), 119(207)
Makisumi, Y. 132(24)
Maksimovic, Z. 199(583)
Malhotra, R. 113(202), 114(202)
Maloney Huss, K. E. 248(721)
Manfredi, A. 285(56)
Manfredini, S. 12(76a), 158(165), 276(17), 294(79, 80)
Mangiapan, S. 11(52), 13(52)
Mannafov, T. G. 26(355), 47(355)
Mannhardt, K. 168(221)
Marbet, R. 106(123)
Marchesini, A. 285(56)
Marchetti, L. 82(114)
Markland, M. P. 160(171)
Markov, V. I. 77(45), 88(181)
Markova, J. G. 12(77)
Markowicz, T. 86(148)
Maron, S. H. 96(7)
Marsden, K. 22(197)
Marti, F. 251(723)
Martin, D. 184(375, 376)
Martin, D. G. 63(69)
Martin, N. 285(54)
Martin, R. J. 253(728)
Martin, S. F. 16(138), 32(138), 33(138), 38(138), 150(141)
Martvon, A. 186(448)
Maruishi, T. 142(114)
Marx, M. 251(723)
Marzin, C. 13(99)
Masamune, T. 86(152)
Mascagni, P. 22(210), 55(2), 76(21a)
Mascherpa, A. 27(289)
Maslen, H. S. 100(55)
Maslina, I. A. 119(210)
Massard, R. 40(318), 254(729)
Massini, R. 28(249)
Masuda, H. 27(362), 47(362)
Masuda, K. 166(199)
Masui, M. 22(205), 23(205), 34(261), 77(24), 82(24), 82(82), 186(444)
Mataka, S. 188(521)
Mathys, G. 27(350), 187(463)
Matsinger, M. 271(3)
Matsubara, S. 157(163)
Matsuda, K. 82(118), 169(229)
Matsukubo, H. 187(461)
Matsumoto, G. 295(90)
Matsumura, Y. 279(26)
Matsuoka, H. 168(216), 171(216)
Matsuura, A. 167(204), 171(260), 211(204)
Matsuura, T. 16(125)
Mattauch, B. 283(44)
Matz, M. J. 157(157)
Mavrodiev, V. K. 77(44)

Mayer, H. 9(35), 84(138a), 88(138a, 172), 137(82)
Mazzanti, G. 187(473, 475, 478, 480)
Mazzu, A. 200(585)
McAlduff, E. J. 39(304)
McCauley, J. P. 42(372)
McCoy, R. E. 105(120), 106(120)
McDouall, J. J. W. 271(114)
McGreer, D. E. 75(6)
McGregor, D. N. 13(84), 145(126), 173(283)
McIntosch, C. L. 172(269)
McKenna, J. I. 11(71c)
McKillop, A. 107(144)
McMurry, J. E. 4(13), 13(13), 79(66), 98(37), 138(86), 139(88, 89), 145(86)
McPhail, A. T. 24(230), 202(590), 273(10)
McVeigh, P. H. 98(24)
Meckler, H. 35(265), 38(265), 44(265), 216(634), 276(18), 288(65)
Medyantseva, E. A. 75(5)
Meek, J. S. 106(124b)
Meier, H. 188(523)
Meier, J. P. 78(48)
Meinke, P. T. 188(507)
Meinwald, J. 22(194)
Meisenheimer, J. 79(67)
Menacherry, M. D. 157(158)
Menard, M. 227(655)
Menke, J. R. 24(240), 197(577)
Menou, C. 132(43)
Merlini, L. 132(23)
Metelli, R. 176(309)
Metz, J. T. 271(2)
Metzger, H. 98(28)
Meunier, J. 227(655)
Meyer, V. 66(118), 98(22, 27)
Meyer, W. L. 132(44)
Meyers, A. I. 3, 4(7)
Micetich, R. G. 11(57), 33(260), 140(94, 110), 141(110)
Michalska, M. 80(69), 173(282)
Michel, G. 88(164)
Michel, R. E. 102(88)
Mickler, W. 168(221)
Middleton, D. M. 66(125)
Middleton, W. J. 55(5), 67(127)
Miescher, K. 186(433)
Migliara, O. 160(168), 178(322), 184(347)
Mihalovic, M. L. 199(583)
Mihaly, E. 168(212), 254(729)
Miki, M. 63(73b)
Milcent, R. 185(407)
Milhart, P. 272(96)
Miller, B. J. 40(316), 63(67), 244(695)
Milliet, P. 84(136)
Milowsky, A. S. 295(110)
Minami, T. 27(364)
Mini, V. 64(83)
Minkin, V. I. 75(5)
Mishchenko, A. I. 77(44, 45), 88(181)

Mitchell, W. R.   55(6), 66(*110, 112*), 67(*110*)
Mitsuhashi, K.   27(*362, 375*), 47(*362*)
Mitsui, H.   88(*175, 176*)
Mitsunobu, O.   106(*124a*)
Mittag, T. W.   22(*200*)
Miyashita, M.   99(*48–50*), 155(*152*)
Miyazaki, M.   11(*69*), 12(*69*), 140(*98, 104*), 145(*104*)
Miyazaki, R.   18(*167, 168*), 164(*188*), 168(*208*)
Mizuguchi, R.   171(*252*), 186(*430*)
Moccia, R.   59(*31*)
Moerli, H.   82(*86*)
Moersch, G. W.   134(*61, 63*), 136(*63*)
Moffat, J. G.   254(*730*)
Mohan, S.   185(*402*)
Moiseenkov, A. M.   169(*242*)
Molina, G.   79(*64*)
Mondelli, R.   188(*499*)
Monk, P.   245(*712*)
Monroe, P. A.   106(*124b*)
Montavon, M.   106(*123*)
Morgenstern, H.   86(*153*)
Morgenstern, K.   60(*35*)
Mori, Y.   77(*33*)
Morita, K.   88(*189*), 191(*540*)
Moroder, F.   144(*117*), 153(*117*), 247(*716*), 294(*79, 80*)
Morozova, N. S.   21(*184*), 104(*105*), 105(*105*), 110(*105, 160*)
Morris, D. G.   80(*72*)
Morris, R.   227(*655*)
Morrocchi, S.   11(*68*), 12(*68*), 13(*83*), 140(*102, 103*), 145(*102, 103*), 184(*360*), 185(*360, 389, 406*), 186(*389*), 187(*360, 460, 464, 465*)
Morrow, S. D.   63(*72*), 102(*81, 82*)
Morton, W. D.   39(*311*)
Moses, S. R.   40(*313*), 239(*679*), 271(*1*)
Mosher, H. S.   108(*155*)
Moskal, J.   272(*96*)
Motoki, S.   81(*77*)
Motoyoshiya, J.   166(*197*)
Moye, A. J.   98(*42*)
Mucchi, L.   294(*79*)
Muckensturm, B.   293(*78*)
Mueller, P. H.   26(*359b*)
Muench, A.   188(*505*)
Mugrage, B. B.   207(*601*)
Mukai, N.   11(*70*), 12(*70*), 140(*100*), 141(*112*), 179(*112*)
Mukai, T.   273(*13*)
Mukaiyama, T.   61(*58*), 68(*58*), 107(*151*), 119(*215*), 185(*386*)
Mukerji, S. K.   15(*134*), 16(*134*), 21(*134*), 113(*187, 190*), 114(*187, 190*), 152(*145*), 154(*148*), 155(*145*), 156(*145*), 158(*145*), 160(*145*), 164(*148*), 179(*145*), 192(*145*), 193(*145*)
Mukhametova, D. Ya.   82(*97*)
Mukherjee, D.   78(*47*)

Mukhopadhyay, T.   102(*86a*), 108(*86a*), 113(*199*), 120(*199*), 129(*4*), 130(*4*)
Mukuta, T.   16(*139*), 113(*191, 192*), 114(*191, 192*), 246(*715*), 247(*717*)
Mullane, M.   64(*77*)
Mullen, G. B.   23(*213, 217*), 216(*631, 632*)
Mumm, O.   10(*45*), 137(*77–80*)
Munchausen, L. L.   26(*359b*), 60(*39*), 78(*46*)
Murahashi, S.-I.   88(*175, 176*), 279(*102*)
Murai, N.   166(*196*)
Murthy, K. S. K.   27(*377*), 289(*69*)
Musante, C.   11(*51, 55*), 132(*29, 46*), 140(*29*), 169(*240, 241*), 170(*241*), 187(*492*)
Mustafa, A.   169(*245*), 170(*245*)
Musumarra, G.   102(*75*)
Müller, E.   57(*22*)
Müller, H.   88(*164*)
Müller, I.   2, 16(*157*), 27(*330*), 147(*673*), 235(*672–674*), 236(*674*), 240(*673*), 241(*673*)
Müller, K.   20(*179*)
Müller, W.   169(*233*)
Münchmeyer, G.   10(*45*), 137(*77*)
Myagi, M. Ya.   116(*205*)
Myszkiewicz, T. M.   83(*122*)
Mzengeza, S.   86(*159*), 244(*708*)

Nadolsky, K.   184(*375*)
Nagagawa, N.   277(*19*)
Nagahara, T.   182(*339, 341, 342*)
Nagy, J. O.   40(*316*), 63(*67*), 244(*695*)
Nair, V.   185(*399*)
Nakamura, H.   106(*133*)
Nakano, A.   168(*215, 216*), 171(*215, 216*)
Nakashima, S.   187(*495*)
Nakayama, A.   182(*337, 340*)
Nakayama, Y.   182(*340*)
Nambu, H.   119(*215*)
Nambu, Y.   12(*76b*), 24(*76b*)
Nametkin, S.   98(*33*)
Napolitano, M.   283(*47*)
Narang, S. C.   113(*202*), 114(*202*)
Narasaka, K.   210(*615*)
Narita, S.   24(*236*)
Nash, R.   76(*14*)
Natale, N. R.   11(*53, 74c*)
Nazarova, E. B.   88(*193*)
Nef, J. U.   57(*19*), 69(*138*), 95(*5*)
Neidlein, R.   186(*436*)
Neill, D. C.   77(*32*), 78(*32*), 81(*74, 75*), 85(*32, 143, 145*), 89(*32*)
Neiman, L. A.   77(*41*), 83(*121*)
Nekrasov, Yu. S.   77(*41*)
Nelson Jr., S. D.   68(*133*)
Nemoto, N.   106(*124a*)
Nenitzescu, C. D.   97(*18*), 105(*119*), 106(*119*), 107(*119*)
Nerdel, F.   88(*180*)
Nesi, R.   276(*15, 98, 115*)
Neuklis, W. A.   134(*61, 63*), 136(*63*)
Newland, R. J.   24(*240*), 197(*577*)

Newton, G. 110(*170*), 111(*170*)
Niazi, U. 271(*114*)
Nicholas, E. S. 66(*111*), 135(*74*)
Nickel, G. W. 57(*15*), 60(*15*)
Nicolaides, D. N. 27(*281*), 187(*462*)
Nielsen, A. T. 2, 95(*1*), 96(*1*), 97(*1*), 99(*46*), 101(*1*), 106(*1, 128*), 107(*1*), 110(*1*)
Niklas, K. 19(*169*), 168(*210*)
Nikonova, L. A. 106(*140*)
Nilsson, N. H. 187(*484*)
Nilsson, T. 98(*31*)
Nimitz, J. S. 108(*155*)
Niou, C.-S. 11(*71c*)
Nishihara, Y. 27(*331*)
Nishitani, T. 34(*264*), 242(*688b*), 254(*688b*)
Nishiyama, H. 209(*604*)
Nishiwaki, T. 3, 166(*201b*), 168(*211, 214–218, 220*), 169(*220, 243*), 171(*215, 216, 220*)
Nitta, M. 12(*76a*), 27(*276, 338*), 177(*312*), 272(*95*), 276(*17*)
Niwa, H. 81(*77*)
Niwayama, S. 27(*365*)
Noble Jr., P. 101(*57*)
Noël, D. 187(*482*)
Noland, W. E. 96(*20*), 98(*24*)
Nold, A. E. 132(*42*)
Nomura, Y. 82(*99*)
Norman, R. O. C. 23(*218*)
Nour-el-din, A. M. 168(*209*), 280(*31*)
Novikov, S. S. 20(*180*), 21(*184, 185–187, 189, 191, 192*), 30(*192*), 96(*8, 10, 12, 13*), 101(*61*), 104(*101–107*), 105(*101–107*), 106(*134–138, 140*), 107(*13*), 110(*101–107, 159, 163*), 111(*166, 173*), 135(*76*), 136(*76*), 147(*76*), 185(*404*)
Novikov, V. M. 119(*212*), 168(*206*)
Nozaki, H. 157(*163*)
Nussberger, G. 174(*291*), 175(*291*)

Oae, S. 79(*61, 62*)
Obayashi, M. 88(*189*), 191(*540*)
Obrecht, J.-P. 42(*329*), 234(*670*)
Ochiai, M. 88(*189*), 100(*51b*), 191(*540*)
Odell, B. G. 76(*8*), 77(*40*)
Oediger, H. 66(*115*)
Oesterloe, T. 113(*195*), 114(*195*)
Ogata, K. 132(*24*)
Ohashi, M. 4(*13*), 13(*13*), 139(*87, 90*), 142(*114*), 176(*306, 307*)
Ohkata, K. 39(*308*), 173(*284*)
Ohki, T. 209(*604*)
Ohloff, G. 157(*163*)
Ohshiro, Y. 166(*196*), 187(*452*)
Ohta, A. 98(*41*)
Ohyama, Y. 173(*284*)
Oinuma, H. 221(*643*)
Ojha, N. D. 24(*240*), 197(*577*)
Oka, O. 244(*702*)
Olah, G. A. 98(*40*), 113(*202, 203*), 114(*202, 203*), 122(*203*)

Olmsted, M. P. 13(*89*)
Olofson, R. A. 9(*35*), 10(*46*), 11(*35, 48, 49*), 137(*81, 82*), 175(*298*), 180(*331*)
Olszewski, W. F. 84(*134*)
Omata, A. 27(*276, 338*)
Omote, Y. 11(*70*), 12(*70*), 141(*112*), 178(*319*), 179(*112*)
Omoto, O. 140(*101*)
Omran, S. M. A. R. 185(*396*), 187(*396*)
Onan, K. D. 24(*230*), 202(*590*)
Onishchenko, A. A. 104(*109*), 105(*109*), 106(*138*), 110(*109, 159, 168*), 111(*168*)
Ono, H. 79(*53*), 85(*53*)
Onomura, S. 168(*217*)
Ooi, N. S. 86(*147, 155*)
Oppolzer, W. 2, 24(*233, 243*), 191(*549*), 193(*556*), 201(*549, 588*), 202(*589*), 223(*649, 650*), 225(*649*), 226(*649*), 227(*657*)
Oravec, P. 27(*376*)
Orlich, I. 80(*69*), 173(*282*)
Orlova, T. 89(*195*)
Orsolini, P. 294(*79*)
Ortega, A. G. 169(*231*)
Osawa, A. 16(*124*)
Oshima, K. 157(*163*)
Osmanski, P. S. 57(*14*), 60(*14*)
Oster, T. A. 244(*707*)
Otomasu, H. 215(*624*)
Otsuji, Y. 187(*459*)
Otsuka, M. 245(*712*)
Ottenheim, H. C. J. 285(*58*)
Overberger, C. G. 184(*379*)
Owens, K. 102(*77*)
Owens, M. 283(*50*)
Øksne, S. 16(*131*)

Paddon-Row, M. N. 39(*306, 307*), 271(*2*)
Padeken, H. G. 98(*26*), 104(*26*), 119(*26*)
Padwa, A. 2, 3, 3(*5*), 7(*5*), 27(*255, 288a, 291*), 43(*288*), 171(*264, 267, 268*), 180(*334*), 193(*555*), 195(*555*), 196(*555*), 199(*585*), 200(*585*), 271(*3, 93*), 279(*27*), 280(*28*), 284(*106*)
Paetzold, P. I. 188(*511*)
Palazzo, C. F. 60(*52*)
Pallie, K. D. 254(*729*)
Palotti, M. 184(*374*)
Pan, Y. 131(*11*), 132(*11*)
Pandey, G. 279(*101*)
Panfil, I. 22(*326*), 24(*246*), 27(*363*), 41(*325*), 42(*326*)
Panizzi, L. 16(*130*), 57(*16*), 60(*16, 50*), 132(*15, 16, 18*), 133(*56*), 147(*129*)
Papadakis, I. 185(*407*)
Papaleo, S. 276(*15a, 115*)
Pappo, R. 98(*26*), 104(*26*), 119(*26*)
Paquette, L. A. 39(*308*)
Parini, C. 27(*348*), 134(*65*), 184(*363*), 185(*413*)
Park, K. P. 14(*108*), 102(*71*)

Park, P.-U. 232(*662*)
Parkash, G. K. S. 98(*40*)
Parker, C. 102(*70*)
Parker, D. R. 77(*34*), 80(*34*)
Parker, W. 157(*162*)
Parpani, P. 271(*92*)
Parsons, P. J. 155(*151*), 283(*105*)
Partis, R. A. 187(*458*)
Patel, R. C. 102(*74*)
Paton, R. M. 55(*6*), 66(*110, 112, 121, 122*), 67(*110*), 184(*354*)
Pattenden, G. 206(*599*)
Paul, K. G. 131(*7*)
Paul, R. 16(*150*), 61(*59*)
Paulus, E. F. 40(*314*), 235(*674*), 236(*674, 676*), 239(*676*)
Pearl, H. K. 184(*361*)
Pearson, D. A. 27(*267*), 38(*267*), 79(*59*), 110(*165*)
Pechet, M. M. 76(*10*), 79(*63*)
Pedrini, P. 187(*475*)
Pedrosa, R. 169(*231*)
Pelosi, P. 60(*47*)
Pennings, M. L. M. 77(*25*), 82(*25, 92–96*), 84(*25*), 182(*343*)
Perekalin, V. V. 102(*80*)
Perez, J. D. 283(*116*)
Perkins, M. J. 89(*194*)
Perold, G. W. 16(*146*), 42(*146*), 167(*203*), 210(*203*)
Perotti, A. 27(*278*)
Perret, F. 40(*319*), 254(*729*)
Perrini, G. 27(*342*), 184(*357*), 185(*357*), 278(*22*)
Perron, Y. G. 227(*655*)
Perry, D. A. 110(*162*), 113(*200*), 114(*200*), 122(*200*), 187(*476*)
Perumattan, J. 284(*106*)
Pesso, J. 82(*107, 108*)
Peters, E.-M. 27(*374*)
Peters, K. 27(*374*)
Petersen, J. W. 132(*36*)
Petrov, A. A. 24(*226, 227*), 27(*366*)
Petrovskii, P. V. 9(*31*), 13(*97*)
Petrus, C. 16(*128*), 185(*417, 454, 455*)
Petrus, F. 16(*127, 128*), 24(*225*), 185(*454, 455*)
Petrusevich, I. I. 147(*135*)
Petruso, S. 13(*87*)
Petrzilka, M. 24(*243*), 189(*530*), 223(*649, 650*), 225(*649*), 226(*649*)
Pettit, G. R. 4(*12*)
Pevarelli, P. 273(*9*)
Pezzanite, J. O. 295(*90*)
Philip, R. 60(*43*)
Piacenti, F. 9(*37*), 13(*79, 95*)
Piancatelli, G. 4(*10*)
Pillay, M. K. 63(*65, 71, 72*), 135(*71, 72*), 136(*71*), 243(*694*)
Pink, P. 103(*93*), 104(*93*)
Pinkey, J. T. 67(*128b*)

Pinkus, J. L. 179(*326*)
Pino, P. 9(*32, 37*), 13(*79, 91*), 60(*47*), 131(*5*)
Pinzauti, S. 11(*65*), 140(*107*)
Pioch, D. 27(*293*)
Piozzi, F. 169(*240*)
Pistikopoulos, P. 27(*344*)
Pivawer, P. M. 99(*43*)
Piyasena, H. P. 276(*99*)
Pizzolato, G. 16(*155*), 42(*155*), 210(*613*), 251(*724*), 252(*725*)
Plate, R. 285(*58*)
Plenkiurce, J. 184(*369*)
Plescia, S. 13(*87*), 160(*168*), 178(*321, 322*), 184(*346, 348*)
Pleshkova, A. P. 77(*44, 45*)
Pless, J. 86(*158*), 245(*713*)
Poliacikova, J. 22(*212*)
Polievktov, M. K. 12(*77*)
Pollini, G. P. 11(*56*), 12(*76a*), 14(*109, 110*), 144(*116–118*), 145(*119*), 153(*116–119*), 158(*165*), 193(*119, 553*), 247(*716*), 292(*74*), 294(*80*)
Polniaszek, R. P. 63(*68*), 244(*701*)
Polo, E. 292(*74*)
Polonski, T. 82(*85*)
Polyanskaya, A. S. 102(*80*)
Poncet, J. 254(*729*)
Ponti, P. P. 188(*499*)
Ponticelli, F. 168(*226*), 169(*226, 247*), 276(*16*)
Ponzio, G. 59(*29*), 65(*89*)
Porter, A. 104(*94*)
Post, M. E. 23(*216*), 195(*570*), 216(*570*)
Považanec, F. 186(*448*)
Pozdeev, A. G. 147(*133*)
Pramanik, B. 273(*10*)
Prandtl, W. 63(*64*)
Prasad, M. 248(*720*)
Prasit, P. 38(*298*), 194(*563–565*)
Pratt, R. N. 166(*193*), 172(*193, 279*)
Preobashchenskii, N. A. 104(*99*)
Preu, L. 82(*90*)
Price, D. T. 42(*372*), 104(*113*), 105(*113*), 110(*113*), 135(*73*), 294(*81, 82*)
Princ, B. 76(*17*)
Prokofiev, E. P. 106(*129*)
Prosyanik, A. V. 77(*44, 45*), 88(*181*)
Prout, K. 245(*712*)
Puar, M. S. 24(*230*), 200(*587*), 201(*587*), 202(*590*), 273(*10*)
Pucci, D. G. 180(*330*)
Pudussery, R. G. 18(*166*), 164(*180*), 168(*180*), 173(*180*)
Puglis, J. M. 287(*60*), 288(*62*)
Pulido, F. J. 11(*71a*), 13(*90*), 19(*170*), 283(*48*)
Purchase, R. 23(*218*)
Purrello, G. 27(*342*), 184(*356, 357*), 185(*356, 357*), 187(*489*)
Pushkov, V. A. 77(*41*)
Putsykin, Yu. G. 89(*195*)
Putt, S. R. 108(*154*)

Pyaterikov, V. F.   113(*204*), 117(*207*), 118(*207*), 119(*207*)

Quang, L. Vo   27(*360*)
Quang, Y. Vo   27(*360*)
Quartara, L.   276(*15a, 115*)
Quartieri, S.   27(*292*)
Querci, A.   176(*308*)
Quilico, A.   2, 3, 9(*29, 33, 36*), 11(*50*), 13(*80, 92*), 16(*148*), 28(*249*), 57(*16, 17*), 60(*16, 17, 49, 50*), 61(*49*), 64(*82*), 68(*134*), 132(*17, 19, 23, 25, 28, 29, 46*), 133(*53, 56*), 140(*29*), 145(*120, 123, 124*), 157(*120*), 178(*120*), 187(*464, 466*)
Quincy, D. A.   11(*53*)
Qureshi, A. K.   18(*166*), 78(*36*), 82(*36*), 83(*123*), 88(*123*), 164(*180*) 168(*180*), 173(*180*), 186(*434*), 187(*434*)

Raasch, M. S.   187(*488*)
Raban, M.   195(*573*), 197(*573*)
Rackow, S.   184(*375*)
Radau, M.   187(*451*)
Rae, J. D.   22(*208*), 27(*208*), 38(*299*)
Raffauf, R. F.   157(*158*)
Rahman, A.   68(*131*)
Rai, M.   185(*393, 402*)
Raimondi, L.   39(*368, 369*), 239(*677, 688*), 271(*91*), 275(*12*), 294(*83*)
Rajagopalan, P.   186(*418*), 188(*508, 514, 515*)
Rajappa, S.   27(*351*)
Rambaud, R.   27(*293*)
Rao, C. N. R.   101(*56, 63*)
Raphael, R. A.   157(*162*)
Rastelli, A.   27(*292*)
Rastetter, W. H.   88(*165*)
Ratcliffe, M. H. G.   188(*520*)
Rathmann, R.   290(*70*)
Reamer, R. A.   22(*211*), 76(*18*)
Reed, W. L.   101(*57*)
Rees, C. W.   111(*175*), 177(*315*), 185(*400*)
Regenass, R. A.   82(*111*), 88(*111*)
Regitz, M.   188(*506*)
Reich, W.   102(*69*)
Reid, E. B.   160(*171*)
Reinhardt, H. G.   281(*33*)
Reinhoudt, D. N.   16(*123*), 77(*25*), 82(*25, 92–96*), 84(*25*), 182(*343*), 280(*30*)
Reisdorff, J.   251(*723*)
Reisenegger, C.   57(*12*), 60(*12*)
Reissig, N.-U.   26(*359a*)
Reiter, L. A.   171(*257*), 285(*117*)
Rembarz, G.   184(*366*)
Renato, C.   27(*348*)
Rennie, R. A. C.   66(*121*)
Renzi, G.   11(*65*), 140(*107*), 169(*244*)
Repke, D. B.   254(*730*)
Restelli, A.   39(*369*), 42(*371*), 150(*142*), 239(*678*), 294(*83*)
Rheinboldt, H.   67(*88*)

Ricca, A.   11(*68*), 12(*68*), 13(*83, 92*), 64(*75*), 140(*102, 103*), 145(*102, 103, 120, 121, 122*), 157(*120, 155*), 178(*120*), 185(*360, 389, 406*), 186(*389*), 187(*360, 460, 464, 465*)
Richter, R.   65(*90, 93*), 186(*425*)
Richtmeyer, N. K.   12(*75*)
Rickett, R. M. W.   104(*100*)
Riddel, F. G.   20(*182*)
Riediker, M.   189(*535*)
Righetti, P. P.   86(*154*)
Rinckenberger, A.   102(*67*)
Risitano, F.   186(*421*), 187(*421*)
Riss, B. P.   293(*78*)
Rivest, P.   227(*655*)
Robb, M. A.   271(*114*)
Roberts, J. D.   59(*32*)
Roberts, S. M.   13(*98*)
Robertson, A. D.   120(*217*)
Robertson, J. E.   88(*182*)
Robinson, G. C.   40(*321*), 294(*87*)
Rochlitz, J.   177(*313*)
Rodina, L. L.   79(*60*), 85(*60*)
Rodriguez, A.   27(*255, 288a*), 43(*288a*), 180(*334*)
Rogers, M. A. T.   22(*197*), 88(*168*)
Rondan, N. G.   38(*302*), 39(*306, 307*), 40(*313*), 239(*679*), 271(*2*)
Ronen-Braunstein, I.   82(*107, 117*)
Roques, R.   27(*293*)
Ros, F.   102(*78*)
Rose, G.   102(*69*)
Rose, J. D.   104(*98*), 105(*98*), 110(*98*)
Rosen, P.   16(*156*), 42(*156*), 98(*32*), 211(*614*), 246(*614*)
Rosen, T.   208(*735*), 292(*76*)
Rosenfeld, B.   69(*142*)
Ross, W. J.   140(*93*)
Ross-Petersen, K. J.   191(*544*)
Rossi, M. L.   27(*361*)
Rösch, N.   188(*506*)
Roush, W. R.   227(*656*)
Ruccia, M.   173(*285*), 175(*302, 303*), 178(*321*)
Rudebaugh, R. K.   194(*566*)
Rukasov, A. F.   89(*195*)
Rundel, W.   2, 8(*20*), 75(*1*), 79(*51*), 81(*1*), 82(*1*), 84(*1*), 86(*1, 51*), 89(*1, 51*)
Rupe, H.   88(*167*)
Ruschig, H.   134(*62*)
Russell, G. A.   98(*42*), 102(*77, 78, 89*)
Russell, M. A.   241(*687*)
Rüegg, R.   106(*123*)
Rühter, G.   187(*481*)
Ryser, G.   106(*123*)
Rzepa, H. S.   16(*160*), 210(*611*)

Saba, G.   69(*143*)
Sachs, F.   88(*178*)
Safir, S. R.   178(*318*)
Saft, M. S.   102(*81*)
Saginova, L. G.   14(*100, 101*)

Sagitdinova, Kh. F. 82(97)
Saimoto, H. 157(163)
Saito, T. 168(211, 217, 218)
Sakai, T. E. 184(362)
Sakamoto, T. 10(44)
Sakurai, H. 211(617)
Sakurai, Y. 24(241)
Sala, A. 8(16), 193(559), 209(559), 283(47)
Salem, L. 33(273), 113(178)
Samokhvalov, G. I. 104(99)
Samuilov, Ya. D. 26(354–356), 27(333), 47(354–356), 184(351)
Sandhu, J. S. 185(402)
Sandmeyer, R. 251(723)
Sarti-Fantoni, P. 60(46), 276(15b, 98)
Sasaki, T. 66(107), 175(301), 184(362, 367, 368), 185(394), 186(419, 424, 438), 187(438, 486), 198(578, 579), 226(652)
Sato, H. 104(96)
Sato, N. 86(152)
Sato, T. 107(148), 168(228)
Sato, Y. 38(296), 198(580), 199(581)
Saucy, S. 106(123)
Savides, C. 101(63)
Savostyanova, I. A. 20(180), 21(185, 188), 104(104, 107), 105(104, 107), 106(135, 137), 110(104, 107)
Sav'yalov, S. I. 132(39)
Sayigh, A. A. R. 175(297)
Scanga, S. A. 135(69), 136(69)
Scarpati, R. 59(31)
Scettri, A. 4(10)
Scevola, L. 38(301)
Schaumann, E. 187(481, 490)
Schechter, H. 98(30), 101(58)
Schechter, M. S. 151(144)
Schenk, C. 86(150), 186(446)
Scherer, W. 57(12), 60(12)
Schiesser, G. A. 295(90)
Schimpf, R. 186(429)
Schlegel, H. B. 271(114)
Schletter, I. 244(699)
Schlewer, G. 244(709)
Schmalle, H. 240(685)
Schmid, H. 168(224), 171(224, 263), 172(275)
Schmidt, E. 63(64)
Schmidt, G. 164(182)
Schmidt, M. 76(21b), 279(23, 24)
Schmitz-Dumont, O. 64(88)
Schohe, R. 2, 16(157), 17(162), 40(313, 314), 147(673), 235(673, 675), 236(675, 676), 239(675, 676, 679), 240(673), 241(673), 294(84)
Scholl, R. 69(139)
Schow, S. R. 248(719), 292(75)
Schöllkopf, U. 11(63), 182(344)
Schramm, D. 169(246), 170(246)
Schroll, G. 77(42), 188(496)
Schröter, D. 2, 294(84)

Schröter, R. 294(84)
Schuierer, E. 57(21), 61(54)
Schultze, O. W. 95(4), 97(4)
Schuster, K. E. 88(192)
Schwab, H. 177(313)
Schwab, W. 15(116, 118), 16(154, 157), 42(116, 154), 147(673), 211(616), 235(673), 240(673), 241(673)
Schwartz, M. A. 204(595, 597), 205(595, 597)
Schwartz, T. R. 38(302)
Schwarz, H. H. 98(26), 104(26), 119(26)
Schwerzel, T. 76(21b)
Scripco, J. G. 166(202b), 253(728)
Sebastian, C. 40(319)
See, C. 39(308)
Seebach, D. 8(26), 99(51a), 102(83–87), 108(83–87), 113(196–199), 114(196–199), 115(197), 116(197), 117(197), 118(197), 119(196–198), 120(199, 216, 217), 122(218), 129(4), 130(4)
Segnitz, A. 98(26), 104(26), 119(26)
Seidl, H. 20(173–175), 24(173), 25(173), 26(174), 27(175, 257), 29(173), 30(257), 34(173), 43(173), 44(164), 110(164), 164(187), 186(428), 187(428)
Seidl, V. 273(10)
Selby, I. A. 164(185)
Seligman, R. 97(16)
Selva, A. 188(500), 187(464, 465)
Semper, L. 68(137)
Senaratne, P. A. 27(286), 35(265), 38(265), 44(265), 181(335), 216(634), 276(18)
Sennewald, K. 56(11), 60(11), 62(61), 63(64)
Senning, A. 187(484)
Seno, M. 187(467)
Seoane, C. 285(54)
Sereno, J. F. 205(598), 292(73)
Sergeeva, T. A. 27(346)
Setsu, M. 27(283)
Sevastyanova, T. K. 82(88)
Severin, E. S. 244(706)
Severin, T. 104(114), 109(114)
Shabarov, Yu. S. 14(100, 101)
Shakhova, M. K. 104(99)
Shalom, E. 17(119), 189(533, 537)
Shapiro, B. L. 88(177)
Sharma, S. C. 8(25), 15(25), 16(25), 21(25), 113(186, 188), 114(188), 116(186), 120(188), 133(58), 136(58), 147(58), 148(58), 151(58), 155(58), 158(58), 159(58), 170(58), 178(58), 192(58)
Sharma, A. H. 89(194)
Sharma, K. K. 15(134), 16(133, 134), 21(134), 113(190), 114(190), 147(137, 138), 148(137, 138), 151(137, 138), 152(137, 145), 155(145), 156(137, 145), 158(138, 145), 160(145), 178(137), 179(145), 192(137, 145), 193(137, 145), 239(137)
Sharma, N. D. 79(58)
Sharp, I. 40(317)

Shashkov, A. S. 118(*208, 209*)
Shatzmiller, S. 17(*119*), 189(*531–534, 536, 537*)
Shaul'skii, Yu. M. 26(*353*), 47(*353*)
Shaw, C. C. 140(*110*), 141(*110*)
Shaw, G. 13(*82*), 178(*317*)
Shaw, E. 283(*49*)
Shcherban, A. I. 79(*60*), 85(*60*)
Shelberg, W. E. 132(*35*)
Shelton, B. R. 65(*96*)
Shemyakin, M. M. 77(*41*)
Shen, Y. 27(*367*)
Shevchuk, M. I. 188(*501*)
Shevelev, S. A. 96(*10, 12, 13*), 101(*61, 64*), 106(*64*), 107(*13*)
Shibafuchi, H. 68(*132*), 277(*21*)
Shimizu, T. 63(*73b*), 66(*123, 124a*), 67(*132*), 106(*124a*), 191(*550*), 277(*20, 21*)
Shinoda, M. 157(*163*)
Shinkai, I. 22(*211*), 76(*18*)
Shiota, T. 88(*176*), 279(*102*)
Shiozawa, A. 10(*44*)
Shiraishi, S. 27(*334, 335, 340*), 64(*86*), 65(*86*), 184(*377*), 187(*467*)
Shiro, M. 59(*33*)
Shishido, K. 219(*638, 640*)
Shitkin, V. M. 8(*27*), 113(*181–183*), 114(*183*), 115(*183*), 116(*205*), 118(*205*)
Shitov, O. P. 119(*211–213*)
Shiue, C.-Y. 14(*108*)
Shlyapochnikov, V. A. 55(*8*), 101(*61*)
Shlykova, N. I. 21(*187*), 104(*103*), 105(*103*), 110(*103*)
Shoji, H. 211(*617*)
Shono, T. 279(*26*)
Shpak, S. T. 188(*501*)
Shriner, R. L. 107(*141*), 108(*141*)
Shvekhgeimer, G. A. 104(*108*), 105(*108*), 110(*108*)
Sigel, C. W. 157(*157*)
Siles, S. 223(*649*), 225(*649*), 226(*649*)
Silverman, R. B. 244(*700*)
Simchen, G. 113(*194, 195*), 114(*194, 195*), 281(*37*)
Simmons, T. 99(*44*)
Simonetta, M. 68(*134*)
Simoni, D. 12(*76a*), 144(*117, 118*), 145(*119*), 153(*117–119*), 158(*165*), 193(*553*), 247(*716*), 276(*17*), 292(*74*), 294(*79, 80*)
Simonyan, L. A. 185(*390*), 187(*390*)
Sims, J. 3(*2, 3*), 27(*2*), 29(*254*), 30(*254*), 33(*2*), 36(*2*), 55(*7*), 60(*39*), 78(*46*), 183(*349*)
Sindona, G. 22(*196*), 155(*150*), 165(*190*), 168(*205*)
Singh, A. 185(*393, 401*)
Singh, B. 168(*213, 219*), 171(*213, 219, 262*)
Singh, J. 185(*401*)
Singh, N. 82(*98*), 185(*402*)
Singh, S. M. 63(*65*), 243(*694*)
Singleton, D. H. 158(*164*)

Sirrenberg, W. 88(*173*), 186(*437*)
Sivanandaiah, K. M. 132(*37*)
Sklarz, B. 18(*166*), 78(*36*), 82(*36*), 83(*123*), 88(*123*), 164(*180*), 168(*180*), 173(*180*), 186(*434*), 187(*434*)
Skolimowski, J. 86(*148*)
Skowronski, R. 86(*148*)
Sletzinger, M. 22(*211*), 76(*18*)
Sliwa, W. 66(*126*)
Slovetskii, V. I. 96(*10, 12, 13*), 101(*61*), 107(*13*), 111(*166*)
Slusarczuk, G. M. J. 22(*201*), 197(*576*)
Smagin, S. S. 21(*191*), 104(*101*), 105(*101*), 110(*101*), 185(*404*)
Smalley, R. K. 132(*33*)
Smets, G. 187(*491*), 188(*502*)
Smith III, A. B. 248(*719*), 292(*75*)
Smith, F. A. 169(*239*)
Smith, K. H. 160(*169*)
Smith, L. I. 106(*127*)
Smith, P. A. S. 77(*26*), 80(*26*), 88(*182*)
Smits, J. M. M. 285(*58*)
Smolinska, J. 11(*64*), 12(*64*), 141(*111*)
Snider, B. B. 224(*651*), 225(*651*)
Snowden, R. L. 223(*649*), 225(*649*), 226(*649*)
Sobhy, M. 185(*396, 397*), 187(*396, 397*)
Sobtsova, N. I. 104(*108*), 105(*108*), 110(*108*)
Sokolov, S. D. 2, 9(*28, 31, 39*), 11(*28, 39, 42, 67*), 12(*77*), 13(*97*), 132(*48*), 140(*106*), 147(*48*)
Solov'eva, S. E. 26(*354–356*), 27(*333*), 47(*354–356*), 184(*351*)
Somanathan, R. 185(*400*)
Sone, K. 82(*118, 119*), 169(*229*)
Soto, J. L. 285(*54*)
Souchay, P. 66(*108*)
Spellmeyer, D. C. 271(*2*)
Spence, G. G. 79(*52*)
Spenser, J. L. 14(*111*)
Spenser, J. T. 188(*507*)
Speroni, G. 11(*50*), 13(*91, 95*), 59(*31*), 60(*49*), 61(*49, 53*), 145(*124*), 184(*364*)
Spevak, P. 140(*110*), 141(*110*)
Spinelli, D. 173(*285*), 175(*303*)
Splitter, J. S. 79(*53*), 85(*53*)
Sportoletti, G. 134(*65*), 184(*363*)
Springer, J. P. 40(*322*), 134(*66*), 135(*66*), 136(*66*), 207(*602*)
Sprio, V. 13(*87*), 160(*167, 168*), 169(*232*), 173(*289*), 178(*320–322*), 184(*346, 347*)
Spurlock, L. A. 23(*215*), 197(*576*)
Sreenivasan, R. 27(*351*)
Srivastava, R. M. 185(*395*), 186(*395*)
Stache, U. 134(*61*)
Stadlwieser, J. 99(*47*)
Stagno D'Alcontres, G. 13(*81*), 57(*17*), 60(*17*), 132(*19*), 133(*53, 54*), 169(*230*), 170(*230*), 187(*466*)
Staib, R. R. 29(*256*), 30(*256*), 43(*256*)
Stamm, H. 16(*158*), 76(*20*), 82(*100, 101, 115, 120*)

Stark, B. P.   188(*520*)
Staudinger, H.   186(*433*)
Stefl, E. P.   107(*142*)
Stein, P. D.   8(*18*), 16(*136*), 194(*567, 568*), 230(*658*)
Stein, S. J.   244(*699*)
Steinkopf, W.   97(*21*), 107(*146*)
Stener, A.   169(*240*)
Stepanyants, A. U.   119(*212*)
Steppan, W.   113(*195*), 114(*195*)
Sternberg, J. A.   281(*34, 35*)
Steudle, H.   16(*158*), 76(*20*), 82(*101, 120*)
Stevens, R. V.   13(*85*), 27(*263*), 34(*263*), 63(*68*), 65(*95*), 181(*336*), 244(*701*), 285(*57*)
Stevens, T. O.   184(*370*)
Stevens, T. S.   88(*174*), 160(*170–175*), 162(*175*)
Stiles, M.   108(*152*)
Stock, R.   9(*34*), 131(*8*), 132(*8*)
Stohr, G.   188(*511*)
Stokes, D. P.   166(*193*), 172(*193, 279, 280*)
Stork, D. G.   98(*34*)
Stork, G.   4(*13*), 13(*13*), 139(*87–89*), 140(*91*), 176(*306, 307*)
Stork, K.   185(*405*)
Storr, R. C.   111(*175, 177*), 185(*400*)
Stournas, S.   180(*331*)
Stracke, H.-U.   164(*181, 182*)
Stradi, R.   30(*262*), 34(*262*)
Strakov, A. J.   169(*242*)
Streith, J.   187(*453*)
Stribanyi, L.   27(*376*)
Strom, L. E.   68(*129*)
Strozier, R. W.   3(*2*), 27(*2*), 33(*2*), 36(*2*)
Stubbs, M. E.   77(*38*), 81(*75*)
Stühmer, W.   210(*608, 609*)
Stytsenko, T. S.   106(*129*)
Su, T.-M.   79(*53*), 85(*53*)
Suda, K.   22(*205*), 23(*205*), 34(*261*), 82(*82*), 186(*444*), 279(*25*)
Suga, S.   277(*19*)
Suginome, H.   86(*152*)
Sugiyama, N.   140(*95*), 141(*95*)
Sugizaki, M.   106(*124a*)
Sugowdz, G.   13(*82*), 178(*317*)
Sultanov, A. Sh.   77(*44*)
Sunada, Y.   244(*703*)
Sunday, B. R.   24(*230*), 202(*590*)
Surov, I.   12(*78*), 16(*78*), 24(*78*)
Sustmann, R.   14(*107*), 27(*269*), 28(*250*), 33(*250*), 33(*269*)
Sutherland, J. K.   83(*131*)
Sutherland, I. O.   82(*79*), 84(*79*), 87(*79*), 88(*79*)
Sutor, D. J.   100(*55*)
Suvorov, N. N.   82(*103*)
Suzuki, H.   77(*27*), 78(*27, 35*), 283(*42*)
Suzuki, M.   215(*625*)
Suzuki, S.   16(*143*), 210(*612*), 240(*680*)
Suzuki, T.   184(*362*), 198(*578, 579*), 226(*652*)
Suzuki, Y.   27(*287*), 175(*301*), 186(*419*)

Svetkin, Ya. V.   26(*353*), 47(*353*)
Swanson, G. C.   204(*595*), 205(*595*)
Swigor, J. E.   13(*84*), 145(*126*), 173(*283*)
Swoboda, P.   60(*38*)
Szeimies, G.   168(*221*)

Tabakovic, J.   12(*78*), 16(*78*), 24(*78*)
Tacconi, G.   86(*154*)
Tagaki, S.   132(*21, 40*)
Takahashi, K.   106(*132, 133*)
Takahashi, S.   27(*285*), 38(*296*), 86(*146*), 184(*350*), 186(*439, 440*), 187(*439*), 198(*580*), 199(*581*), 240(*681*)
Takano, S.   219(*638*)
Takano, Y.   167(*204*), 211(*204*)
Takasuka, M.   59(*30*)
Takatsu, N.   215(*624*)
Takeda, T.   285(*57*)
Takei, H.   16(*139*), 113(*191*), 114(*191*), 246(*715*), 247(*717, 718*), 295(*111*)
Takeuchi, Y.   3, 20(*171*), 23(*171*), 71(*155*), 82(*99*), 147(*128*)
Talaty, E. R.   98(*42*)
Tamagaki, S.   79(*61, 62*)
Tamura, O.   296(*113*)
Tamura, Y.   296(*113*)
Tanaka, I.   59(*30*), 65(*94*), 85(*141*)
Tanaka, J.   226(*652*)
Tanaka, K.   27(*362, 375*), 47(*362*), 219(*640*)
Tanaka, M.   216(*626*), 219(*639*)
Tanaka, T.   11(*69*), 12(*69*), 140(*98, 104*), 145(*104*)
Tane, J. P.   88(*165*)
Tangthongkum, A.   88(*191*), 191(*547*)
Taniguchi, T.   66(*124a*)
Tartakovskii, V. A.   2, 8(*27, 159*), 16(*159*), 20(*180*), 21(*184–189, 191, 192*), 30(*192*), 101(*61*), 104(*101–107, 109*), 105(*101–107, 109*), 106(*134–138, 140*), 110(*101–107, 109, 159, 160, 163, 168*), 111(*166, 168*), 113(*179, 181–184, 204*), 114(*183*), 115(*183, 185, 206*), 116(*205*), 117(*207*), 118(*205–210*), 119(*179, 207, 211–213*), 135(*76*), 136(*76*), 147(*76*), 168(*206*), 185(*404*), 210(*610*), 281(*36*)
Tartkovski, E.   17(*119*)
Tashchi, V. P.   89(*195*)
Tashiro, M.   27(*331*), 186(*430*), 188(*521*)
Tauer, E.   172(*276*)
Taya, N.   140(*98*)
Taylor, E. C.   79(*52*), 177(*314, 316*)
Taylor, G. A.   58(*27b*), 166(*193, 198*), 172(*193, 278–281*)
Taylor, H. J.   102(*73*)
Taylor, W. J.   101(*58*)
Tchelitcheff, S.   16(*150*), 61(*59*)
Tedeschi, P.   168(*226*), 169(*226*), 276(*15b, 16, 98*)

Author Index

Tegeler, J. J. 23(*213*), 216(*632, 633*), 218(*636*)
Teles, J. H. 278(*100*)
Teramura, K. 63(*73b*), 66(*123, 124a*), 68(*132*), 191(*550*), 277(*20, 21*)
Tette, J. P. 24(*229*), 212(*618*)
Thakur, R. S. 82(*112*), 83(*132*), 88(*132*)
Thal, C. 272(*5*)
Thesing, J. 84(*138a*), 88(*138a, 164, 169, 172, 173*), 186(*437*)
Thestrup-Pedersen, K. 70(*153*)
Thianpantangul, S. 191(*548*)
Thiele, J. 12(*73*), 57(*12*), 60(*12*)
Thijs, L. 187(*479, 480*)
Thomas, A. 66(*126*)
Thomas, C. B. 23(*218*)
Thomas, E. 113(*199*), 120(*199*), 129(*4*), 130(*4*)
Thomas, E. J. 15(*120*), 27(*288b*), 40(*321*), 42(*288b*), 252(*726*), 294(*87*)
Thomas, T. 88(*191*)
Thompson, A. S. 248(*719*), 292(*75*)
Thompson, N. J. 77(*38*)
Thompson, W. J. 242(*688a*), 254(*688a*)
Thomsen, I. 65(*158*), 159(*166*), 227(*166*), 272(*6*), 285(*107*), 294(*109*), 296(*6*)
Thomson, R. H. 83(*125*)
Thorndal-Jaccard, S. 40(*318, 319*)
Thorneycroft, F. J. 75(*4*)
Thuc, L. V. 200(*586*)
Thurston, J. T. 107(*141*), 108(*141*)
Tiberio, T. 11(*60*), 132(*47*), 169(*238*)
Tice, C. M. 23(*219*), 242(*689*)
Ticozzi, C. 188(*499*)
Tiernan, P. L. 77(*31*)
Tikhonov, A. Ya. 82(*83*)
Tilley, J. N. 175(*297*)
Timms, G. H. 16(*144*), 147(*130*)
Tischler, S. A. 142(*113*)
Tobe, S. 140(*95*), 141(*95*)
Tobias, R. S. 60(*38*)
Todd, A. 76(*8*), 79(*57*), 82(*79*), 84(*79*), 87(*79*), 88(*79*)
Todd, Lord 82(*112*), 83(*130, 132*), 88(*132*)
Toi, N. 185(*394*)
Toma, L. 272(*7*)
Tomioka, Y. 271(*93*)
Tomoda, S. 82(*99*)
Tonnard, F. 27(*290*), 104(*111*), 110(*111*)
Toppet, S. 175(*304*)
Tori, K. 77(*33*)
Torigoe, M. 24(*241*)
Tornetta, B. 175(*299*)
Torssell, K. B. G. 8(*24, 25*), 14(*114*), 15(*25, 114, 134*), 16(*24, 25, 114, 132–134, 145*), 17(*24, 25, 114, 132*), 21(*24, 25, 114, 134, 145, 190*), 22(*206*), 24(*24, 206*), 30(*24, 114, 206*), 39(*132, 312*), 40(*15, 312*), 61(*60*), 64(*97*), 65(*97, 98, 158*), 75(*7*), 76(*7*), 78(*37*), 86(*188*), 88(*7, 188*), 98(*31*), 106(*139*), 110(*139, 158*), 111(*174*),
113(*158, 174, 186–190, 193*), 114(*158, 187–190, 193*), 115(*158*), 116(*186*), 119(*158*), 120(*158, 188, 193*), 129(*1*), 133(*57–60*), 136(*58*), 147(*1, 58, 59, 136–138*), 148(*1, 58, 136–138*), 150(*59, 60*), 151(*57–59, 137, 138*), 152(*137, 145, 146*), 154(*148*), 155(*58, 145*), 156(*137, 145*), 158(*58, 59, 138, 145*), 159(*58, 59, 166*), 160(*145*), 164(*57, 148*), 170(*250*), 178(*58, 137*), 179(*145*), 191(*545*), 192(*58, 136, 137, 145*), 193(*136, 137, 145*), 227(*166*), 238(*60*), 239(*59, 60, 137, 146*), 272(*6*), 276(*6, 85*), 280(*29*), 281(*40*), 285(*107*), 294(*29, 85, 86, 109*), 296(*6*)
Tosolini, G. 185(*388*)
Toy, A. 242(*688a*), 254(*688a*)
Trager, W. F. 68(*133*)
Treibs, A. 88(*170*)
Trickes, G. 188(*523*)
Trimarco, P. 27(*361*)
Trimmer, R. W. 172(*270, 272*)
Tronchet, J. 40(*319*)
Tronchet, J. M. J. 40(*318, 319, 320*), 168(*212*), 254(*729*)
Trybulski, E. J. 23(*213*), 24(*242*), 216(*630, 632*), 222(*646*), 223(*630*)
Tsai, J. H. 254(*732*)
Tschaen, D. M. 273(*11*)
Tschang, K. T. 88(*183*)
Tsuda, Y. 11(*66, 70*), 12(*70*), 140(*96, 97, 100, 101*), 141(*112*), 144(*97*), 153(*97*), 179(*112*)
Tsuge, O. 27(*331*), 82(*118, 119*), 166(*199*), 169(*229*), 186(*430*), 188(*521*), 202(*591*), 235(*591*), 277(*19*)
Tsuji, A. 16(*124*)
Tsujii, Y. 187(*459*)
Tsujimoto, T. 279(*25*)
Tufariello, J. J. 2, 8(*23*), 22(*195*), 23(*213, 217*), 24(*229, 242, 244*), 27(*286*), 30(*274*), 31(*274*), 35(*265*), 38(*265*), 44(*265*), 75(*3*), 79(*3*), 80(*3*), 82(*3*), 86(*3*), 165(*191*), 166(*191*), 181(*335*), 185(*384*), 188(*384*), 212(*618*), 213(*384, 619*), 214(*621, 622*), 215(*623*), 216(*630–634*), 217(*384, 623*), 218(*384, 636, 637*), 222(*646*), 223(*630*), 276(*18*), 287(*60*) 288(*62–65*), 289(*118*), 295(*110*)
Tulegenova, N. K. 188(*524*)
Turnbull, D. 96(*7*)
Turnovska-Rubaszewska, W. 101(*60*)
Turs, V. E. 55(*8*)
Turyan, Ya. J. 96(*9*)
Tyler, P. C. 194(*564*)
Tyler, P. J. 38(*298*)
Tylor, C. M. B. 22(*197*)
Tyurin, Yu. M. 96(*9*)

Uccella, N. 22(*196*), 155(*150*), 165(*189, 190*), 168(*205*)
Uchida, H. 98(*41*)

Uchida, Y.   66(*116*), 67(*116*)
Uemori, M.   140(*101*)
Ueno, K.   8(*17*), 202(*591*), 235(*591*)
Ueyanagi, J.   244(*702, 703*)
Ugarov, B. N.   188(*518*)
Ukaji, Y.   210(*615*)
Ullman, E. F.   168(*213*), 171(*213, 262*)
Ulrich, H.   175(*297*)
Umani-Ronchi, A.   188(*499, 500*)
Umezawa, B.   187(*495*)
Umezawa, S.   254(*734*)
Uno, K.   27(*283*), 64(*86*), 65(*86*), 283(*42*)
Urano, S.   82(*118, 119*), 169(*229*)
Urbanski, T.   101(*62*), 107(*143, 145*), 108(*143*)
Urrea, M.   227(*657*)
Urry, G. W.   102(*92*)
Uskokovic, M. R.   16(*155*), 42(*155, 327*), 205(*598*), 210(*613*), 232(*664*), 251(*724*), 252(*725*), 292(*73*)
Utsino, S.   57(*13*), 60(*13*)
Utzinger, G. E.   82(*111*), 88(*111, 162*)
Uzawe, J.   140(*98*)

Vagurtova, N. M.   9(*39*), 11(*39*)
Vail, P. D.   63(*72*), 135(*72*)
Vajna de Pava, O.   32(*275*), 145(*122*), 168(*222*)
Van Eijk, P. J. S. S.   280(*30*)
Valentini, F.   64(*85*), 66(*109*)
Van den Heufel, W. J. A.   79(*65*)
Van der Drift, J. A. M.   86(*150*), 186(*446*)
Vander Meer, R. K.   180(*331*)
Van der Plas, H. C.   3, 173(*286*)
Vangehr, K.   27(*373*)
Van Hummel, G. J.   77(*25*), 82(*25, 92, 93*), 84(*25*), 182(*343*)
Van Meerssche, M.   64(*76*)
Van Tamelen, E. E.   4(*12*), 16(*147*)
Vasella, A.   24(*220, 221, 237, 238*), 42(*328, 329*), 86(*157, 158*), 194(*562*), 234(*667-670*), 245(*713, 714*)
Vasilev, A. F.   88(*193*)
Vasileva, L. I.   27(*341*)
Vaughan, W. R.   14(*111*)
Vecchia, F. D.   294(*79*)
Vecchietti, V.   185(*388*)
Vecchio, G.   27(*348*), 184(*364*), 185(*413*)
Vedejs, E.   110(*161, 162*), 113(*200*), 114(*200*), 122(*200*), 187(*476, 477*)
Vederas, J. C.   12(*72*), 142(*115*), 153(*115*)
Veenstra, G. E.   187(*480*)
Veglio, C.   189(*513*)
Veil, P. D.   102(*81*)
Veit, A.   97(*15*)
Velezhava, V. S.   14(*112*), 82(*103*)
Velo, L.   184(*360*), 185(*360, 389*), 186(*389*), 187(*360*)
Venkataraman, M. K.   271(*3, 93*)
Verducci, J.   16(*127*), 24(*225*)

Veronesi, P.   235(*671*)
Vialle, J.   187(*482, 483*)
Vicentini, C. B.   14(*110*), 153(*116*), 193(*553*)
Villareal, J. A.   80(*71, 73*)
Villaume, M. L.   24(*224*), 88(*190*), 191(*541, 542*)
Vinick, F. J.   194(*566*), 131(*11*), 132(*11*)
Vintani, C.   188(*512*)
Visser, R.   280(*30*)
Vita Finzi, P.   14(*104*), 32(*275*), 66(*106*), 145(*120*), 157(*120*), 178(*120*), 184(*380*), 187(*380*)
Vitali, D.   38(*302*)
Vivona, N.   173(*285*), 175(*302, 303, 305*), 178(*321*)
Voeffray, R.   24(*221, 238*), 42(*329*), 86(*158*), 234(*667*), 245(*713, 714*)
Volodarskii, L. B.   82(*83, 88*)
Von Auwers, K.   132(*34, 42*)
Von Reiche, F. V. K.   16(*146*), 42(*146*), 167(*203*), 210(*203*)
Von Schickh, O.   98(*26*), 104(*26*), 119(*26*)
Von Schnering, H. G.   27(*374*)
Voznesenskii, V. N.   77(*44, 45*)
Vulfson, N. S.   77(*41*)
Vyas, D. M.   244(*696*)

Wada, M.   254(*734*)
Wade, P. A.   42(*372*), 63(*65, 70-72*), 102(*81, 82*), 104(*113*), 105(*113*), 110(*113*), 135(*70-73*), 136(*71*), 243(*694*), 294(*81, 82*)
Wagner, G. W.   108(*153*)
Wagner, J.   20(*182*)
Wakefield, B. J.   3, 9(*58*), 11(*58*), 64(*84*), 71(*156*), 171(*266*), 173(*266*)
Waldner, A.   285(*57*)
Walentowski, R.   88(*192*)
Walser, H.   174(*294*)
Walts, A. E.   227(*656*)
Walz, K.   57(*23*)
Wamhoff, H.   169(*246*), 170(*246*), 171(*258*)
Wang, B. C.   207(*601, 602*), 230(*661*)
Waring, L. C.   85(*143*)
Warren, J. P.   59(*32*)
Wasylichen, R. E.   77(*38*)
Watanabe, H.   166(*199*)
Watanabe, N.   283(*42*)
Watanabe, T.   88(*175*)
Watanabe, Y.   27(*285*), 86(*146*), 216(*627, 628*), 240(*681*)
Waters, J.   102(*70*)
Waters, W. A.   24(*228*)
Watson, K. G.   186(*431*), 187(*474, 494*)
Watts, C. R.   3(*3*), 55(*7*), 183(*349*)
Weber, A.   60(*44*)
Weber, H. P.   201(*588*)
Weber, W. P.   113(*180*)
Wegler, R.   187(*487*)
Wegmann, H.   227(*657*)
Wehrli, A.   22(*209*)

Wehrli, H.   221(*645*)
Weidner, M.   188(*496*)
Weiler, L.   142(*113*)
Weinreb, S. M.   29(*256*), 30(*256*), 43(*256*), 273(*11*)
Weinstein, B.   132(*41*)
Weinstein, F.   79(*67*)
Weintraub, P. M.   77(*31*)
Weise, A.   184(*376*)
Weisler, L.   102(*66*)
Weller, T.   120(*216*), 122(*218*)
Wenis, E.   169(*239*)
Wentland, M. P.   160(*172*)
Wentrup, C.   61(*57*), 163(*177*)
Werner, A.   63(*65*), 64(*74*)
West, W.   113(*195*), 114(*195*)
Whang, J. J.   8(*21*), 193(*554*), 195(*554, 570*), 216(*570*)
Wheeler, O. H.   77(*28*)
White, E. H.   107(*149*), 108(*149*)
White, J. D.   295(*90*)
Whitney, R. A.   66(*111*), 86(*159*), 135(*74*), 244(*708*)
Whittle, R. R.   273(*11*)
Wicks, G. E.   271(*941*)
Widdowson, D. A.   83(*131*)
Wieland, H.   11(*61*), 57(*12, 13, 24*), 59(*28*), 60(*12, 13*), 61(*56*), 64(*81*), 68(*137*), 69(*81, 142*), 107(*147*)
Wierenga, W.   244(*699*)
Wildschmidt, E.   16(*144*), 147(*130*)
Wiley, R. H.   64(*84*)
Wilk, M.   177(*313*)
Wilkinson, D. J.   157(*162*)
Will, J.   57(*12*), 60(*12*)
Willard, P. G.   285(*57*)
Willbrand, A. M.   205(*597*)
Williams, D. J.   241(*687*)
Williams Jr., F. T.   98(*30*), 101(*58*)
Williams, M. J.   287(*61*)
Williams, W. M.   185(*409*)
Wilson, D. A.   81(*76*), 86(*147, 155, 160, 161*)
Wilson, J. D.   222(*647*)
Wilson, V. E.   79(*58*)
Wilson, W.   102(*69*)
Winnewisser, B. P.   60(*37*)
Winnewisser, M.   60(*34, 36, 37*)
Winterfeldt, E.   164(*181, 182*), 191(*543*)
Winternitz, F.   132(*43*)
Winther, F.   60(*37*)
Winzenberg, K. N.   248(*719*), 288(*64, 65*), 292(*75*)
Wislicenus, W.   97(*17*)
Witkop, B.   221(*642*)
Wittle, E. L.   134(*63*), 136(*63*)
Wittmann, D. K.   102(*76*)
Wittwer, R.   88(*167*)
Wolfart, P.   140(*110*), 141(*110*)
Wolff, G.   187(*453*)
Wollweber, H.-J.   163(*177*)

Wollenberg, R. H.   193(*560*), 206(*560*)
Wong, C. M.   98(*38*)
Wong, D.   166(*195*)
Wong, S. C.   23(*213*), 216(*632, 633*)
Wong, G. S. K.   27(*255*), 279(*27*)
Woodman, D. J.   10(*47*), 138(*83*)
Woodward, D. R.   39(*311*)
Woodward, R. B.   3(*1*), 9(*35*), 10(*46*), 11(*35, 48, 49*), 137(*81, 82*), 138(*83, 84*), 175(*298*), 179(*328*)
Woolhause, A. D.   185(*400*)
Wovkulich, P. M.   42(*327*), 205(*598*), 232(*664*)
Wöhler, L.   60(*44*)
Wragg, A. H.   88(*174*)
Wren, D.   98(*35*)
Wright, D. J.   3, 9(*58*), 11(*58*), 71(*156*), 171(*266*), 173(*266*)
Wright, J. J.   24(*230*), 202(*590*)
Wright, R.   22(*197*)
Wu, Y.-D.   38(*302*), 40(*313*), 239(*679*), 271(*1, 2*)
Wulff, J.   82(*104*), 186(*423*), 188(*423, 503*)
Wunderlin, D. A.   283(*116*)
Wunsche, K.-H.   179(*327*)
Wurster, C.   98(*22*)
Wünsch, K.-H.   132(*32*)

Xu, Z.-B.   207(*601, 602*), 230(*661*)

Yajima, Y.   27(*276, 338*)
Yakovlev, J. P.   106(*129*)
Yamaguchi, K.   25(*358*), 27(*358*), 33(*358*)
Yamaguchi, M.   34(*261*), 186(*444*)
Yamakawa, M.   59(*30, 33*), 65(*94*)
Yamamoto, I.   166(*197*)
Yamamoto, K.   168(*228*)
Yamamoto, M.   140(*95*), 141(*95*)
Yamamoto, Y.   11(*66, 70*), 12(*70*), 27(*283*), 140(*97*), 141(*112*), 144(*97*), 153(*97*), 173(*284*), 178(*319*), 179(*112*)
Yamanaka, Y.   10(*44*)
Yamauchi, M.   22(*205*), 23(*205*), 82(*82*), 279(*25*)
Yamazaki, E.   175(*295, 296*)
Yanami, T.   99(*48, 50*), 155(*152*)
Yankelevich, A. Z.   8(*27*), 113(*181*)
Yaroslavski, I. S.   82(*103*)
Yasuda, H.   132(*20–22, 40*)
Yeh, H. J. C.   77(*38*)
Yelland, L. J.   69(*145*), 70(*145, 151*), 113(*201*), 114(*201*), 122(*201*)
Yen, H.-K.   63(*72*), 104(*113*), 105(*113*), 110(*113*), 135(*72, 73*), 294(*81*)
Yeung Lam Co, Y. Y. C.   188(*505*)
Yijima, C.   22(*205*), 23(*205*), 34(*261*), 77(*24*), 82(*24*), 82(*82*), 186(*444*), 279(*25*)
Ykman, P.   188(*502*)
Yokohama, S.   27(*287*)
Yokoyama, A.   132(*40*)
Yoshida, A.   187(*459*)

Yoshii, E.   24(*222*), 41(*324*)
Yoshikoshi, A.   99(*48–50*), 155(*152*)
Yoshimura, Y.   77(*33*)
Yoshioka, T.   66(*107*), 175(*301*), 184(*367, 368*), 186(*419, 424*), 187(*486*)
Younas, M.   68(*131*)
Young, A.   111(*167*)
Yousif, M. M.   166(*199*)
Yudintseva, I. M.   9(*31*), 13(*97*)
Yuen, P.-W.   224(*653*)
Yusuf, H. M.   111(*172*)

Zabrodena, A.   98(*33*)
Zabrodina, K. S.   98(*36*)
Zaitsev, P. M.   96(*9*)
Zamkanei, M.   187(*485*)
Zanarotti, A.   187(*460, 465*)
Zapulsky, P.   186(*448*)
Zawalski, R. C.   82(*87*)
Zähner, H.   240(*683*), 290(*70*)
Zbaida, S.   82(*102, 106–108*), 188(*504*)
Zecchi, G.   8(*16*), 28(*252*), 66(*99, 100*), 193(*559*), 209(*559, 605*), 271(*92*)
Zeeh, B.   57(*22*)
Zehetner, W.   66(*102*)
Zeifman, U. V.   185(*390*), 187(*390*)
Zeller, P.   106(*123*)
Zen, S.   68(*130*), 102(*79*), 106(*130–133*), 107(*150*), 166(*200*), 184(*355*), 185(*355*), 187(*355*)
Zenki, S.-i.   88(*175, 176*)
Zenou, J.-L.   189(*533*)
Zerner, B.   283(*50*)
Zeuthen, O.   8(*24*), 16(*24*), 21(*24*), 24(*24*), 30(*24*), 61(*60*), 75(*7*), 76(*7*), 88(*7*), 110(*158*), 113(*158*), 114(*158*), 115(*158*), 119(*158*), 120(*158*), 129(*1*), 133(*57*), 147(*1*), 148(*1*), 151(*57*), 164(*57*), 280(*29*), 294(*29*)
Zheng, J.   27(*367*)
Zhukova, S. V.   83(*121*)
Zhvirblis, V. E.   10(*42*), 11(*42*)
Zibuck, R.   79(*65*)
Zimmerman, G.   240(*685*)
Zinner, G.   186(*441–443, 447, 449, 450*)
Zutter, U.   285(*57*)
Zwanenburg, B.   187(*475, 479, 480*)
Zweig, A.   168(*219*), 171(*219*)

Subject Index

Acivicin, 242
L-Acosamine, 232
Acrolein, β-acylated, 151, 152
Acrylic esters, β-acylated, 151
Acrylonitrile, 154, 164
   β-silyloxy, 283
Adalia bipunctata, 221
($\pm$)-Adaline, 221
Agroclavine I, 227
Aldehydes, homologation, 154
Aldols, 17, 147–151
Alkaloids, 209, 212–232, 285–290
Alkyl nitronates
   cyclic, 106
   cycloaddition, 107, 110, 137
   N,O-dialkyl-N-nitronates, 110–112
   fragmentation, 105
   geometric isomerism, 105
   IR spectra, 111
   mechanism of cycloaddition, 25
   NMR spectra, 105
   physicochemical properties, 109
   reactions, 105, 109
   reactivity, 110
   reduction, 110
   regioselectivity of cycloaddition, 27, 30
   stability, 105, 109
   synthesis, 101, 104, 107, 281
   UV spectra, 111
Allethrolone, 192
Allosedamine, 217
Amanita species, 242
Amaryllis alkaloids, 222
Amines
   allylic, 283
   homoallylic, 212, 283
Amino acids, 239–246, 290, 291
β-Amino acids, 245
α-Amino-3-chloro-4,5-dihydroisoxazole-5-acetic acid (AT-125), 242, 243
5-Amino-5-deoxyidose, 235
1,2-Aminoalcohols, 129-131, 284
1,3-Aminoalcohols, 17, 24, 209–212
Aminoisonitrile, N,N-dialkylated, 57
2-Amino-oxime-O-trialkylsilyl ether, 281

Anabaena flos aquae, 216
Anatoxin-a, 216
Anilines, 145
Antibiotic ($\pm$)-A 26771 B, 247
L-Asparagine analogues, 245
L-Aspartic acid analogues, 245
Asymmetric induction, 25, 39–44, 233, 235, 237, 238, 245, 271, 276
Avermectins, 248
Azetidines, 180, 276
Aziridines, 18, 82, 83, 165, 167–169
   N-chloro, 20
   N-methoxy, 20
Azirines, 167–169

Bacillus species, 233
Beckmann's rearrangement, 79, 172
Benzenesulfonylnitrile oxide, 63
Benzisoxazoles, 172, 174
Biheteroaromatics, 285
Bikaverin, 142
Biotin, 251
α-Bisabolol, 204
Blastmycinone, 254
Bromonitrile oxide, 63, 244, 283
Bullatenone, 157
Bullvalene, 273
Butadiene,
   acylation, 151
   cyanation, 151
Di-t-Butylacetonitrile oxide, 56

Calythrone, 192
Captopril analogues, 245
Carbocyclic systems, *see* Cycloalkanes
Carbohydrates, 232–239, 283, 291, 296
β-Carbolines, 285
Carbonyl compounds
   α,β-unsaturated, 17, 19, 144, 151–156, 234, 284
   β,γ-unsaturated, 284
1,3-Carbonyl transposition, 142, 153
Carboxy-hydroxylation, 134–137, 283
Chanoclavine I, II, 227
Chloronitrile oxide, 63, 243

α-Chloronitrones, 189
Claisen's reaction, 17
Claviceps alkaloids, 227–231
Cocaine, 216
Coleus forscolii, 292
Compactin, 208, 292
Coriolin, 203
Corrins, 160–162, 286
Costaclavine, 227
Crispatic acid, 292
(±)-Croalbinecine, 213, 288
Crotalaria species, 289
(±)-Cryptoleurine, 218
Cumulated double bonds, 172, 271, 287
Curcumin, 12, 141
Cyanates, 55
2-Cyanoalcohols, 15, 131, 134–136, 150, 283
Cyanohydroxylation, 15, 131, 134–136, 150, 283
α-Cyanoketones, 11, 131–134, 170, 283
Cycloalkanes
  bridged, 192–209
  condensed, 192–209
Cyclopentane derivatives, 192–209
  chiral, 194
Cyclopropanes, 14, 106, 276
Cycloreversion, 15, 16, 22, 23, 216, 272
Cycloserine, 242

α- and β-Damascones, 142
Darlingia darlingiana, 287
L-Daunosamine, 232, 296
(±)-Decaline, 219
Dehydrofukinone, 139
Deoxyaldoses, 232–239
2-Deoxyaldoses, 237
D,L-2-Deoxygalactose, 237
Deoxyhexoses, 239
Deoxyketoses, 237
D-2-Deoxyribose, 237
D,L-2-Deoxyribose, 238, 283
Detoxinine, 290
D,L-1,3-Dideoxyfructose, 238
1,5-Dienes, 295
D,L-Digitoxose, 238
Dihydrocinerolone, 192
3,4-Dihydroisoquinoline N-oxide, 274
Dihydrojasmone, 144, 155, 192
Dihydromuscimol, 283
1-α-Dihydroxyergocalciferol, 292
β,β'-Dihydroxyketones, 150
2,6-Dimethylbenzonitrile oxide, 56
1,2-Diones, 16
1,3-Diones, 14, 138, 145, 157
1,4-Diones, 17, 151, 155
1,5-Diones, 155
1,6-Diones, 159
1,4,2,5-Dioxadiazine, 55, 58, 187, 278
1,3,4-Dioxazolidines, 187

1,3-Dipolar cycloadditions
  asymmetric, 39–41, 271
  cumulated bonds, 172, 271, 287
  cycloreversion, 16, 22, 187, 195, 216
  direction of approach, 38, 235, 271
  heterodipolarophiles, 183, 188
  intramolecular, 8, 23, 66, 192–209, 216, 217, 221–233, 246–253, 275, 281, 282, 287, 290, 292, 293
  mechanism, 25–38, 271
  nitrile oxides and acetylenes, 28, 31
  nitrile oxides and olefins, 28, 31
  nitronates and olefins, 29
  nitronates and acetylenes, 29
  nitrones and acetylenes, 29
  nitrones and olefins, 29, 34
  oximes, 191
  reactivity, 44
  regiochemistry, 25–39, 271–274
  relative rates, 26, 44
  stereochemistry, 25, 40–44, 235–239, 271
Discorea dumentorum, 289
N,N-Disilyloxyenamines, 282
Dumentorine, 289

(±)-Elaeocarpine, 214
Elaeocarpus species, 214
(±)-Elaeokanine-A and -C, 214
β-Enaminoketones, 12, 13, 139, 144, 146, 153, 160, 161, 173, 176–178, 183
2-Ene-1,4-diones, 151, 152
(±)-Epilupinine, 218
α,β-Epoxynitrones, 189
Eremantholides, 157
Ergot alkaloids, 227–231
Erythroxylum coca, 216
β-Eudesmol, 205

Ferruginol, 142
Festuca arundianacea, 289
Ficus septica, 215
Flavonoids, 294
Formhydroxyimoyl iodide, 60
Formonitrile oxide, *see* Fulminic acid
Fulminates, 61, 62, 69
Fulminic acid
  dimerization, 57
  halogen, 62, 243, 283
  oligomerization, 56
  polymerization, 57, 60
  properties, 56
  structure, 57
  sulfur substituted, 62
  synthesis, 60–62
Fulminotrimethylsilane, 61
3-(2H)-Furanones, 157–159
Furans, 157–159, 285, 286
Furoxans, 55, 56, 66, 67
  cycloreversion, 66
  pyrolysis, 66

Subject Index

Geiparvarin, 158
Gephyrotoxin, 232
Gingerol, 144, 150, 153
C-Glucosides, 254, 296
Glucosinolates, 69, 122
Glutarimide antibiotics, 146, 147

Henry's reaction, 102, 104, 119–121, 129
  stereoselectivity, 130
Heterocycles, see also Alkaloids
  condensed systems, 13, 183, 192–209
  miscellaneous, 183–189
  in synthesis, general 3–9
Hirsutene, 203
Histrionicotoxines, 221
L-Homoserine analogues, 245
Hydroazulene ring system, 199, 207
Hydroximoyl chlorides, 64–66
2-Hydroxy-1,4-diketones, 155
α-Hydroxyacetals, 152, 239
α-Hydroxyaldehydes, 76, 152, 156
β-Hydroxyaldehydes, 14, 147, 151, 238
Hydroxycotinine, 227
α-Hydroxyesters, 151
γ-Hydroxyglutamic acids, 240
α-Hydroxyketones, 76, 151, 156
β-Hydroxyketones, 14, 147–151
Hydroxylamines
  condensation with carbonyl compounds, 86
  dehydrogenation, 88
4-Hydroxyproline, 240
Hydroxyputresine, 242
Hypusine, 242

Ibotenic acid, 242–244
Imidazoles, 82, 169, 171, 172, 277, 285
Imidazolinone N-oxides, 87
Imines (Schiff's bases), N-oxidation, 89
Indoles, 160, 166, 167
Indolizidine alkaloids, 214–216, 289
β-Ionone, 142
(±)-Ipalbidine, 216
Ipomoea species, 216
Iridodial ring skeleton, 206
Isatins, 166
Isochanoclavine, 227
Isocyanates, 55, 69
Isocyanilic acid, 57
Isodiazomethane, 57
(±)-Isoelaeocarpine, 214
Isonitramine, 224
(±)-Isoretronecanol, 212, 213
α-Isospartein, 221
Isoxazoles
  alkylation, 10, 11, 137–139, 296
  basicity, 10
  3-chloroisoxazoles, 244
  cleavage, 11, 131–134, 137
  condensations in the side chain, 11, 12, 138–142, 153, 276
  halogenation, 9
  IR spectra, 13
  isoxazolium salts, 11, 137–139
  masked functionality, 9, 14
  mercuration, 9
  nitration, 9
  NMR spectra, 13
  oxidation, 11
  photolysis, 168, 171
  physicochemical properties, 9
  physiological effects, 71
  quaternization, 10
  reactions, 9
  rearrangements, 171, 285
  reduction, 12, 276
  stability, 11
  sulfonation, 9
  synthesis of α-cyanoketones, 11, 131–134, 170, 283
  tautomerism, 14
  thermolysis, 168, 171
  unmasking, 12
  UV spectra, 13
Isoxazolidines
  N-alkoxy, 21, 107, 110
  basicity, 24
  chiral, 41–43, 232–235, 246
  cleavage, 21–24, 155, 205
  cycloreversion, 22, 23, 216, 272
  inversion barrier, 20, 21
  masked functionality, 20, 24
  NMR spectra, 24
  oxidation, 23
  physicochemical properties, 20
  reduction, 24
  stability, 22
  synthesis, 20, 279
  N-trimethylsilyloxy, 9, 21, 00
Isoxazoline route, 17
2-Isoxazolines
  alkylation, 15, 16, 42, 276
  basicity, 16
  cleavage, 15, 16, 131, 133–135, 152, 154, 273
  cycloreversion, 15, 16
  decarboxylation, 15, 134
  epoxidation, 14
  IR spectra, 17
  masked functionality, 6, 17, 19, 147
  NMR spectra, 17
  oxidation, 14
  3-α-oxygenated, 234–239, 293
  photocycloaddition, 276
  physicochemical properties, 14
  quaternization, 16, 18
  reduction, 16, 17, 135, 147–149, 276, 290
  route to aldols, 17, 147
  stability, 15
  synthesis, 14, 21
  unmasking, 17
3-Isoxazolines

physicochemical properties, 18
reactions, 18
synthesis, 16, 18
4-Isoxazolines
IR spectra, 19
NMR spectra, 19
reactions, 18, 164, 168, 272, 279
synthesis, 18, 164, 272, 279
Isoxazolium salts, cleavage, 12, 137, 283

Jatrophone, 157
($\pm$)-Julandine, 218

5-Keto-3-hydroxyesters, 156
γ-Ketoaldehydes, 156
γ- and δ-Ketoesters, 151, 152, 156
Ketone annelation, 138
Kröhnke's reaction, 89

β-Lactams, 180, 274
β-Lactones, 180
Lancacidin C, 252
($\pm$)-Lasubine I, II, 219, 294
Leguminosae species, 221
D and D,L-Lividosamine, 237, 239
Loline, 288
Lolium cuneatum, 289
Lossen's rearrangement, 58
Luciduline, 223
($\pm$)-Lupinine, 218
Lupinus alkaloids, 218
Lycopodium-alkaloids, 223
Lyngbyatoxin A, 166

Macrocycles, 246–252
Maytansine, 249
Maytenus serrata, 250
Metafulminuric acid, 57
4-Methoxy-2,6-dimethylbenzonitrile oxide, structure, 59
α-Methylene-γ-lactones, 208
γ-Methylhomoserine, 240
Milbemycins, 247, 292
Mukaiyama–Hoshino's procedure, 61, 68
Muscimol, 242–244, 283
Muscone, 247
($\pm$)-Myritine, 219

Naphthalene derivatives, 142
Nef's reaction, 96
Negamycins, 291
Nikkomycin antibiotics, 240, 290
Nitramine, 224
Nitrile oxides, *see also* 1,3-Dipolar cycloadditions
allergenic properties, 70
chiral, 41, 58
dimerization, 55, 67, 278
half lives, 56

hydrolysis, 58, 69
IR spectra, 59
mechanism of cycloaddition, 25 *et seq.*
NMR spectra, 59
nucleophilic addition, 69
as oxidant, 59
phosphorous functionalized, 277
physicochemical properties, 55 *et seq.*
polymerization, 55, 58
reactivity, 44, 45, 55
rearrangement, 58, 278
regioselectivity, 27, 28, 31, 272–274
stability, 55–57
stereoselectivity, 38–41
synthesis, 60, 62, 64–69, 277
x-ray structure, 59
2-Nitroalcohols, 102, 104, 119–121, 129
Nitro compounds
acylation, 68, 107, 108, 280
aliphatic, 95 *et seq.*
alkylation, 101–109
basicity, 96
Diels–Alder reaction, 281
double deprotonation, 108
inorganic nitronates (B, Sn, Pb, P), 119, 281
IR spectra, 101
Michael addition, 102, 104
nitro-aldol reaction (Henry's reaction), 102, 104, 119, 129
2-nitroalcohols, 129
aci-nitro form, 95–101, 280–283
UV spectra, 101
Nitrolic acids, 68, 98
Nitronates and nitronic esters, *see* Alkyl nitronates *and* Silyl nitronates
Nitrones, 75 *et seq.*, *see also* 1,3-Dipolar cycloadditions
activation energy for Z-E conversion, 77, 85
acylation, 76, 80, 243
basicity, 75
Behrend's rearrangement, 80
blocking, 280
chiral, 42, 233–235, 246, 291, 296
cleavage, 81, 83
cyclic, 18, 23, 76, 83, 87, 181, 212–222, 235, 243, 245, 274, 280, 287, 288, 295
dimerization, 77, 278
geometric isomerism, 75, 77
IR spectra, 76
mechanism of cycloaddition, 25 *et seq.*
MS spectra, 78
nitron-amide rearrangement, 79
nitrone-oxime-*O*-ether rearrangement, 80
nitrone-*N*-hydroxyenamine tautomerism, 76, 278
NMR spectra, 77, 78, 278, 279
oxaziridine rearrangement, 79
oxidations, 83, 279
photoelectron spectra, 78
physicochemical properties, 75

radical scavenger, 84
reactions, 79
reactions with nucleophiles, 81
reductions, 84
regioselectivity of cycloaddition, 27, 29, 34, 271
relative rates in cycloaddition, 45
silylation, 76
synthesis, 86–89, 278, 279
thermolytic alkene elimination, 81
UV spectra, 76
x-ray structure, 77
Z-E-Isomerization, 85
Nitronic acids
  acylation, 68, 107, 108, 280
  alkylation, 101–109
  cycloaddition, 100
  dimerization, 98
  half-life, 97
  ionization constants, 96
  IR spectra, 99, 101
  NMR spectra, 99, 283
  oxidation, 98
  physicochemical properties, 95
  reactions, 95–100
  salts, 98, 280
  stability, 97
  structure, 99–101
  UV spectra, 101
Nitroso compounds, 89, 154
Nitroxide radicals, 84, 278
Nojirimycin, 233, 235
5-epi-Nojirimycin, 235
Norloline, 288
Nornicotyrine, 227
Nuphar alkaloids, 289

Ocimenones, 297
D,L-Oleose, 238
Olivanic acid, 273
1,2,4- and 1,2,5-Oxadiazoles, 55, 174, 185–187
Oxalodinitrile oxide, 55
1,4,2-Oxathiazoles, 187
Oxazines, 23, 179, 190, 200
1,2-Oxazines, 99, 190, 281
Oxaziridines, 79
Oxazoles, 169, 172
Oxazoline N-oxides, 87
Oxime O-ethers, 75, 79, 80
Oximes
  alkylation, 88
  cycloaddition, 191
  halogenation, 64, 65
  homolytic fragmentation, 209
  oxidation, 64, 281
  tautomerism, 75
  α,β-unsaturated, 15, 153, 296
7-Oxopyrazolo[1,5a]pyrimidine, 184

Paliclavine, 227
5-epi-Paliclavine, 231
Penicillium vermiculatum, 246
Phenols, 145
γ-Phenylhomoserine, 240
Phytosphingosin, 211
Piperidine alkaloids, 217–224, 287, 289
Polyketones, 12, 145
Poranthera corymbosa, 220
($\pm$)-Porantheridine, 220
($\pm$)-Porantherilidine, 220
β-Prodine, 227
Proline analogues, 245
Prostanoids, 144, 153, 192–194
Ptilocaulin, 227, 289
Ptilocaulis aff. P. spiculifer, 227
($\pm$)-Pumiliotoxin-C, 223
Pyrrolidine alkaloids, 217, 287, 288
Pyrans, 285
Pyrazoles, 132, 169, 170, 174
Pyrazolo[4,3-b]pyridine, 183
Pyrenophora avenae, 246
Pyridazines, 178, 184
Pyridazones, 178
Pyridines, 176, 183, 227, 286
Pyrimidines, 176, 178, 184
γ-Pyrones, 157
Pyrroles, 18, 160–167, 279, 286
Pyrrolidine alkaloids, 217, 287, 288
Pyrrolidones, 163, 165, 166, 227, 295
Pyrrolizidine alkaloids, 212, 213, 287

Quadrone, 206
Quinolines, 177, 178
Quinolizidine alkaloids, 217–220, 289, 294
Quinoxalines, 152, 179

Rethrolones, 192
($\pm$)-Retronecine, 213
($\pm$)-Rhodinose, 233

Salicylaldoxime, chlorination 64, 65, 294
Sarcomycin, 194
Seco-corrin skeleton, 160
6,7-Secoagroclavine, 227
7,12-Secoishwaran-12-ol, 203
Sedridine, 217
Sedum alkaloids, 217
Senecioaldoxime, 296
Senecio alkaloids, 212
($\pm$)-Septicine, 215
Sibirine, 224
Silylating agents, 113, 282
Silyl nitronates
  N-alkyl-O-silyl nitronate, 118
  cycloaddition, 9, 42, 62, 120, 134, 135, 152, 154, 157, 192, 238, 246, 291, 294
  geometric isomerism, 119
  hydrolysis, 119
  IR spectra, 115

mechanism of cycloaddition, 25
NMR spectra, 116–118
oxidation, 122
physicochemical properties, 114
reactions, 119–122
reduction, 122
stability, 116
structure, 118
N-sulfonyl-O-silyl nitronate, 118
synthesis of glucosinolates, 69, 122
synthesis, 8, 61, 114, 153, 156, 281
UV spectra, 115
Solenopsis ants, 217, 287
Spiro compounds, 22, 23, 198, 199, 202–204, 221, 222, 224, 253, 254, 274, 281, 289
Streptomyces species, 233, 248
Streptomyces sviceus, 242
Streptomyces tendae, 240
Streptomyces viridochromogenes, 232
($\pm$)-Supinidine, 212

Talaromyces stipitatus, 253
Talaromycin B, 253
Terpenoids, 203–208, 292, 297
α-Terpineol, 205
Tetracyclins, 140, 253
Tetrahydrocinnolones, 178
Tetrahydrorethrolone, 192
Tetrazoles, 174

Tetronic acids, 295
1,2,4-Thiadiazoles, 174
Thiazoles, 173
Thiohydroxamic acids, 69, 120
Thioimidate N-oxide, 180
Thiophenealdoxime, chlorination, 64, 65
($\pm$)-Trachelantamidine, 212
1,2,3 and 1,2,4-Triazoles, 174
S-Triazolo[4–3b]pyridine-4-one, 184
Tricholoma muscarium, 242
Tricholomic acid, 242–244
Trifluoroacetonitrile oxide, 55
1-α,25S,26-Trihydroxycholecalciferol, 205
2,4,6-Trimethylbenzonitrile oxide, 59
Triphenylacetonitrile oxide, 69
Tropane alkaloids, 216
Tropones, 272
Tylophorine, 215

α,β-Unsaturated aldoximes, chlorination, 296

($-$)-Vermiculine, 211, 246
($\pm$)-Vertaline, 219

Woodward-Olofson's peptide synthesis, 137, 283

Xanthons, 142

/547.2T698N>C1/